ELEMENTARY
SYMBOLIC
DYNAMICS

and Chaos in Dissipative Systems

ELEMENTARY SYMBOLIC DYNAMICS

and Chaos in Dissipative Systems

Hao Bai-lin

Institute of Theoretical Physics, Beijing

World Scientific

Singapore • New Jersey • London • Hong Kong

Published by

World Scientific Publishing Co. Pte. Ltd.,
P O Box 128, Farrer Road, Singapore 9128
USA office: 687 Hartwell Street, Teaneck, NJ 07666
UK office: 73 Lynton Mead, Totteridge, London N20 8DH

Library of Congress Cataloging-in-Publication Data

Hao, Bai-Lin.
    Elementary symbolic dynamics and chaos in dissipative systems/
Hao Bai-lin.
    p. cm.
Bibliography: p.
Includes index.
ISBN 9971-50-682-3. — ISBN 9971-50-698-X

1. Chaotic behavior in systems.   2. Differentiable dynamical systems.
I. Title.
Q172.5.C45H36   1989
003 — dc20
                                        89-14766

Printed in Singapore by Utopia Press.

Preface

Perhaps the first question to be touched on in this foreword is why write yet another book on chaos, since many monographs have appeared in recent years (see, e.g., Part I of the References at the end of this book, where many monographs, conference proceedings and collections of papers are listed). While working on the manuscript I have skimmed through the available literature with the hope that someone else would have released me from the task. However, as always I could not get rid of the thought that there exist some gaps in the literature on chaos. Many monographs authored by pure or applied mathematicians were competently written, but to a practitioner in physical sciences, he often feels the "language barrier". A few excellent books were written by our fellow physicists, but most were on an introductory level. Many topics such as the renormalization group approach, the role of external noise, transient behaviour and characterization of the attractors in laboratory and computer experiments, and especially, symbolic dynamics as a practical tool in studying chaos, have not been exposed in sufficient detail.

This book, however, is still an introduction to the study of chaotic behaviour in simple mathematical models, written by a physicist for those working in the physical sciences. The models considered are well-known ones, ranging from one-dimensional mappings to systems of a few ordinary differential equations. Throughout the book emphasis is laid on the use of symbolic dynamics in an elementary way, but we touch on a few other methods to study such models. Symbolic dynamics itself has been a long-studied mathematical topic and is still largely wrapped in fairly abstract (at least, for non-mathematicians) form. Most of the experts would agree that for the time being symbolic dynamics might be the only rigorous way to define chaos and perhaps everyone should start the study of chaos from learning symbolic dynamics, but few practitioners can carry out this programme properly. We ourselves are learners. What we have been doing is to make use of a small part of this beautiful theory in a down-to-earth manner and we are glad to share the experience with colleagues from other branches of physical science.

The reader may find this book written quite inhomogeneously. Some parts contain only qualitative discussions, while other parts are furnished with detailed derivations. The author tried to use elementary mathematics and calculus, and, whenever possible, to rely on physical intuition. Substantial attention has been paid to numerical techniques in studying chaos, but the use of algebraic manipulation languages is also mentioned. I should admit that it is our intention to present the materials in this way in order to put those topics that deserve more scrutiny in depth on a more general background.

This book would not be possible without the interactions and communication with colleagues at home and abroad. There are too many names to be listed. At least, I would like to express my gratitude to the following scientists: P. Bak, T. Bohr, E. Brézin, Chen Shi-gang, R. Conte, P. Cvitanovic, Vic. Dotzenko, M. J. Feigenbaum, J. Ford, L. Glass, Gu Yan, G. Gunaratne, H. Haken, B. Hasslacher, Bambi Hu, Hu Gang, D. K. Kondepudi, C. Y. Liaw, P. Limcharoen, Liu Ji-xing, A. J. Mandell, G. Mayer-Kress, Ni Wan-sun, G. Nicolis, E. Ott, G. Parisi, Peng Shou-li, I. Percival, Y. Pomeau, I. Prigogine, L. Reichl, G. Schmidt, Ya. G. Sinai, H. E. Stanley, B. H. Stewart, H. L. Swinney, C. Tsallis, M. G. Velarde, Wang Guang-rui, Wang Xiao-jing, K. Wiesenfeld, K. M. Khanin, Xu Jing-hua, J. A. Yorke, K. Young, J. M. Yuan, Zhang Hong-jun, and Zheng Wei-mou, for discussions or hospitality or both. I would like to thank Drs. A. Arneodo, J. Bélair, Ping Chen, J. D. Farmer, G. Casati, C. Grebogi, He Da-ren, B. Hess, B. A. Huberman, J. L. Hudson, M. Inoue, K. Kaneko, H. Mori, K. Nakamura, V. I. Oseledes, R. E. Rapp, G. Riela, D. Ruelle, T. Tél, C. Tresser, K. Tomita, Y. Ueda, F. Vivaldi, and many others for providing their publications.

I would like to emphasize that discussions with Dr. Zheng Wei-mou, who has returned to our team recently, have deepened my understanding of symbolic dynamics and changed significantly the presentation of Chapters 3 and 4. Still

I have not been able to take into account all his comments, since a cut-off point must be introduced otherwise the manuscript would never be finished. In addition, I must mention my students Zeng Wan-zhen, Ding Ming-zhou, Li Jia-nan, Lu Li-sha, Zou Chuan-ming, and Yang Wei-ming for I have certainly learnt more from them than what I could offer.

Special thanks go to Zhang Shu-yu who has taken the trouble with all the logistics of my research, from the implementation of algebraic manipulation languages and TeX, to assistance in numerical calculations.

About the reference convention in this book, references to the References at the end of the book are given as, e.g., Poincaré (B1899), the capital B indicating the first part of the References on "Books, Conference Proceedings and Collections of Papers", or Feigenbaum (1980a), addressing a paper in the second part of the References. A few citations to papers not included in the References are given in the footnotes. No efforts have been made to clarify the chronology of one or other statements. In a rapidly expanding and interdisciplinary field like *Chaos* there must have been many rediscoveries of important facts. It is better to leave these to the historians of science.

During the writing of this book I have been constantly perturbed by the limitation of using a foreign language. I apologize for any inconvenience caused by my broken English, although I have honestly tried to meet the definition of the latter, as given by H. B. G. Casimir[1]. Prof. Rainer Radok kindly read the manuscript and corrected the English. Nevertheless, it is my sole responsibility for all the mistakes and inexactitudes that still remain in the text.

Our work on chaos has been partially supported by the Division of Mathematics and Physics, Academia Sinica (1983-1985), and by the Chinese Natural Science Foundation (1986-1988).

The text was typeset by the author using LaTeX of Leslie Lamport with indispensable help from the staff of World Scientific. In particular, I would like to thank Dr. K. K. Phua, the Editor-in-Chief, and Misses P. H. Tham and K. Tan, the editors, for their patience and advice.

---

[1] H. B. G. Casimir, *Scientific American*, **194**(1956) 96.

Contents

Chapter 1

Mathematical Models

Exhibiting Chaos

Throughout this book we shall use the term "physics" in a broad sense, understanding it as a synonym for the physical sciences. In this chapter a few physical systems will be analyzed and modelled by nonlinear equations which may exhibit chaotic behaviour in a suitable range of the control parameters. We shall use some of these models in subsequent chapters to show one or other aspects of a mathematical theory or a certain numerical method, and to gain deeper insight into the phenomenon of chaos. We start with a question which has been asked by everyone who comes across the somewhat fuzzy word *chaos* in the context of an exact science like physics.

1.1   What Is Chaos?

We should consider the question at least in a threefold way: the phenomenon, the word, and the science. To begin with the phenomenon, let us recall an old experiment that has been redone in view of the new understanding of chaos.

In 1831, Faraday observed shallow water waves in a container vibrating vertically with a frequency ω and discovered the occurrence of a subharmonic component of frequency $\omega/2$. Later Lord Rayleigh repeated and discussed this

experiment from the viewpoint of parametric resonance in his famous trea-
tise *The Theory of Sound* (first edition, 1877). Why did the appearance of
subharmonics deserve special attention?

Many experimental apparatus in physics may be viewed as frequency trans-
formers. A system is linear if the frequency components of the input and output
signals coincide, and nonlinear otherwise. In nonlinear systems, higher harmonics as well as sums and differences of the input frequencies appear naturally,
but it is not trivial to have subharmonics. In fact, if the output quantity Q_{out}
depends on the input Q_{in} nonlinearly, e.g.,

$$Q_{out} = a\,Q_{in} + b\,Q_{in}^2$$

and Q_{in} contains a single frequency component, say, $Q_{in} \propto \cos(\omega t)$, then Q_{out}
will contain frequency components 2ω, ω, and 0. If Q_{in} is a linear combination
of $\cos(\omega_1 t)$ and $\cos(\omega_2 t)$, then one can observe in Q_{out} also $\omega_1 \pm \omega_2$. All these
are a simple consequence of basic relations among trigonometric functions. Fur-
thermore, the appearance of frequency harmonics, sums and differences does
not exhibit a threshold. They present themselves inevitably, perhaps, with
very small amplitude, no matter how weak the nonlinearity may be. The subharmonic observed by Faraday and Rayleigh, however, was a kind of threshold
phenomenon: it appeared suddenly when the nonlinearity reached a certain
magnitude. The appearance of both the threshold and the subharmonic calls
for a highly nontrivial explanation.

In 1981, the Faraday experiment was repeated with modern data acquisi-
tion and analysing systems (Keolian *et al.*, 1981; it contains also a few historical
references). Not only the second subharmonic, but also a sequence of subharmonics such as 1/2, 1/4, 1/12, 1/14, 1/16,..., each with its own threshold,
were found. This sequence then turned into a noise-like output with a continuous frequency spectrum. This is just what today people call chaos. 150 years
after Faraday, scientists returned to his experiment, because it had become
clear then that subharmonics usually appear as the first tone in the overture
to chaos. In fact, in many nonlinear systems, various chaotic states of motion
may be reached via a series of finite or infinite numbers of sharp transitions
(sometimes referred to as scenario or route to chaos) and the final chaotic states
are characterized by a number of quantities that distinguish them from pure
randomness.

Broadly speaking, to classify various types of time evolution, one may think of the following possibilities:

1. Purely random processes, e.g., coin tossing.

2. Entirely deterministic processes, e.g., the two body problem in classical mechanics.

3. Deterministic processes subject to random fluctuations, e.g., Brownian motion of a pollen particle in water.

4. Seemingly random behaviour in deterministic systems without any external stochastic source. It is this category of phenomena that now comes under the term chaos.

5. Chaotic processes subject to external noise. Any experimental study of chaos, in laboratories or on computers, must deal with the inevitable effect of external noise and one must be able to distinguish chaos from noise.

It is curious enough that chaotic phenomena have been overlooked for centuries. Once they had been recognized, people began to see chaos everywhere, in Nature and in laboratories. Indeed, chaotic phenomena are ubiquitous whereas purely random or deterministic processes occur only as exceptions. The great time lag in recognizing chaotic phenomena might be attributed partially to the dominance of the deterministic point of view in science ever since the time of Newton, partially, until the 1960s, to the lack of means to tackle and visualize the complicated dynamics, i.e., the modern computers with their graphics.

The word *chaos* appeared as a scientific term in L. Boltzmann's assumption on molecular chaos in his derivation of the famous H-theorem more than a hundred years ago. N. Wiener used the word chaos in the titles of several papers[1]. Both scientists, however, used it to denote disorder caused by or closely related to stochastic processes. The modern usage of the word for intrinsic stochastic behaviour in deterministic systems seems to have appeared

[1]N. Wiener, "The homogeneous chaos"(1938) and "The discrete chaos"(1943), both reprinted in *Collected Works*, ed. by P. Masani, MIT Press, 1976. We thank Dr. P. Limcharoen for telling us these references.

for the first time in the title "Period 3 implies chaos" of Li and Yorke's 1975 paper.

For the time being, there is still no generally accepted definition of chaos. Indeed, there have been some nice and rigorous mathematical definitions for stochastic behaviour in dynamical systems, however, usually it is very difficult to fit a realistic system or model into the mathematical framework. Therefore, one tends to use a working or operational definition for chaos. If seemingly random motion occurs in a system, without applying any external stochastic forces, and the individual output depends on the initial conditions sensitively, but, at the same time, some global characteristics, (e.g., a positive Lyapunov exponent or entropy, fractal attractor dimension, etc., see Chapter 6) turn out to be quite independent of the initial conditions, then one may well be dealing with chaos. Fortunately, physicists rarely disagree on what they observe as being chaos. We shall see that symbolic dynamics provides us with a rigorous way to define chaos. For the time being, however, it is better to look at concrete models instead of playing with definitions.

1.2   The Concept of Universality and the Role of Models

It is appropriate to precede the models by a brief discussion of the deep change of our attitude towards the role of mathematical models in physics that has taken place during the last two decades. In order to grasp the essence of a physical phenomenon, one has to put aside all the secondary factors and to construct simple, yet nontrivial, mathematical models. Since few models can be solved rigorously, it is often necessary to resort to further approximations. It is quite natural, however, to raise questions on the conclusions of a model study as to what are the peculiarities introduced by the particular model, and what are the artifacts introduced by the adopted approximation. For many years, most scientists have paid respect to rigorously solved models, at least for their mathematical beauty, but have viewed with scepticism approximate model studies.

Successes in the theory of phase transitions and critical phenomena have infused new meaning into the notion of universality. Many natural phenomena and mathematical models are grouped into classes characterized by similar behaviour in their parameter dependence, especially, when these parameters are close to some critical values where abrupt change takes place. As a rule, a great number of degrees of freedom are required to describe the routine evolution of a more or less complex physical system. The sudden change of the system at a certain transition point, however, may be characterized by only a few variables. Generally speaking, it is sharp transition rather than smooth development that reveals the universal nature of systems. In fact, these remarks are backed by some weighty mathematical cornerstones which we shall not touch on in this book. Nevertheless, we shall study models consciously as representatives of relevant universality classes.

Another factor that has influenced our viewpoint on models is the use of high-speed digital computers. No matter how difficult a model may be, we are now able to simulate most of its behaviour on computers, provided, of course, necessary precautions are taken. The barrier between analytically solvable and unsolvable models has been diminishing since the introduction of computers. Nowadays, even analytical manipulation of models benefits from using computers. In this book, we shall pay due attention to both numerical and analytical aspects of the use of computers.

Historically, there have been a few models that have greatly stimulated the development of science. Over the decades, they have served as paradigms or touchstones for the development of many important theories. In the first place, we have in mind the two-body problem, starting from the classic Kepler problem of celestial motion, passing through the explanation of Mercurian perihelion procession in relativistic theory and hydrogen atom spectra in quantum mechanics, culminating in the impact of the understanding of the Lamb shift on the development of nonrelativistic and relativistic quantum field theory. As a second example, one may recall Brownian motion, which has been the source of inspiration for the whole stochastic approach in physical sciences, from Langevin and Fokker-Planck equations, path integral representation of Wiener to the Onsager-Machlup functional. In studying chaotic behaviour in nonlinear systems, we are lucky enough to have another such paradigm, namely, one-dimensional mappings of the interval, with which we start our study.

1.3   Insect Population and the Logistic Map

The simplest example of a nonlinear dynamical system comes from ecological models. Suppose there is a seasonally breeding insect population in which generations do not overlap. Then the average (or total, or maximal, depending on which quantity is measured) population Y_{n+1} of the next, i.e., $(n+1)$-th, generation will be determined entirely by the population Y_n of the present generation, i.e.,

$$Y_{n+1} = \Phi(Y_n).$$

This is a first order difference equation whose simplest possible form is a linear relation

$$Y_{n+1} = A Y_n. \tag{1.1}$$

The linear difference equation (1.1) can be easily solved to yield

$$Y_{n+1} = Y_0 A^n,$$

which states that, if, on an average, each insect lays A eggs and all eggs hatch, then the population will grow exponentially, provided $A > 1$. If so, it would take only a few tens of generations before the globe would be overpopulated solely by this single species. However, new phenomena come into play when Y_{n+1} gets large enough: the insects will fight and kill each other for limited food, a contagious epidemic may sweep through the population, etc. Either fighting or touching requires the contact of at least two insects, and the number of such events is proportional to Y_{n+1}^2 (or, more pedantically, to $\frac{1}{2}Y_{n+1}(Y_{n+1} - 1)$, that makes no difference when Y_{n+1} is large). Taking into account this suppressing factor, we can modify Eq. (1.1) into

$$Y_{n+1} = A Y_n - B Y_n^2. \tag{1.2}$$

Despite its apparently simple form, Equation (1.2) may exhibit a quite complicated dynamical behaviour, as we are going to learn in this book (see especially Chapter 2). Obviously, one of the two parameters A and B in (1.2) can be scaled out. Usually, one normalizes Y_{n+1} as well and writes (1.2) in one of the two following forms,

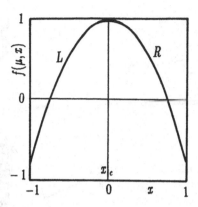

Figure 1.1: The logistic map.

$$y_{n+1} = 4\lambda y_n(1 - y_n), \quad y \in (0,1), \quad \lambda \in (0,1) \tag{1.3}$$

or

$$x_{n+1} = 1 - \mu x_n^2, \quad x \in (-1,+1), \quad \mu \in (0,2). \tag{1.4}$$

Since both expressions appear frequently in the literature, we will write down the corresponding transformation for quick reference. In order to go from (1.3) to (1.4), let

$$y = (\lambda - 1/2)x + 1/2,$$
$$\mu = 4\lambda(\lambda - 1/2).$$

The reverse transformation reads

$$x = \frac{1-k}{\mu}y + \frac{k-1}{2\mu},$$
$$\lambda = (1-k)/4 \quad \text{where} \quad k = \pm\sqrt{1+4\mu}.$$

Note that the parameter range $\lambda \in (0,1)$ corresponds nonmonotonically to $\mu \in (-1/4, 2)$. This difference does not matter as long as we are concerned with chaotic regimes and the associated periodicities. Sometimes (1.4) is modified to read

$$x_{n+1} = \mu - x_{n+1}^2, \quad x \in (-\mu, +\mu), \quad \mu \in (0,2). \tag{1.5}$$

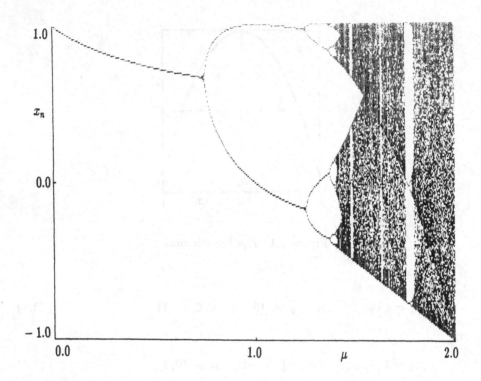

Figure 1.2: The bifurcation diagram of the logistic map.

Throughout this book we shall adhere to (1.4). This first order difference equation describes the time evolution of the normalized population x_n. Starting from a number x_n belonging to the interval $I = (-1, +1)$, it generates in a deterministic manner the next number x_{n+1} belonging to the same interval I, i.e., the nonlinear transformation $f(\mu, x) \equiv 1 - \mu\, x^2$ "maps" the interval I into itself. In the literature Equations (1.3) or (1.4) are often referred to as the *logistic map*. The function (1.4) is shown in Fig. 1.1. The letters R and L denote the *R*ight and *L*eft monotone branches of the map. These letters are the symbols which we shall be playing with throughout this book (see Chapter 3).

The simplest way to get a feeling regarding the logistic map is to visualize it on a personal computer with graphics. One covers the parameter range $\mu \in (0, 2)$ by small steps, and calculates, at each parameter value, the iterates of (1.4) starting from some initial value, say, $x_0 = 0.5$. After throwing away a few hundred points as transients (more about transient regimes in Chapter 7), one displays the remaining points on the screen. In this way we get the *bifurcation diagram* in $x - \mu$ coordinates (see Fig. 1.2). At every given parameter value, plotted along the x direction, one has the "limiting set" of the mapping. For $\mu \leq 0.75$, the limiting set consists of one point. This is the *fixed point* of the mapping. For $\mu = 0.75$ to 1.25 the limiting set comprises two points, giving rise to a 2-cycles or period 2 orbit. Then there come consecutively the 4, 8, ..., 2^n, ... cycles, forming a *period-doubling bifurcation sequence*. If one looks downwards along the μ axis, starting from $\mu = 2$, then one encounters first a one-band chaotic region, followed by two, four, ... and 2^n-band chaotic regions. Both sequences meet at a certain $\mu_\infty = 1.40115...$. Within a chaotic region, there are many *periodic windows*, i.e., lucid intervals where only periodic orbits exist instead of chaotic output. Moreover, we see many dark lines, either going through chaotic regions or becoming boundaries of the latter.

A bifurcation diagram is essentially a diagram of attractors, because almost all initial points are attracted to the points shown in the figure, provided a sufficient number of transients has been thrown away. Fixed points and periodic points are trivial attractors, while the darkened vertical segments are chaotic attractors.

One can "zoom" into the details of the bifurcation diagram by changing to smaller and smaller scales both in x and in μ. Figure 1.3 is a blow-up of a small part of Figure 1.2, namely, for $\mu = 1.78$ to 1.79 and $x_n = -0.16$ to 0.16. Figure 1.4 is a further blow-up of a small part of Figure 1.3, namely, for $\mu = 1.78632$ to 1.78650 and $x_n = -0.02$ to 0.02. This process can be repeated *ad infinitum*, and the self-similar structure is obvious. If all that we have described occurs only for some particular map, then we would be dealing with no more than a rare species in the nonlinear zoo. It is remarkable that all these features happen to be shared by many nonlinear systems. Both the global structure of the bifurcation diagram and the numerical characteristics of many local transitions within the diagram are universal properties. We shall

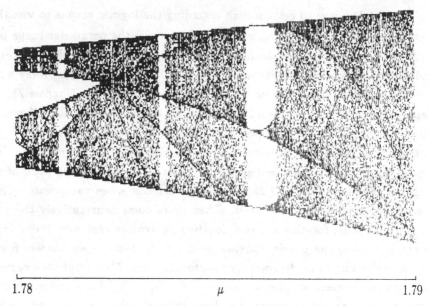

1.78   μ   1.79

Figure 1.3: Blow-up of a small part of the preceding Figure 1.2.

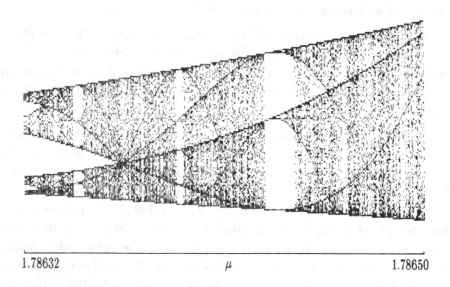

1.78632   μ   1.78650

Figure 1.4: Further blow-up of a small part of the preceding Figure 1.3.

study this bifurcation structure in detail in Chapters 2 and 3.

At this stage, we point out that the logistic map serves as a representative of a wide class of one-dimensional mappings, i.e., of functions with only one hump. Those universal properties that depend on there being only one local maximum, but not on the nature of the maximum, are sometimes called *structural universality* (Derrida, Gervois and Pomeau, 1978 and 1979). Close to the local maximum at $x = x_c$, if the nonlinear function f may be expanded in the form

$$f(x) = f_{max} - A\,(x - x_c)^z + \cdots,$$

those properties, which are shared by maps with one and the same value of z, are classified as *metric universality*. Since z values other than 2 require additional conditions to be satisfied, e.g., $z = 4$ necessitates simultaneous vanishing of the second and third derivatives of f at x_c, the $z = 2$ maps, like the logistic map, represent the most generic case. They are sometimes called *unimodal maps* because they have only one hump. The necessary conditions that these unimodal maps must satisfy in order to enjoy the universal properties, will be formulated below in Chapter 2. The apparent simplicity of these maps may be deceitful, since their dynamics appears to be so rich, that one cannot say even now that everything related to these maps has been fully understood. In fact, a substantial part of this book will be devoted to the scrutiny of these maps.

1.4 Thermal Convection, the Lorenz Model and the Antisymmetric Cubic Map

Many mathematical models have been devised to simulate the turbulent processes in the earth's atmosphere; obviously, this is a problem of great concern to mankind. A simplified model treats the problem as thermal convection of a fluid between two infinite plates, subject to a temperature gradient. If one or both of the horizontal boundaries are considered to be free surfaces, one arrives at the Bénard problem. If both, the upper and lower, interfaces obey rigid boundary conditions, the problem reduces to the case first studied by Lord Rayleigh. In the Boussinesq approximation, the two-dimensional Rayleigh-Bénard problem leads to a pair of partial differential equations (Saltzman, 1962):

$$\frac{\partial \Delta \psi}{\partial t} = -[\frac{\partial \psi}{\partial x}\frac{\partial \Delta \psi}{\partial z} - \frac{\partial \psi}{\partial z}\frac{\partial \Delta \psi}{\partial x}] + \sigma\frac{\partial \theta}{\partial x} + \sigma\Delta^2\psi,$$
$$\frac{\partial \theta}{\partial t} = -[\frac{\partial \psi}{\partial x}\frac{\partial \theta}{\partial z} - \frac{\partial \psi}{\partial z}\frac{\partial \theta}{\partial x}] + R\frac{\partial \psi}{\partial x} + \Delta\theta, \qquad (1.6)$$

where ψ is the stream function, θ is the deviation of the temperature from the linear profile, established by thermal conduction only, and the dimensionless parameters σ and R are the Prandtl and Rayleigh numbers, respectively.

If the boundary conditions are taken to be periodic in the horizontal directions, and free in the vertical direction, one has

$$\psi = \Delta\psi = \theta = 0, \qquad (z = 0, \pi),$$
$$\psi = \Delta\psi = \frac{\partial \theta}{\partial x} = 0, \quad (x = 0, \pi/a),$$

where $1/a$ is the aspect ratio. In this form Equations (1.6) could only be solved by numerical calculation. Such a brute force approach, however, would keep us from gaining deeper insight into the mechanism by which the motion becomes more and more chaotic through a series of transitions. A better way to look into this type of physics involves truncation of the system and its transformation into a system of ordinary differential equations, describing the evolution of a finite number of "modes". In this manner, Saltzman (1962) and Lorenz (1963) arrived at their celebrated model of three ordinary differential equations. As hundreds of papers and a monograph (Sparrow, B1982) have been devoted to the Lorenz model, instead of going into details, we confine ourselves only to an outline of how the Lorenz system appears as the first one in an hierarchy of systems (Curry, 1978; Zhong and Yang, 1986).

A systematic way of truncation of a system of partial differential equations is the Galërlkin expansion that satisfies the boundary conditions from the outset. In our case, a choice of M sine modes and $M + 1$ cosine modes in the horizontal directions and N sine modes in the vertical direction, leads to the expansions:

$$\psi(x, z, t) = \sum_{m=1}^{M} \sum_{n=1}^{N} \psi_{m,n} \sin(amx) \sin(nz),$$
$$\theta(x, z, t) = \sum_{m=0}^{M} \sum_{n=1}^{N} \theta_{m,n} \cos(amx) \sin(nz). \qquad (1.7)$$

On substitution of the above result into (1.6), all spatial derivatives disappear and one obtains a system of ordinary differential equations for the coefficient functions $\psi_{m,n}$ and $\theta_{m,n}$. If one takes initially only one nonzero mode (n_0, m_0), then the truncated system will "excite" additional nonzero modes, according to the scheme (Zou, Yang and Zhou, 1986)

$$
\begin{aligned}
(m_0, n_0) \quad &\rightarrow \quad (m_0, n_0), (0, 2n_0) \\
&\rightarrow \quad (m_0, n_0), (0, 2n_0), (m_0, 3n_0) \\
&\rightarrow \quad (m_0, n_0), (0, 2n_0), (m_0, 3n_0), (0, 4n_0), \ldots
\end{aligned}
\tag{1.8}
$$

On stopping at $(m_0, n_0), (0, 2n_0)$ and taking $M = 1$, $N = 2$, one finds the 3-mode Lorenz model

$$
\begin{aligned}
\dot{\psi}_{11} &= -\sigma(a^2 + 1)\psi_{11} + \frac{\sigma a}{a^2 + 1}\theta_{11} \\
\dot{\theta}_{11} &= a\psi_{11}\theta_{02} + Ra\psi_{11} - (a^2 + 1)\theta_{11} \\
\dot{\theta}_{02} &= -\frac{a}{2}\psi_{11}\theta_{11} - 4\theta_{02}
\end{aligned}
$$

which, after redefinition of the variables and parameters, acquires the standard form

$$
\begin{aligned}
\dot{x} &= \sigma(y - x), \\
\dot{y} &= rx - xz - y, \\
\dot{z} &= xy - bz.
\end{aligned}
\tag{1.9}
$$

We shall present some new results for this much-studied system in Chapter 5, using the method of symbolic dynamics (see Section 5.8).

The next choice of $M = 3$, $N = 4$ leads to the 14-mode system studied by Curry (1978), while that of $M = 5$, $N = 6$ results in a 33-mode system (Zhong and Yang, 1986). Each of these larger systems introduces some qualitatively new features in their transitions to chaos; the question as to how to truncate systematically the original system (1.6), with preservation of its essential physics, remains still open.

One might have started with two initial modes, say, (m_0, n_0) and (m'_0, n_0), and then climbed up a different ladder, other than that represented by (1.8). The next step would give a 5-mode double Lorenz model (Zou, 1986; Zou, Yang and Zhou, 1986; for other 5-mode double-diffusive models, see, e.g., Velarde, 1981, Knobloch and Weiss, 1981). The truncation of the hydrodynamical partial differential equations has become a minor industry (see, e.g., Franceschini and Tebaldi, 1979 and 1981; Riela, 1982; Tedeschini-Lalli, 1982; Franceschini,

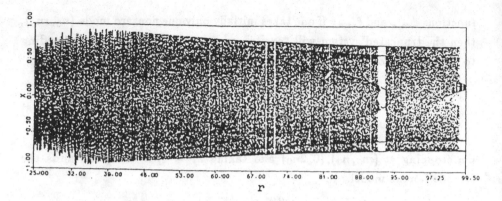

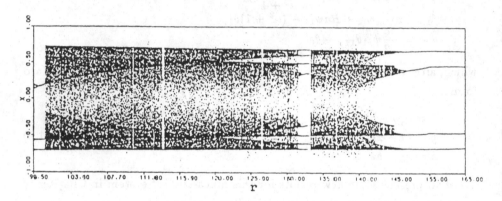

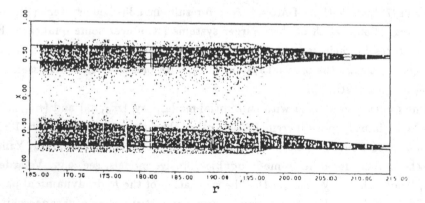

Figure 1.5: The $x - r$ bifurcation diagram for the Lorenz model.

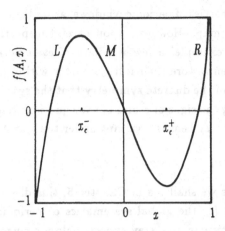

Figure 1.6: The antisymmetric cubic map.

1983; Ahlers and Lücke, 1987, etc.) We refrain from listing these systems and return to the Lorenz equations (1.9).

The Lorenz system possesses only quadratic nonlinearities which appear naturally in many contexts. Therefore, it is not surprising that (1.9) can be derived from other physical problems. Among the direct derivations, we refer only to its relevance to the single mode laser model (Haken, 1975) and to a mechanical model (see Appendix B in Sparrow, B1982).

In order to obtain a feeling regarding the Lorenz equations, we repeat our old trick of presenting a bifurcation diagram, as we did with the logistic map. However, it is a much more time-consuming job, that simply goes beyond the capacity of personal computers. Figure 1.5 shows a long bifurcation diagram cut into three pieces (Ding and Hao, 1988). Two parameters in (1.9) are fixed at $\sigma = 10$ and $b = 8/3$, while the parameter r varies over a wide range from 25 to 215. The values of x, obtained after throwing away sufficiently long transients, are plotted along the ordinate against the r-abscissa. Note that, in order to show some details, the abscissa has a varying scale, i.e., it is stretched where necessary.

First of all, unless told, one would take Fig. 1.5 as the output of some one-dimensional mapping: there are the same kind of dark lines and sharp boundaries and similar series of periodic windows, as we have seen in Figures 1.2 to 1.4 for the logistic map. However, upon careful inspection, one discovers some difference. For example, a few windows of even periods show a half-broken feature and then restore their full symmetry within the chaotic region. This is a consequence of the discrete symmetry that the system (1.9) possesses. Indeed, upon changing the signs of x and y, the equations remain the same. In other words, the Lorenz system is invariant under the transformation

$$(x, y) \rightarrow (-x, -y)$$

with z unchanged. As we shall see in Chapter 5, this discrete symmetry has profound implications for the global systematics of periodic windows in the Lorenz model. In particular, one may expect a close connection between this system and the simplest one-dimensional mapping that shares the same discrete symmetry, namely, the antisymmetric cubic map (May, 1979)

$$x_{n+1} = A x_n^3 + (A - 1) x_n, \qquad x \in (-1, +1). \tag{1.10}$$

Its graph is shown in Fig. 1.6; the meaning of the letters R, M and L will become clear in Chapter 3. We shall study this map and explore its links to the Lorenz model in subsequent chapters. To conclude this section, we show the bifurcation diagram of map (1.10) in Figure 1.7. This diagram demonstrates some of the peculiarities seen in Figure 1.5, e.g., the symmetry-breaking in the main period 2 region and its restoration in the chaotic band. A symbolic dynamics analysis of symmetry breakings and restorations in nonlinear mappings will be developed in Section 3.8.

1.5 Chemical Kinetics Models and Forced Limit Cycle Oscillators

Generally speaking, chemical kinetics provides a good language for the formulation of nonequilibrium and nonlinear problems, since chemical reactions inevitably involve interactions, i.e., nonlinear terms, and can be realized under stationary as well as under far from equilibrium conditions. Unlike simple experiments with nonlinear circuits, which are almost designed to fit the theo-

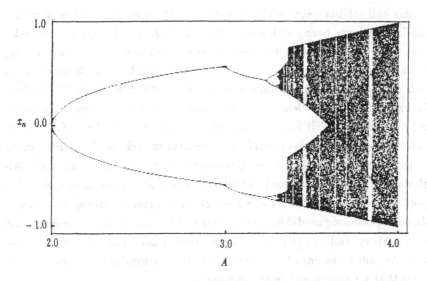

Figure 1.7: The bifurcation diagram of the antisymmetric cubic map.

retical equations, well-controlled chemical reactions show a rich variety of dynamical behaviour and provide grist for the theorist's mills.

In a complete setting, a chemical reaction must be described by a system of reaction-diffusion equations of the form

$$\frac{\partial n_i}{\partial t} = f_i(\mathbf{n}) + D_i \triangle n_i,$$

where n_i is the suitably normalized concentration of the i-th component and the nonlinear functions f_i represent the reaction kinetics. The diffusion coefficient D_i appears in a linear manner, but it is responsible for the formation of spatial patterns, etc. If the reactor is being well-stirred, one can neglect the spatial variation of the concentrations and the diffusion term drops out. In this case we shall have a system of nonlinear ordinary differential equations which describe only the temporal behaviour of the reaction.

Since the discovery, and the recognition, which, by the way, came much later than the discovery, of oscillatory chemical reactions (chemical clocks), the observation of chaotic chemical reactions (or chemical turbulence, as some researchers call it) has been in the foreground. Most earlier studies of chemical oscillations have been carried out on the so-called Belousov-Zhabotinskii reaction, a reduction reaction of bromate ions by an organic compound, e.g., malonic acid; it is not surprising that the first observation of chemical chaos occurred on the same reaction (Schmitz, Graziani and Hudson, 1977; Hudson and Mankin, 1981). The Belousov-Zhabotinskii reaction involves some 20 intermediate species, and it has been modelled by systems of three, four, seven or more ordinary differential equations. The first models of this kind carried the name *Oregonator*, owing to the university where they were conceived. Although some of these models could simulate the periodic behaviour, they failed to reproduce the experimentally observed chaotic regime. Only quite recently a modified seven-mode model has turned out to be good enough to mimic most of the laboratory findings (Richetti, Roux, Argoul and Arneodo, 1987). This system looks much like one of the seven-mode truncations of the Navier-Stokes equations that we mentioned in the last section:

$$\dot{X}_1 = -a_1 X_1 - a_2 X_1 X_2 - a_3 X_1 X_3 + k_8 X_5 X_6 + k_0(X_1^0 - X_1),$$
$$\dot{X}_2 = a_1 X_1 - a_2 X_1 X_2 - a_4 X_2 + a_5 X_4^2 + a_6 X_4 - k_0 X_2,$$
$$\dot{X}_3 = a_1 X_1 + 2a_2 X_{+1} X_2 - a_3 X_1 X_3 + k_5 X_2^2 - a_7 X_3 - k_0 X_3,$$
$$\dot{X}_4 = 2a_4 X_2 - 2a_5 X_4^2 - a_6 X_4 - k_0 X_4,$$
$$\dot{X}_5 = a_6 X_4 - k_8 X_5 X_6 - k_9 X_5 X_7 - k_0 X_5,$$
$$\dot{X}_6 = a_7 X_3 - k_8 X_5 X_6 - k_0 X_6,$$
$$\dot{X}_7 = k_8 X_5 X_6 - k_9 X_5 X_7 - k_0 X_7.$$

The rich dynamical behaviour of such a system with so many adjustable parameters and seven-dimensional phase space can hardly be exhausted in numerical simulations. One has to be content with the study of certain restricted sections of the phase and parameter spaces.

In order to reveal the typical behaviour of chemical kinetics models, however, it is possible to look at simpler systems. In a sense, the choice of model does not mean very much, as chaotic behaviour in physically reasonable systems is expected, to a certain extent, to be "universal". Among the rich variety of chemical models that have been studied, a special class has attracted long-lasting attention, namely, chemical oscillators, coupled to each other or affected by periodical changes in the ambient condition or controlled input. The simplest model reduces to a single nonlinear oscillator, driven externally

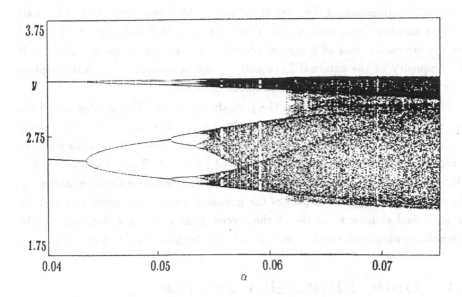

Figure 1.8: A bifurcation diagram of the forced Brusselator.

by a periodic force, or, equivalently, coupled to a linear oscillator.

In this book, we shall frequently refer to a model of trimolecular kinetics under far from equilibrium conditions, namely, the so-called *Brusselator*. The name was coined by Tyson (1973). With a diffusion term added, it exhibits a variety of spatial and temporal patterns and has been described in detail in the literature[2]. We shall deal with a simpler case when there is no spatial inhomogeneity and diffusion, but the system is disturbed periodically. The model has the form

$$\dot{x} = A - (B+1)x + x^2y + \alpha\cos(\omega t),$$
$$\dot{y} = Bx - x^2y. \tag{1.11}$$

This is a system of non-autonomous differential equations, since there occurs an explicitly time-dependent term on the right-hand side. It can be transformed into an apparently autonomous form (see Section 5.7). We note that the phase

[2]See, for example, G. Nicolis, and I. Prigogine, *Self-organization in Nonequilibrium Systems*, Wiley, 1978.

space is two-dimensional, i.e., the (x, y) plane. One can include the time axis t to get an extended phase space. There are 4 control parameters: A and B are the concentrations of some chemicals that are kept at given values, ω is the frequency of the external force and α may be thought of as the coupling constant between the nonlinear and linear oscillators.

In order to demonstrate that the periodically forced Brusselator may have something in common with one-dimensional mappings, we show a bifurcation diagram of Eqs. (1.11) in Fig. 1.8 (Ni, Pei and Hao, 1988). This is the y versus α plot with the other three parameters fixed ($A = 1.2$, $B = 0.4$ and $\omega = 0.85$). The similarity to Fig. 1.2 is obvious, although there exists a definite difference. We shall see that in a large part of the parameter space the systematics of the periodic and chaotic solutions of the forced Brusselator are decribed by the symbolic dynamics of two letters, just like the logistic map (Chapter 5).

1.6   Optical Bistability and the Sine-Square Map

In recent years, optical bistability has become a fascinating subject from both the applied and fundamental points of view. A persistent effort has been made to explore the potential of optical bistability in order to realize optical logical circuits (see, e.g., the book by Gibbs[3]). Theoretically, optical bistability provides us with a good example of bifurcations and transitions to chaos in nonlinear physical systems.

We first describe an optical bistability device briefly. When a light ray passes through a cavity filled with a suitably chosen nonlinear medium, e.g., a saturable absorber, the output intensity I_{out} may depend on the input intensity I_{in} in a highly nonlinear fashion. At low input power, I_{out} changes with I_{in} more or less linearly. At higher input power, say, when the absorber gets saturated and the cavity becomes more transparent, I_{out} may jump to a higher value. If one lowers now the input intensity, the output, though decreasing, will follow another pattern up to a certain point when it jumps back to the linear regime (see Fig. 1.9), thus forming a hysteresis loop. In order to achieve

---

[3]H. M. Gibbs, *Optical Bistability: Controlling Light with Light*, Academic Press, 1985.

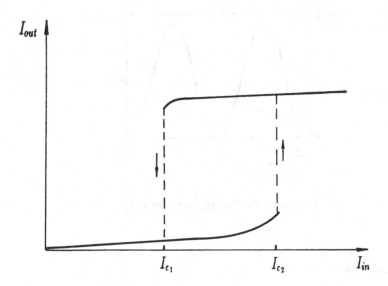

Figure 1.9: Optical bistable state (schematic).

bistable behaviour, a feedback mechanism is required. This may be accomplished by using a ring cavity, an electro-optical or an acoustic-optical loop. The latter type of devices are known as hybrid optical bistable systems. Ikeda and coworkers (1979, 1980) were the first to show that the presence of a time delay in the feedback loop might cause the device to undergo a transition to a chaotic regime.

The equations governing a hybrid optical bistable system with time delay may be derived from the well-known Maxwell-Bloch equations. We skip the details and only point out that in an abstract dimensionless form such a device is described by a time-delayed differential equation of the type

$$\tau \frac{dx(t)}{dt} + x(t) = f(\mu, x(t - T_R)), \tag{1.12}$$

where $f(\mu, x)$ is a nonlinear function which depends on the parameter μ. In (1.12), τ is the relaxation time of the system, and T_R the time delay introduced by the feedback loop. Several choices of the nonlinear function $f(\mu, x)$ have been studied to date:

    1. The cosine model

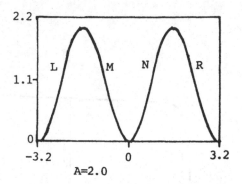

Figure 1.10: The sine-square map.

$$f(\mu, x) = A\,[1 - B\,\cos(x - x_B)]$$

where here and below μ stands for the set of the three parameters A, B and x_B . This is the model first suggested by K. Ikeda (1979).

2. Let $f(\mu, x)$ to be the logistic map (1.4). This is an apparently simple, but, in fact, rather complicated model (Li and Hao, 1985).

3. The sine-square model

$$f(\mu, x) = A\sin(x - B)^2. \qquad (1.13)$$

This case corresponds to a hybrid optical bistable device using a twisted nematic liquid crystal as the nonlinear medium[4], because the transmittence of the liquid crystal may be well approximated by

$$T(V) = H\sin(\frac{\pi V}{2H_V})^2,$$

where V is the voltage applied to the cell, which in turn is the sum of a constant bias and a component proportional to the feedback signal. H is the maximal transmittence at $V = V_H$. The parameter A in (1.13) is proportional to the input light intensity, while the phase B is determined by the bias voltage. A and B are the adjustable control parameters.

[4]H.-J. Zhang *et al.*, *Acta Physica Sinica* **30**(1981), 810; **33**(1984), 1024; **34**(1985), 992; *J. Opt. Soc. Am.* **B3**(1986), No. 2.

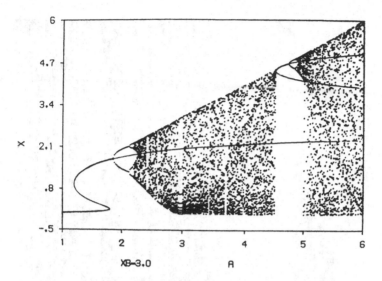

Figure 1.11: The $x - A$ bifurcation diagram for $B = 3.0$ with superimposed steady state solution x^* (solid lines).

In the limit of a very long delay $T_R \ll \tau$, one can neglect the time derivative on the left-hand side of (1.12) and measure the time in units of T_R to get a mapping

$$x_{n+1} = f(\mu, x_n),$$

i.e., exactly the type of one-dimensional mappings considered in Section 1.3. In the last case, we are led to the *sine-square map* (Zhang *et al.*, 1985, 1987)

$$x_{n+1} = A\sin(x - B)^2. \qquad (1.14)$$

The map (1.14) is drawn in Fig. 1.10 for $A = 2.0$, the abscissa being $x - B$. Here again the letters R, M, N and L will be used in the symbolic description of orbits. A different choice of B corresponds to parallel shift of the bisectrix. This is why the bisectrix in Fig. 1.10 does not pass through the origin. Since there are two independent parameters A and B, the bifurcation diagrams must be drawn with one parameter fixed. Fig. 1.11 gives the $x - A$ diagram for fixed

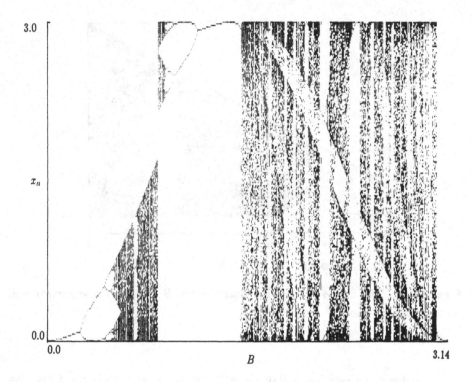

3.0

x_n

0.0

0.0 B 3.14

Figure 1.12: The $x - B$ bifurcation diagram for $A = 3.0$.

$B = 3.0$. Superimposed on this diagram (the solid lines in Fig. 1.11) is the steady state solution x^* of (1.14), i.e.,

$$x^* = A \sin(x^* - B)^2.$$

This multivalued function of A gives rise to bistable and multistable behaviour of the system and may be compared with the schematic drawing in Fig. 1.9.

Figure 1.12 shows the $x - B$ bifurcation diagrams for various values of A. We again see various limiting sets in these bifurcation diagrams: fixed points, periodic cycles and chaotic regimes. There occur also dark lines and lucid windows. We shall see in subsequent chapters that this map can be studied in a manner similar to the logistic map (1.4).

Chapter 2

One-Dimensional Mappings

In the last chapter we have seen a number of simple one-dimensional mappings which may exhibit quite complex dynamical behaviour. Unlike many other one-dimensional models in physics which are either too specific or too trivial to have any practical other than pedagogical meaning, one-dimensional mappings are very rich in content and, at the same time, simple enough to be accessible to analytical means and not very time-consuming in numerical studies. For this reason we shall concentrate in this book on a few typical one-dimensional mappings. Although one-dimensional mappings cannot cover all the properties of higher dimensional nonlinear systems, they do have many features in common. Therefore, one-dimensional mappings may provide a good beachhead to enter the vast Chaosland, and it is quite rewarding to return to them when studying more complicated dynamical systems.

2.1   The Skeleton of Bifurcation Diagrams

All the mappings introduced in Chapter 1 have the general form

$$x_{n+1} = f(\mu, x_n), \qquad n = 0, 1, \ldots, \tag{2.1}$$

where x_n belongs to a certain interval I and μ denotes the set of control parameters. The simplest way to get a feeling for the behaviour of these mappings is to draw the bifurcation diagrams on a computer screen. We have shown a few in Figs. 1.2, 1.11 and 1.12. Two characteristic features immediately stand out on inspection of these diagrams. Firstly, there are many dark lines which either pass through the chaotic regions or become their boundaries. In a sense, they form the skeleton of bifurcation diagrams. Secondly, there are many, and, in fact, enumerably infinite, "windows" embedded in the chaotic regions. These are the locations of stable periodic orbits which exist at the corresponding parameter values.

Can one write down the equations for all the dark lines and band boundaries seen in these diagrams? Is there a way to enumerate all the periodic windows and to determine their locations on the parameter axis? The answer to both of these questions is in the affirmative. We give the answer to the first question in this section (Zheng, Hao, Wang, and Chen, 1984; cf.: Jensen and Myers, 1985, Eidson, Flynn, Holm, Weeks and Fox, 1986). The study of the second question leads to symbolic dynamics to be treated in the next chapter.

The mapping (2.1) is a play of numbers: a seed x_0 produces a long sequence $\{x_i, \ i = 0, 1, 2, \ldots\}$. Now, let us "lift" this iteration of numbers to a recursive definition for a family of composite functions, using the same nonlinear transformation $f(\mu, x)$. As a function of x, $f(\mu, x)$ may have one or more extremes at, say, $x = x_c$, etc., where its first derivative vanishes, i.e., $f'(x_c) = 0$. Starting from the critical point x_c, we define recursively a set of functions $\{P_n(\mu)\}$:

$$
\begin{aligned}
P_0(\mu) &= x_c, \\
P_{n+1}(\mu) &= f(\mu, P_n(\mu)), \qquad n = 0, 1, \ldots.
\end{aligned}
\tag{2.2}
$$

If the map has more than one critical point x_{c_i}, then we start the recursion from different critical points and obtain several sets of composite functions. One might attach a superscript i to the functions P_n^i to indicate the starting point.

Our main assertion is:

These functions $\{P_n^i(\mu)\}$ give all the dark lines and chaotic zone boundaries seen in the bifurcation diagrams; their intersections and tangent points determine the locations of an important class of chaotic orbits as well as the locations of periodic windows.

We will first of all look at two examples and then explain why this occurs in Section 2.1.3.

2.1.1 The Logistic Map

In the case of the logistic map

$$f(\mu, x) = 1 - \mu x^2, \tag{2.3}$$

we have $x_c = 0$ and the functions P_n are polynomials of μ:

$$
\begin{aligned}
P_0(\mu) &= 0, \\
P_1(\mu) &= 1, \\
P_2(\mu) &= 1 - \mu, \\
P_3(\mu) &= 1 - \mu + 2\mu^2 - \mu^3, \\
P_4(\mu) &= 1 - \mu + 2\mu^2 - 5\mu^3 + 6\mu^4 - 6\mu^5 + 4\mu^6 - \mu^7, \\
P_5(\mu) &= 1 - \mu + 2\mu^2 - 5\mu^3 + 14\mu^4 - 26\mu^5 + 44\mu^6 - 69\mu^7 + 94\mu^8 \\
&\quad -114\mu^9 + 116\mu^{10} - 94\mu^{11} + 60\mu^{12} - 28\mu^{13} + 8\mu^{14} - \mu^{15},
\end{aligned}
$$

. . .

The curves P_n vs. μ for $n = 0$ to 8 are shown in Fig. 2.1. We see that they do give the skeleton of the bifurcation diagram Fig. 1.2. It is remarkable that these functions also determine the locations of all band-merging points, i.e., the parameter value where a 2^n chaotic band merges into a 2^{n-1} band. In particular, the merging of the two main chaotic bands into a single band occurs at the intersection point of all but the first two functions, i.e., at $\bar{\mu}$, where

$$P_3(\bar{\mu}) = P_4(\bar{\mu}) = P_5(\bar{\mu}) = \dots . \tag{2.4}$$

To the left of $\bar{\mu}$, these functions are grouped by the parity of their subscripts, and the merging point of four bands into two is determined by either of the two sets of equalities (excluding the first four P_n's, if we do not count $n = 0$):

$$
\begin{aligned}
P_5 &= P_7 = P_9 = \dots, \\
P_6 &= P_8 = P_{10} = \dots .
\end{aligned}
\tag{2.5}
$$

Similarly, to the left of the above band-merging point, the P_n's are grouped according to their subscripts (mod 4) excluding the first eight functions which have become the outer boundaries of the bands, and the next merging takes place at the intersections

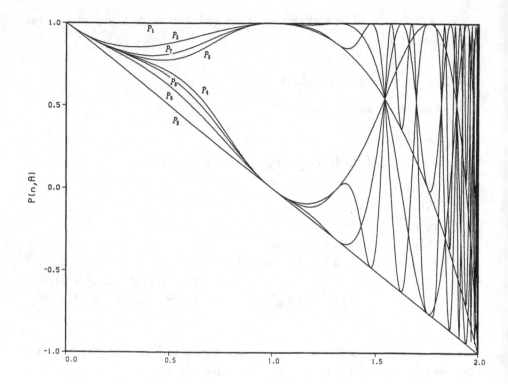

Figure 2.1: Skeleton of the bifurcation diagram for the logistic map.

$$P_9 = P_{13} = P_{17} = \ldots,$$
$$P_{10} = P_{14} = P_{18} = \ldots,$$
$$P_{11} = P_{15} = P_{19} = \ldots,$$
$$P_{12} = P_{16} = P_{20} = \ldots. \tag{2.6}$$

These relations hold for all unimodal maps. Moreover, the locus of the unstable fixed points (see, e.g., (2.32) below) passes through the first band-merging point (2.4); the loci of the two unstable period 2 orbits (see, e.g., (2.39)) intersects the next band-merging point (2.5), etc. In other words, the drastic changes of the structure of the chaotic bands may be viewed to be due to the "collision" of unstable objects in the phase space with attractors. These phenomena have been called the *crisis* of chaotic attractors [Grebogi, Ott and Yorke (1982a and b)]. We shall give a description of the crisis in the general setting of symbolic dynamics in Section 3.6.7.

Continuing with the logistic map (2.3), more useful properties of the P_n may be derived. First of all, it follows from the recursion formula

$$P_{n+1}(\mu) = 1 - \mu P_n(\mu)^2 \tag{2.7}$$

that at those values of μ where $P_n(\mu) = 0$ one must have $P_{n+1}(\mu) = 1$. Next, the above equation may be viewed as the first of the recursion relations

$$P_n = P_k - M_{n-k,k} P_{n-k}^2, \quad k = 1, 2, \ldots.$$

Repeated use of this relation leads to

$$M_{n-k,k+1} = -\mu(P_n + P_k) M_{n-k,k}.$$

Therefore, if $P_n + P_k = 0$ at a certain point $\bar{\mu}$, then one must have $P_{n+1} = P_{k+1}$ at the same point.

Replacing n by $m + k$ in the above formulae, we rewrite them in the form

$$M_{m,k+1} = -\mu(P_{m+k} + P_k) M_{m,k},$$

and

$$P_{m+k} = P_k - M_{m,k} P_m^2.$$

If at certain $\bar{\mu}$ and some k we have $M_{m,k} = 0$, then it follows that

$$M_{m,j} = 0 \quad \forall j \geq k,$$

and at this point we are led to

$$P_{m+j}(\bar{\mu}) = P_j(\bar{\mu}) \quad \forall j \geq k. \tag{2.8}$$

We shall see that this kind of relations determines various band-merging points, including the point where a period 2^{n+1} chaotic band merges into a 2^n band. For example, a nontrivial root of

$$M_{1,3}(\mu) = 0$$

is given by the real root of

$$P_2(\mu) + P_3(\mu) = 0$$

i.e.,

$$2 - 2\mu + 2\mu^2 - \mu^3 = 0 \qquad (2.9)$$

at $\bar{\mu} = 1.543689013\ldots$. Other band-merging points can be calculated in the ·
same way. However, the solution of the above equation and other similar
equations encounters the same kind of numerical difficulties as that associated
with the calculation of supercritical parameter values (see Section 3.2). We
shall show a much better way out in Section 3.6.7 when we learn more about
symbolic dynamics.

2.1.2   The Sine-Square Map

We already know that if there are more than one critical point, one has to define
several sets of composite functions, starting from different critical points. In
the case of the sine-square map

$$x_{n+1} = A \sin^2(x_n - B), \qquad (2.10)$$

four critical points only lead to two sets of composite trigonometric functions,
namely,

$$
\begin{aligned}
P_0(A, B) &= B \pm \pi/2, \\
P_{n+1}(A, B) &= A \sin(P_n(A, B) - B)^2, \\
Q_0(A, B) &= B \text{ or } B + \pi, \\
Q_{n+1}(A, B) &= A \sin(Q_n(A, B) - B)^2.
\end{aligned}
\qquad (2.11)
$$

The first few functions $P_n(A, B)$ and $Q_n(A, B)$ are shown in Figures 2.2 and 2.3,
respectively. These figures should be compared with the corresponding bifurcation diagrams (Figs. 1.11 and 1.12). The equations (2.11) do yield the dark
lines and chaotic band boundaries in these diagrams. In particular, the almost
imperceptible dark lines in the lower part of Fig. 1.11 are just the lines Q_2, Q_3,
and Q_4 shown in Fig. 2.2; $Q_1 = 0$ coincides with the lower frame of the figure.

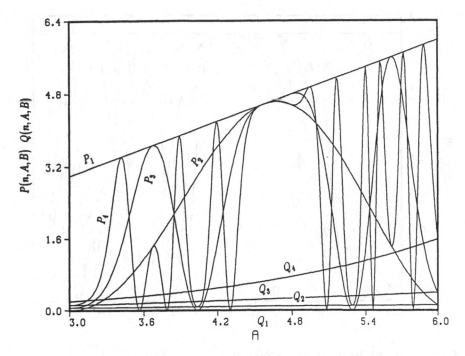

Figure 2.2: Skeleton of the bifurcation diagram of the sine-square map. $P_n(A, 3.0)$ is denoted as P_n, and $Q_n(A, 3.0)$ — as Q_n.

2.1.3 The Rainbow: Explanation of the Dark Lines

In order to understand why these recursively defined functions describe all the dark lines and band boundaries in the chaotic regions, we will digress a little to a middle-school physics problem: the rainbow. If asked to explain the origin of the rainbow, a good student would draw a picture like Fig. 2.4, and then would point out how the droplets work as prisms due to double refraction and single reflection of the light rays at the water-air interfaces. This might be an excellent answer for a middle-school student, but certainly not a complete answer for a graduate student in physics. The point is that the angle of deflection θ of the outgoing ray depends not only on the refraction index n of water, but also on the sight distance (also called the impact parameter) δ of the incoming ray.

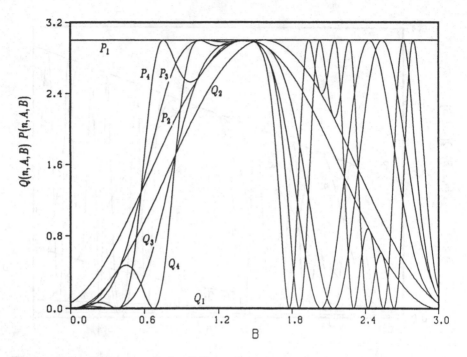

Figure 2.3: Skeleton of the bifurcation diagram of the sine-square map. $P_n(3.0, B)$ is denoted by Q_n, and $Q_n(3.0, B)$ — by Q_n.

Since all sight distances from 0 to R (the radius of the droplet) are present in the sunlight, outgoing rays of the same wavelength will spread out, thus different colors will overlap and mix again. Hence it has still to be explained why we can see at all the rainbow. The correct answer follows from an inspection of the quantitative dependence of θ on n and δ. It is an elementary exercise to derive the formula

$$\theta = 2 \arcsin\left\{ x \left[\frac{2}{n}\sqrt{(1 - x^2)(1 - \frac{x^2}{n^2})} - 1 + \frac{2x^2}{n^2} \right] \right\}, \tag{2.12}$$

where $x = \delta/R$. The $R - \theta$ dependence at fixed n happens to be a unimodal function (Fig. 2.5 is drawn for the refraction index $n = 1.3$, approximately that for water). If we take the incoming rays at equally spaced discrete values of the

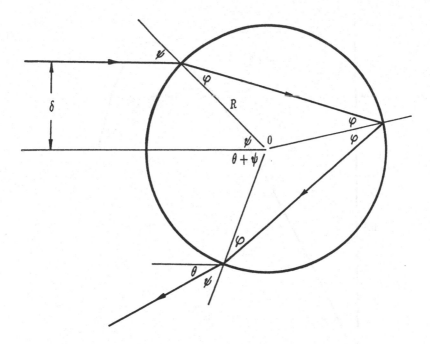

Figure 2.4: Refraction and reflection of a light ray in a water droplet.

sight distance δ, then the resulting rays will come out more densely at angles closer to that corresponding to the critical value x_c. If the incoming rays are homogeneously distributed over the interval $x \in (0, 1)$, then the outcoming distribution will display a singularity, i.e., an infinite peak at the maximum of (2.12). In fact, this result gives the rainbow as we see it [1].

In one-dimensional mappings, the amplification near x_c occurs at each iteration, because, being in the chaotic region, we can look at the iterations from an "ensemble" point of view, taking a continuous distribution as input instead of following the orbit of a single initial point and invoking the "rainbow" argument at each iteration. Old singularities in the input distribution

[1]For more on the physics of the rainbow, see, e.g., H. M. Nussenzveig (1977), Scientific American, April, p.116.

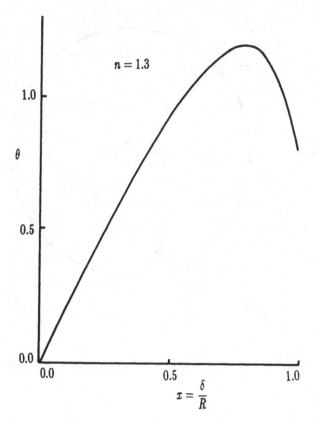

Figure 2.5: The $\theta - x$ dependence for $n = 1.3$.

will be retained in the output, a smooth distribution near x_c gives rise to a new singularity. Furthermore, since x_c corresponds to the maximum of the map, the iterates from both sides of x_c will assemble on the same side of the singularity generated. This explains why it is sufficient to follow the iterates of the critical point x_c in order to find the location of the dark lines and band boundaries, as well as why there exists one-sided "shadowing" near these lines and boundaries.

We note that the "amplification effect" of a single variable function near its extremes is a common phenomenon. Consider another simple example: a free electron gas. In order to calculate the density of states, one needs merely express the momentum space volume element in terms of the energy

$$d^D \mathbf{p} = N(E)dE,$$

where D denotes the dimension of the momentum space. Using the energy-momentum relation $E = p^2/2m$, one readily obtains

$$N(E) \propto E^{\frac{D-2}{2}}.$$

For one-dimensional electron gas, a singularity in the density of states occurs at $E = 0$:

$$N(E) \propto \frac{1}{\sqrt{E}}.$$

When applied to one-dimensional crystal, one has in the effective mass approximation

$$N(E) \propto \frac{1}{\sqrt{|E - E_c|}},$$

where E_c is the band edge. This is just the same kind of singularity that we have seen in the logistic map close to the dark lines and zone boundaries (see Section 2.5.4)[2]. By the way, an interesting question arises for one-dimensional crystals when the gap between the valence and conduction bands shrinks to zero: what happens to, say, the specific heat or magnetic susceptibility when the singularities at the two energy band edges touch each other?[3]. Other examples of the "amplification effect" are the steepest descent method to approximate an integral, the caustics in light or sound propagation, and even the relation between statistical mechanics and thermodynamics.

The results of this section apply to any mapping whenever it has an explicit expression. It illustrates eloquently the deterministic nature of the "skeleton" underlying the seemingly chaotic iterates.

[2]Cf. J. V. José (1983), *J. Phys.*, **A16**, L206.
[3]The answer was given in B.-L. Hao (1961), *Acta Physica Sinica*, **17**, 505.

2.2   Mathematical Preliminaries

In order to avoid interruption in the discussion of the subsequent sections, we introduce here a few simple mathematical concepts and notation.

2.2.1   Composite Functions

Repeated or nested use of several functions leads to a composite function, for example, $f(g(h(x)))$, where $f(x)$, $g(x)$ and $h(x)$ are functions of their own arguments. There exist various notations for composite functions, the most convenient one being the $\circ$ product:

$$f \circ g \circ h(x) \equiv f(g(h(x))). \tag{2.13}$$

If a composite function consists of n nested applications of one and the same function f, we will sometimes use the simple notation F, e.g.,

$$F(n, \mu, x) \equiv \underbrace{f(\mu, f(\mu, \cdots f(\mu, x) \cdots))}_{n \ times} \equiv \underbrace{f \circ f \circ \cdots \circ f(\mu, x)}_{n \ times}. \tag{2.14}$$

An alternative notation for this function is

$$f^n(\mu, x) \equiv F(n, \mu, x). \tag{2.15}$$

Sometimes, for the sake of clarity, we will write $f^{(2^m)}$ instead of f^{2^m}, i.e., adding a pair of parentheses. The reader should not confuse this notation with higher order derivatives. By the way, throughout this book, we will never present derivatives in this form, except that occasionally we will denote the first three derivatives by f', f'' and f'''.

2.2.2   The Chain Rule of Differentiation

Next, we recall the chain rule for the differentiation of composite functions. For three nested functions, it has the form

$$\frac{d}{dx} f(g(h(x))) = \frac{df(y)}{dy}\bigg|_{y=g(h(x))} \frac{dg(z)}{dz}\bigg|_{z=h(x)} \frac{dh(x)}{dx}. \tag{2.16}$$

In particular, when f, g, h ... are one and the same function, we have

$$\frac{d}{dx} f^{(n)}(x) = \prod_{i=0}^{n} f'(f^{(i)}(x)),$$ (2.17)

where $f^{(0)}(x) \equiv x$. If the derivative $\frac{d}{dx} f^n(x)$ is taken at the point x_i, x_i being one of the n periodic points of an n-cycle, where

$$\begin{aligned}
x_2 &= f(x_1), \\
x_3 &= f(x_2), \\
&\cdots \\
x_n &= f(x_{n-1}), \\
x_1 &= f(x_n),
\end{aligned}$$ (2.18)

then

$$\frac{d}{dx} f^n(x) \mid_{x=x_i} = \prod_{j=1}^{n} f'(x_j) \quad \text{for any } i = 1 \text{ to } n.$$ (2.19)

Obviously, this expression is symmetric with respect to permutations of the x_j's.

2.2.3 The Implicit Function Theorem

The number of solutions of a linear problem, whether given by a system of algebraic or differential equations, is fixed and does not change when the parameters in the problem are varied. In contrast, the number of solutions of a nonlinear problem may vary when the values of the parameters are changed. The mathematical theory of the mode in which the solutions change at certain parameter values is called the bifurcation theory. In general, chaotic motion emerges as the result of a series of bifurcations. Therefore, one requires some elements of bifurcation theory in order to understand the appearance of chaos.

A basic tool of bifurcation theory is provided by various extensions of the implicit function theorem of calculus. Since only the simplest form of this theorem will be needed, we merely recall the ideas, skipping the proof which can be found in almost every textbook on advanced calculus. Let there be given an equation $G(x, y) = 0$; is it possible to determine the explicit functional dependence $y = h(x)$ or $x = k(y)$ from this implicit relation between x and y? Consider first of all the differential

$$dG = \frac{\partial G}{\partial x}dx + \frac{\partial G}{\partial y}dy = 0. \tag{2.20}$$

Obviously, if $\partial G/\partial x \neq 0$ then one can solve the equation for dx/dy, and, if $\partial G/\partial y \neq 0$, solve it for dy/dx. Subsequently, there might be some hope to find the function $x = k(y)$ or $y = h(x)$ its derivatives. However, the last step can be accomplished only in particular cases. In general, one can only make a statement on the existence of $x = k(y)$ or $y = h(x)$ from the non-vanishing condition of the corresponding derivative. Usually the above discussion is carried out locally near a point (x_0, y_0), where we know the implicit relation $G(x_0, y_0) = 0$ holds; we look then for the existence condition for $y = h(x)$ such that $G(x, h(x)) = 0$, or for $x = k(y)$ such that $G(k(y), y) = 0$ in a neighbourhood of (x_0, y_0). Hence we require only the following formulation of

The Implicit Function Theorem. Let $G(x_0, y_0) = 0$ and $G(x, y)$ be differentiable near the point (x_0, y_0) in the $x - y$ plane. If $\frac{\partial}{\partial y}G(x_0, y_0) \neq 0$, then there exists a unique function $y = h(x)$ such that

1. $G(x, h(x)) = 0$ in the vicinity of (x_0, y_0), and

2. $$\frac{dh(x)}{dx} = -\frac{\frac{\partial G(x,y)}{\partial x}}{\frac{\partial G(x,y)}{\partial y}}\bigg|_{y=h(x)}.$$

This theorem can also be formulated for $x = k(y)$ and $dk(y)/dy$ provided $\frac{\partial}{\partial x}G(x_0, y_0) \neq 0$.

Actually, the power of the implicit function theorem manifests itself when one or another partial derivative of G vanishes, since it then gives birth to multiple solutions of the implicit equation $G(x, y) = 0$. This situation is called a *bifurcation*. In Sections 2.6.2 and 2.6.4, we shall apply the implicit function theorem to the study of the basic types of bifurcations in one-dimensional mappings.

2.2.4   The Schwarzian Derivative of a Function

The implicit function theorem relies only on local properties of the derivatives. However, some other features of the mapping, e.g., the maximal number of stable periods at a fixed parameter value, may depend on global property of the function through the Schwarzian derivative being negative on the whole interval. The Schwarzian derivative $Sf(x)$ of a function $f(x)$ is an old notion in analysis, introduced in 1869 by H. A. Schwarz[4] and defined by

$$S(f,x) = \frac{f'''(x)}{f'(x)} - \frac{3}{2}\left(\frac{f''(x)}{f'(x)}\right)^2 = \left(\frac{f''(x)}{f'(x)}\right)' - \frac{1}{2}\left(\frac{f''(x)}{f'(x)}\right)^2, \quad (2.21)$$

where a prime denotes a single differentiation. Definition (2.21) requires the existence of f', f'' and f''', a requirement which will not be stated explicitly hereafter. It is readily seen that both the logistic and the sine-square maps have negative Schwarzian derivatives.

The Schwarzian derivative has a number of important properties:

1. The Schwarzian derivative of a linear rational fraction behaves like a constant equal to unity in ordinary differentiation; namely, if $f(x) = (ax + b)/(cx + d)$ and $g(x)$ is another function, then

$$S(f,x) = 0 \quad (2.22)$$

and

$$S(f \circ g, x) = S(g, x). \quad (2.23)$$

2. If both f and g are smooth functions, then direct calculation shows that

$$S(f \circ g, x) = S(f, g(x))[g'(x)]^2 + S(g, x). \quad (2.24)$$

Equation (2.23) is a consequence of (2.24) and (2.22).

3. An important corollary of (2.24) is: if both $S(f, x) < 0$ and $S(g, x) < 0$ for all values of x, then also $S(f \circ g, x) < 0$. This assertion holds also when $<$ is replaced by $>$. In particular, if f and g are one and the same function, then it follows from $S(f, x) < 0$ for all x that

$$S(f^{(n)}, x) < 0 \quad \forall \ x. \quad (2.25)$$

[4]Futher details of the Schwarzian derivative are given, e.g., in E. Hille, *Ordinary Differential Equations in the Complex Domain*, Chapter 10, Wiley, 1976. Schwarzian derivatives also occur naturally in the Liouville-Green, or WKB approximation of differential equations, see, e.g., C. E. Pearson, ed., *Handbook of Applied Mathematics*, Van Norstrand, 1974, Section 12.6.

This inequality is the key relation through which the Schwarzian derivative becomes so useful in the study of one-dimensional mappings.

The fact that a function $f(x)$ has a negative Schwarzian derivative for all x imposes a strong constraint on its shape. Suppose f' has a critical point c, i.e., $f''(c) = 0$, then at this point the second term in (2.21) drops out. A negative Schwarzian derivative simply means that f' and f''' must have opposite signs. Therefore, if the critical point corresponds to a local maximum of f', its second derivative, i.e., f''', is negative, the maximum itself must have a positive value, i.e., be located above the abscissa. Conversely, a local minimum of f' must be located below the abscissa. In fact, it follows thus that for unimodal maps with negative Schwarzian derivatives there exists at most one stable periodic orbit at each parameter value (Singer, 1978). More generally, a map of the interval with n critical points may have at most $n + 2$ stable periodic orbits at a given parameter value (Devaney, B1986). The number 2 comes from a cautious consideration of the possible role of end points, and it usually does not appear. In many cases, we can safely say that n critical points may generate at most n different attracting orbits . In conclusion of this section, we note that all maps considered in this book do have negative Schwarzian derivatives.

2.3   Stable and Superstable Orbits

Next, we turn to a closer study of one-dimensional mappings. A general mapping (2.1) can be considered from different points of view, and hence be given different names. Usually, the parameter μ and the function f are chosen and normalized in such a way that if x_n is taken from a finite interval I, e.g., $(0,1)$, $(-1,+1)$, or (a,b), then the iterate x_{n+1} belongs to the same interval. Therefore, the function f "maps" or "transforms" the interval I into itself, whence follows the name "map (or mapping) of the interval". We say "into itself", because, even when x_n takes all possible values from I, x_{n+1} may not fill up the entire interval I. The mathematical term "endomorphism" describes this situation; hence originates the term "one-dimensional endomorphism". However, a more natural point of view is to look at it as an evolutionary process, measured (or sampled) at discrete time instances t_n, taking the subscript n as an abbreviation for t_n.

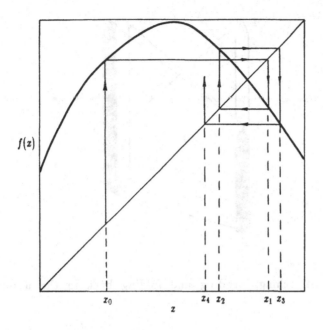

Figure 2.6: Graphic iteration of a map.

Given explicitly the function $f(\mu, x)$ and an initial value x_0, one can compute a sequence

$$x_0, \ x_1, \ x_2, \ x_3, \ \cdots \tag{2.26}$$

by iterating (2.1), using a desk calculator. It is quite helpful to present this iterating process graphically. In Figure 2.6 $f(\mu, x)$ is plotted against x for a general f with fixed μ. At the same time, this is a plot of $x_{n+1} - x_n$. The thin straight line at 45° is the bisector that divides the first quadrant equally. Given an initial x_0, one goes up vertically to find the intersection with f. This is x_1. From there one goes horizontally to the intersection with the bisector, then projects down to locate x_1 on the abscissa. The process is then repeated with x_1 to find x_2, etc. This construction being mastered, one forgets about the dashed lines in the figure and follows only the solid lines in between the bisector and the mapping function. We shall use this type of graphic constructions throughout the book.

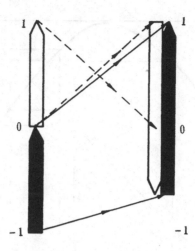

Figure 2.7: Stretching, bending and folding back in the unimodal map (schematic).

What happens to the sequence (2.26) if one keeps on iterating? There may be several possibilities. Usually, there may be a "transient" in the beginning and then the sequence settles down to a "stationary" regime. The values of a certain number of transient points depend more or less on the initial value x_0, while the stationary points do not. The stationary regimes sometimes show a clear, regular or even periodic, pattern, sometimes they may look quite "chaotic". Figure 2.7 may help us to understand the possibility that chaotic orbit might arise. In this figure, the "phase space" for the logistic map (2.3), i.e., the interval $(-1, +1)$, has been drawn vertically twice. The left side represents the initial points, the right one — the points after one iteration. We have to juxtapose a black and a white segment on the right side, because points from the black lower half of the initial interval map into the black part, while points from the white upper half (coreresponding to the monotone decreasing branch of the nonlinear function f) map into the white part which overlays the black part. At the parameter value $\mu = 2$, the right vertical is of the same length as the left, otherwise it may be shorter. Thus a subinterval in the phase space may be stretched, bent and folded back during iterations. This process provides a mechanism for mixing up of initially ordered points, and it may give rise to very complicated orbits. Of course, what has just been described requires more precise formulation. We shall look at this problem from various angles.

Returning to the general structure of bifurcation diagrams, two problems arise naturally: how to find the law governing the appearance of the regular regimes and how to characterize the irregular orbits. In a sense, this whole book, in particular, the present and the next chapters, are devoted to these two problems.

2.3.1 Fixed Point and Its Stability

We start from the simplest case when the x_i repeat themselves without any change, i.e., when, after the transients die away, one reaches a situation where $x_i = x^*$ for all i greater than a certain N:

$$x^* = f(\mu, x^*). \tag{2.27}$$

One says then that the mapping has reached a *fixed point*, or one has found a solution for the (generally) nonlinear equation (2.27). A useful dogma to remember is that, whenever one has obtained a solution for some nonlinear equation (be it an algebraic, difference, differential, or functional equation), the next question to ask is the stability of the solution: what happens to the solution if we perturb it slightly?

In the case of (2.1) stability simply means that on approaching the fixed point x^* the difference between the nth iterate x_n and x^* must keep decreasing. In other words, letting

$$\begin{aligned} x_n &= x^* + \epsilon_n, \\ x_{n+1} &= x^* + \epsilon_{n+1}, \end{aligned} \tag{2.28}$$

then one must have

$$\left| \frac{\epsilon_n}{\epsilon_{n+1}} \right| < 1. \tag{2.29}$$

Substituting (2.28) into both sides of Eq. (2.1) and expanding f to the first order in ϵ, we get

$$x^* + \epsilon_{n+1} = f(\mu, x^*) + f'(\mu, x^*)\epsilon_n + \ldots \tag{2.30}$$

Using the fixed point condition (2.27), Equations (2.29) and (2.30) lead to the stability condition for the fixed point x^*:

$$|f'(\mu, x^*)| < 1. \tag{2.31}$$

What we have just done is the simplest example of a *linear stability analysis* for a nonlinear problem. It is quite instructive, because the equations may become more complicated, but the essence of such an analysis remains the same, with the derivative f' changed to Jacobians, etc. For the logistic map, there are two fixed points, i.e., two solutions of (2.27) and the condition (2.31) yields

$$\begin{array}{ll} x_1^* = (-1 + \sqrt{1+4\mu})/2\mu, & \text{stable for } \mu < 3/4, \\ x_2^* = (-1 - \sqrt{1+4\mu})/2\mu, & \text{unstable for } \mu \in (0, 2). \end{array} \tag{2.32}$$

In the subsequent sections, when we refer to the stable fixed point of the logistic map, we shall write it simply as x^* dropping the subscript 1.

2.3.2 Periodic Orbits and Their Stability

A period n sequence emerges when the iteration loops among n points: $x_{i+n} = x_i$ for all i. Written down explicitly, we have

$$\begin{aligned} x_2 &= f(\mu, x_1), \\ x_3 &= f(\mu, x_2), \\ &\cdots \\ x_n &= f(\mu, x_{n-1}), \\ x_1 &= f(\mu, x_n). \end{aligned} \tag{2.33}$$

These n points form a period n orbit or an n-cycle. Using the composite function notation (2.14), we can present each point of the n-cycle as a fixed point of F:

$$x_i = F(n, \mu, x_i) \qquad i = 1, 2, \ldots n.$$

The last statement allows us to carry over the stability condition (2.31) for a fixed point to a period n orbit. In order to do so, we recall the chain rule of differentiation (2.17). Using the iterations (2.33), the condition

$$| F'(n, \mu, x_i) | < 1, \quad i = 1, 2, \dots, n$$

becomes

$$\left| \prod_{i=1}^{n} f'(\mu, x_i) \right| < 1. \tag{2.34}$$

Obviously, this condition remains the same for all points x_i in the n-cycle, i.e., they acquire or lose stability at the same time. A fixed point of f corresponds to the case $n = 1$; it is called sometimes a period 1 or $1P$ orbit.

The most favourable case for stability occurs when

$$\left| \prod_{i=1}^{n} f'(\mu, x_i) \right| = 0, \tag{2.35}$$

and the convergence of $\{x_i\}$ happens to be quadratic instead of linear. This is called a *superstable orbit* or period. On the other hand, the marginal condition

$$\left| \frac{\epsilon_n}{\epsilon_{n+1}} \right| = 1$$

detremines the boundary of stability, i.e.,

$$\left| \prod_{i=1}^{n} f'(\mu, x_i) \right| = 1.$$

The events at the two limits

$$\prod_{i=1}^{n} f'(\mu, x_i) = +1 \tag{2.36}$$

and

$$\prod_{i=1}^{n} f'(\mu, x_i) = -1 \tag{2.37}$$

cannot be analyzed in the linear approximation, i.e., by keeping only the first derivatives in expanding $F(n, \mu, x^* + \epsilon_n)$. We shall see that interesting phenomena occur at these two extremes, i.e., that tangent bifurcation and period-doubling bifurcation occur at the $+1$ and -1 limits (2.36) and (2.37), respectively. The corresponding discussion invoking higher derivatives will be given in Section 2.6. In the meantime we note that the product of the first derivatives $f'(\mu, x_i)$ changes from $+1$ to -1 on an interval of μ, passing through 0 somewhere in the middle where superstability takes place. (For the time being we consider only good smooth functions f.) On this whole interval of μ, the period n orbit is stable and gives rise to a *periodic window* on the parameter μ axis. For most purposes, the superstable orbit may be taken as the representative of the whole window.

Since it is sufficient to have only one null factor to cause the product $\prod f'(\mu, x_i)$ to vanish, and since this happens at any critical point x_c of the map, we can define a superstable orbit as one that contains at least one of the critical points of the map. We shall exploit this definition later on (see Section 3.2.1).

2.4   The Period-doubling Cascade

Now let us return to the logistic map and see what happens when the parameter μ goes beyond the stability limit of the fixed point x^*, i.e., when $\mu > 0.75$ (see Equation (2.32)). We will give a description of the phenomenon in this section and postpone the detailed analysis, based on bifurcation theory, to Section 2.6.

First of all, we note that right at the $\mu = 0.75$ point, the derivative f' equals -1. A few strokes on a desk calculator will show that it enters into a 2-cycle, i.e., the period has doubled, from 1 (a fixed point) to 2:

$$x_2^* = 1 - \mu x_1^{*2},$$
$$x_1^* = 1 - \mu x_2^{*2}. \tag{2.38}$$

In terms of the language of the original insect population model (see Section 1.3), this is quite a meaningful result: the number of insects oscillates between two levels from generation to generation, an empirically well-known fact.

For the logistic map, the dependence of the periodic points x_1^* and x_2^* can be written down explicitly. Since any fixed point of $f(x) = x$ will also be a periodic point of $f^2(x) = x$, it is better to divide off the trivial factor $(f - x)$ and to solve the equation

$$\frac{f^2(\mu, x) - x}{f(\mu, x) - x} = \mu^2 x^2 - \mu x - \mu + 1 = 0,$$

which yields

$$x_1^* = (1 - \sqrt{4\mu - 3})/2\mu,$$
$$x_2^* = (1 + \sqrt{4\mu - 3})/2\mu. \tag{2.39}$$

It is now clear that these two solutions exist (as observable real roots) only for $\mu \geq 3/4$. When $\mu < 3/4$, they form a complex pair of roots and cannot be observed in iterations of real numbers. On the other hand, the stability condition for this 2-cycle

$$\left| \frac{d}{dx} f^2(\mu, x) \right|_{x=x_i^*} = 4\mu^2 |x_1^* x_2^*| = 1 \tag{2.40}$$

yields

$$-1/4 < 1 - \mu < 1/4,$$

which leads to the stability limits and the superstable parameter:

$$\begin{aligned} \text{stable}: && 3/4 < \mu < 5/4, \\ \text{superstable}: && \mu = 1. \end{aligned}$$

We see that this 2-cycle exists in the interval $(0.75, 1.25)$ of the parameter axis with the superstable orbit just at the center. The derivative

$$\frac{d}{dx} f^2(\mu, x) = 4 - 4\mu$$

(see (2.40)) assumes the value $+1$ at $\mu = 3/4$ and the value -1 at $\mu = 5/4$. Accordingly, both periodic points (2.39) are stable in between. We shall put all these observations in a more general framework in Section 2.6 after an analysis of the tangent bifurcation.

If we take the twice nested function $F(2, \mu, x) \equiv f \circ f(\mu, x)$ as a new map, then each point of the 2-cycle [Eq. (2.38)] becomes a fixed point of F, and the above story repeats itself to yield a 2-cycle for F at some suitable value of μ, which is a 2^2-cycle for the original map f. This process continues *ad infinitum* and leads to a sequence of periods 2^n, $n = 0, 1, \ldots, \infty$. The stability interval of each period diminishes quickly with increasing n, so that infinitely many periods will accumulate at a limiting parameter value μ_∞, for the logistic map $\mu_\infty = 1.40115518909205\ldots$. Later on, we shall study the convergence of the

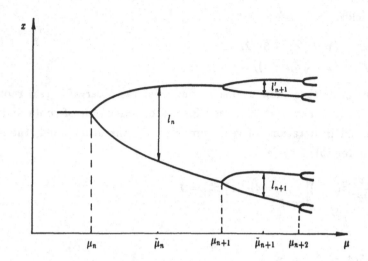

Figure 2.8: The successive bifurcation and superstable points (schematic).

bifurcation points to μ_∞. For the time being, we simply summarize a few "experimental" facts that can be verified by means of a desk calculator.

1. **The convergence rate** of the successive bifurcation points. We already know that for the logistic map the bifurcation from a fixed point to a 2-cycle occurs at $\mu_1 = 0.75$, the bifurcation from the 2-cycle to a 4-cycle takes place at $\mu_2 = 1.25$, etc. One can proceed further to determine the following bifurcation points μ_n numerically and then explore the mode of their convergence to μ_∞. In fact, the bifurcation parameters μ_n are not the best quantities to start with, because the iterations slow down significantly in the vicinity of these points (for this "critical slowing down", cf. Section 7.1). However, included in between two successive bifurcation points, there always exists a superstable periodic point (Section 2.3) $\tilde{\mu}_n$, where the iterations converge much more quickly (see Figure 2.8). These superstable $\tilde{\mu}_n$ must converge in the same manner as the μ_n. M. J. Feigenbaum (1978) discovered that they converge geometrically, i.e.,

$$\tilde{\mu}_n = \mu_\infty - \frac{\text{const}}{\delta^n} \, . \tag{2.41}$$

It is remarkable that

$$\delta = 4.66920160910299067\ldots \tag{2.42}$$

happens to be a "universal" constant for a large class of mappings. The convergence law (2.41) shows that δ may be estimated by looking at the ratio

$$\delta_n = \frac{\tilde{\mu}_{n+1} - \tilde{\mu}_n}{\tilde{\mu}_n - \tilde{\mu}_{n-1}}$$

for several consecutive values of $\tilde{\mu}_n$. Actually, we shall see that δ is only the first, the simplest, and perhaps the most important one among an infinite number of "universal" constants (see, e.g., Section 3.4).

2. **The scaling factor** of typical phase space lengths. We have seen the self-similar structure of the bifurcation diagrams Fig. 1.2 through Fig. 1.4. Along the parameter axis this structure is characterized by the convergence rate δ and other not readily seen constants. If we take some typical lengths along the x direction, i.e., in phase space, for example, the separation of two periodic points at superstable parameters, denoted by l_n, l_{n+1} and l'_{n+1} in Fig. 2.8, then their ratios also approach a certain limit, i.e., both ratios

$$\frac{l_n}{l_{n+1}} \quad \text{and} \quad \frac{l_{n+1}}{l'_{n+1}}$$

approach the constant value

$$\alpha = 2.5029078750958928\ldots. \tag{2.43}$$

[The values of α and δ have been taken from Mao and Hu (1988).] We shall see that α also happens to be the most important of an infinite family of scaling factors (Section 3.4).

The most significant contribution of M. J. Feigenbaum (1978, 1979) comprises not only the discovery of these universal constants, but also his device of a universal, renormalization group approach, which reveals the physical meaning of these constants and opens up a way to the calculation of α and δ to high precisions, independently of any particular model. We shall discuss the renormalization equations in Sections 2.7, 2.9 and 3.4.2. In fact, there are infinitely many renormalization group equations, associated with the unimodal mapping, and symbolic dynamics turns out to be a nice way to tell the basic functional form of these equations.

Usually, μ_∞ is considered as the borderline of chaos beyond that one enters the chaotic regime. However, this statement is not very precise. Firstly, the motion right at μ_∞ is not chaotic, but rather "quasiperiodic", as will be explained later on when we know more about symbolic dynamics (see Section 3.6.7). Secondly, as we have seen in the bifurcation diagrams, many, in fact, infinitely many periodic windows exist beyond the accumulation point μ_∞. Actually, this value μ_∞ is only the first of an infinite number of similar accumulation points, because there are infinitely many converging sequences of periods in the mapping. Only after picking out all periodic windows, is there a hope of seeing the chaotic orbits at the remaining parameter values. Do we still have some chaotic values left on the parameter axis? It is believed that the set of parameter values that correspond to aperiodic orbits, including chaotic as well as quasiperiodic ones, has a positive measure on the parameter axis. There has been some progress towards a complete proof of this statement (Jakobson, 1981; Benedicks and Carleson, 1985; note the remark made by Eckmann and Ruelle, 1985, p.633). Nevertheless, there is at least one parameter value where genuine chaotic orbits do exist. This corresponds to the rightmost boundary of the bifurcation diagram in Figure 1.2 ($\mu = 2$ for the logistic map), where it becomes a *surjective map*, i.e., it maps the whole interval I onto itself. In fact, we shall often refer to this point as a standard for chaos. Namely, whenever we can put another orbit in correspondence with this standard (by means of symbolic dynamics, see Section 3.6), we shall also call it a chaotic orbit. Although this discussion does not exhaust all possible types of chaotic orbits, it certainly presents a best-studied example of chaotic orbits in one-dimensional mappings. Therefore, we should know more about the nature of the orbit at this reference point.

2.5   Chaotic Nature of the Surjective Logistic Map

In this section, we shall study the logistic map at $\mu = 2$

$$x_{n+1} = f(x_n) \equiv 1 - 2\,x_n^2, \tag{2.44}$$

when it maps the entire interval $(-1, +1)$ onto itself. Historically, this was the first chaotic map studied by Ulam and von Neumann (1947), see also Stein and Ulam (1964). In studying this map we shall learn a few more useful notions such as *topological conjugacy*, the *tent map*, and the *invariant distribution* of the iterates. We start by recollecting a little elementary mathematics.

2.5.1 Principal Branch of the Arccosine Function

Define a simple function

$$x = h(\theta) \equiv -\cos(\pi\theta), \tag{2.45}$$

where the minus sign is inserted for convenience. The function h maps $\theta \in (0,1)$ to $x \in (-1,+1)$ monotonically. Therefore, the inverse function is simply

$$\theta = h^{-1}(x) = \frac{1}{\pi}\arccos(-x) = 1 - \frac{1}{\pi}\arccos(x). \tag{2.46}$$

However, for the same range of θ, the function

$$h(2\theta) = -\cos(2\pi\theta) \tag{2.47}$$

has a hump and maps $\theta \in (0,1)$ twice onto $x \in (-1,+1)$ (see Fig. 2.9). Consequently, in taking its inverse, one must be careful with the definition of the principal branch of $\arccos(x)$. For instance, instead of simply annihilating h^{-1} with h, we should write (the new notation $T(\theta)$ will be used below):

$$T(\theta) \equiv h^{-1} \circ h(2\theta) = \begin{cases} 2\theta, & \text{if } 0 \le \theta < 1/2, \\ 2 - 2\theta, & \text{if } 1/2 \le \theta < 1, \end{cases} \tag{2.48}$$

where, if we wish to express the right-hand side of (2.48) in terms of x, θ should be replaced by $1 - \arccos(x)/\pi$ [see (2.46)].

2.5.2 The Surjective Tent Map

We pause to look at the function $T(\theta)$ defined by (2.48). T maps θ from the interval $(0,1)$ onto itself. It consists of two straight lines and looks like a tent, hence the name tent map. Its derivative

$$\left|\frac{dT(\theta)}{d\theta}\right| = 2$$

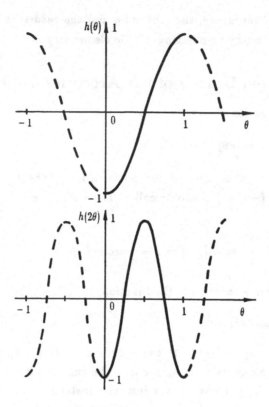

Figure 2.9: The functions $h(\theta)$ and $h(2\theta)$.

is bigger than 1 everywhere, so it does not have any stable periods. Most initial values θ_0 (except for such rational values as $0, 2/3, \ldots$, which, taken together, form a set of measure zero) will lead to chaotic sequences $\{\theta_n, n = 0, 1, \ldots\}$. One can now ask a question: how are these points distributed on the interval $(0, 1)$? The answer defines the density of points

$$\rho(\theta) = \lim_{\Delta\theta \to 0} \lim_{n \to \infty} \frac{\text{number of } \theta_n \in (\theta, \theta + \Delta\theta)}{n\Delta\theta}. \tag{2.49}$$

For piecewise linear maps, $\rho(\theta)$ can be calculated exactly (see Grossmann and Thomae, 1977). In the case of the tent map (2.48), the result can be anticipated by simple counting arguments to be

$$\rho(\theta) = 1. \tag{2.50}$$

Clearly, this distribution is invariant under the map $T(\theta)$, i.e., it remains the same when θ is replaced by another $\theta' = T(\theta)$. We add an asterisk to denote the invariant distribution: $\rho^*(\theta)$.

Now we return to the surjective logistic map (2.44). Let us replace x by (2.45) and look at the result

$$f(x) = f(h(\theta)) = 1 - 2\cos^2(\pi\theta) = -\cos(2\pi\theta) = h(2\theta).$$

Applying h^{-1} to both sides of the above relation and taking into account (2.47), we get

$$h^{-1} \circ f \circ h(\theta) = T(\theta), \tag{2.51}$$

or, equivalently, after inserting $\theta = h^{-1}(x)$,

$$f(x) = h \circ T \circ h^{-1}(x). \tag{2.52}$$

If two maps are related by this type of transformation involving a good enough, i.e., continuous and invertible, function h, they are said to be *topologically conjugate*. So the surjective logistic map is topologically conjugate to the surjective tent map. We shall return to the tent map again in Chapter 3.

2.5.3 Topological Conjugacy of Maps

The concept of topological conjugacy is important for the study of maps, because it divides various maps into equivalent classes. We present some more properties of the topological conjugate relation.

Suppose we have two maps: $f(x)$ maps the interval I into itself, and $g(\theta)$ maps J into itself, where I and J may be different intervals. If there exists a continuous invertible function h such that h transforms J into I and h^{-1} transforms I back into J, i.e.,

$$\begin{aligned} f(x) &= h \circ g \circ h^{-1}(x), \\ g(\theta) &= h^{-1} \circ f \circ h(\theta), \end{aligned} \tag{2.53}$$

then f and g are topologically conjugate.

It is easy to verify that the conjugacy relation extends also to the iterates $f^{(n)}$ and $g^{(n)}$:

$$f^{(n)}(x) = h \circ g^{(n)} \circ h^{-1}(x),$$
$$g^{(n)}(\theta) = h^{-1} \circ f^{(n)} \circ h(\theta). \qquad (2.54)$$

Therefore, an n-cycle of g corresponds to an n-cycle of f and vice versa. Moreover, these cycles enjoy the same stability property. In order to show this, we write

$$f^{(n)} \circ h(\theta) = h \circ g^{(n)}(\theta)$$

and differentiate both sides with respect to θ:

$$\frac{df^{(n)}}{dx}\bigg|_{x=h}\frac{dh}{d\theta} = \frac{dh}{d\theta'}\bigg|_{\theta'=g^{(n)}(\theta)}\frac{dg^{(n)}(\theta)}{d\theta}.$$

Now we see that only when θ belongs to an n-cycle of g, i.e., $g^{(n)}(\theta) = \theta$, we have

$$\frac{df^{(n)}(x)}{dx} = \frac{dg^{(n)}(\theta)}{d\theta}.$$

Therefore, from the fact that the tent map does not have any stable periods follows the same conclusion for the logistic map at $\mu = 2$.

Topological conjugacy provides not only a global relation between two maps, but the function $x = h(\theta)$ also establishes a local correspodence between the points in the two intervals. The conservation of probability (or the conservation of number of points) yields a local relation for the invariant distributions

$$\rho_f^*(x)dx = \rho_g^*(\theta)d\theta \qquad (2.55)$$

(a subscript has been attached to the distribution in order to indicate to which map it belongs). By differentiating the identities

$$h \circ h^{-1}(x) = x h^{-1} \circ h(\theta) = \theta \ ,$$

we get

$$\begin{aligned} h'(h^{-1}(x)) \times \frac{dh^{-1}(x)}{dx} &= 1 \\ h^{-1'}(h(\theta)) \times h'(\theta) &= 1 \end{aligned} \ .$$

The last result and (2.55) permit us to calculate the invariant distribution of one map from that of the other map:

$$\rho_f^*(x) = \rho_g^*(h^{-1}(x)) \left| \frac{dh^{-1}(x)}{dx} \right|. \tag{2.56}$$

Next, we will apply (2.56) to the surjective logistic map (2.44).

2.5.4 The Invariant Distribution for the Surjective Logistic Map

Let g be the tent map $T(\theta)$ in (2.56). We know that $\rho_T^* = 1$ [cf. (2.50)] and (2.46) for $h^{-1}(x)$. Differentiating the arccos(x), we find

$$\rho_f^*(x) = \frac{1}{\pi\sqrt{1-x^2}}. \tag{2.57}$$

The shape of this function is shown in Fig. 2.10. It is sometimes called the Chebyshev distribution, since it is the weight function in the definition of the Chebyshev polynomials. In fact, the Chebyshev polynomials provide the topological conjugacy for a class of piecewise linear maps (Grossmann and Thomae, 1977). The $\mu = 2$ logistic map corresponds to the simplest case, namely, the Chebyshev polynomial of order 2.

In passing, we mention that the invariant density (2.57) may be obtained as the $n \to \infty$ limit of the uniform density (2.50) via a discrete time evolution, generated by the map. All intermediate time densities may be written down analytically (Falk, 1984). Figure 2.10 also gives an idea of the density of the electronic states in a one-dimensional crystal, which we have discussed in connection with the dark lines in the bifurcation diagrams (see Section 2.1).

We take this opportunity to say a few more words on the deterministic nature of chaotic iterations. In principle, the high order iterates of the tent map or the surjective logistic map can all be written down analytically as functions of the initial point. For the tent map, they will be given in Section 2.9; the iterates of the $\mu = 2$ logistic map may be expressed via the Chebyshev polynomials (Erber, Johnson and Everett, 1981). If one could fix the initial value and retain the intermediate points with infinitely many digits, an orbit would be determined entirely by the initial value and might be reproduced whenever one wishes. However, due to the instability of all cycles, an orbit will be extremely

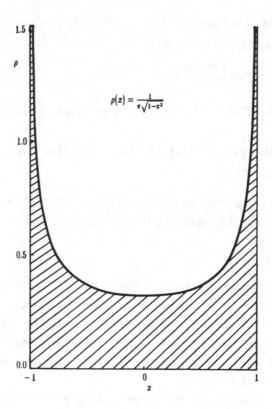

Figure 2.10: The invariant distribution of the surjective logistic map.

sensitive to tiny changes in the initial value. On the other hand, it is just this extreme sensitivity to the initial values that guarantees the existence of the invariant distribution. Since one is unable to retain, in practice, the initial point with infinite precision and it is unnecessary in theory, to bother about single orbits, the invariant distribution (plus some other statistical characteristics) provides us with a satisfactory description of the chaotic motion. In this sense, the output of the $\mu = 2$ logistic map is just as random as any stochastic process with a continuous distribution.

How many chaotic maps are there in a small interval $\mu \in (2 - \epsilon, 2)$ close to $\mu = 2$? It has been conjectured and partially proved (Collet and Eckmann, B1980, p.32) that the ratio

$$\frac{\text{the measure of aperiodic parameters}}{\epsilon} \rightarrow 1$$

as $\epsilon \rightarrow 0$. Note the word *aperiodic* in the last expression. It includes chaotic as well as quasiperiodic (see the end of Section 3.6.7) orbits. It is just in this sense that the $\mu = 2$ point and its vicinity from below may be considered as a real and observable chaotic region on the parameter axis. We shall take the $\mu = 2$ logistic map as a prototype for chaos in studying the orbits at other parameter values. In fact, there are infinitely many such points in the chaotic region of the logistic map. This statement will be made more precise by using the notation of symbolic dynamics (see Section 3.6).

2.5.5 Homoclinic Points in the Logistic Map

Perhaps this is a suitable place to mention for the first time in this book an important notion in chaotic dynamics. We have in mind the *homoclinic intersections* of the stable and unstable manifolds of hyperbolic fixed points and unstable periods, introduced by H. Poincaré in the 1890's. Geometrically speaking, these points are the organizing centers of chaotic motion. These geometrical objects, however, become fully fledged only in two and more dimensions, when a distinction may be made between stable and unstable directions. Compared with the long history of the study of homoclinic points in the general theory of dynamical systems since the time of Poincaré, the analogue of homoclinic points in one-dimensional mappings has been described much later (Block, 1978). Nevertheless, the idea is very simple and it may help us to understand more general cases. In addition, it is possible to make use of the composite functions $P_n(\mu)$, introduced in Section 2.1, for the determination of these points.

We begin with the surjective logistic map (2.44). Since there exist no stable cycles whatsoever, the whole interval $(-1, 0)$ belongs to the unstable set of the unstable fixed point $x^* = -1$, i.e., any point in this interval will get away from the fixed point upon iteration. There is, however, a point at the boundary of the interval, namely, the critical point $x_c = 0$, which is in the "stable" set of the unstable fixed point x^*. Taking $x_0 = x_c$ as the initial point, the iterations lead to a well defined sequence of numbers:

$$0,\ 1,\ -1,\ -1,\ -1,\ -1,\ \ldots \qquad (2.58)$$

i.e., it leads simply to the unstable fixed point itself. Therefore, the points $x_c = 0$ and $x_1 = 1$ are very specific, as their forward iterations fall into the unstable fixed point in a finite number (here two and one) of steps, while their backward iterations, i.e., their pre-images $f^{(-k)}(x)$ return to the same fixed point x^*. The infinite sequence of backward iterates plus the finite forward iterates are called a homoclinic orbit, and the point $x_c = 0$ or $x_1 = 1$ are termed homoclinic points with respect to x^*, or simply, they are homoclinic to x^*.

More generally, if there is an unstable fixed point

$$f(\mu, x^*) = x^*$$

and another point $x \neq x^*$ such that some finite iterates of x settle exactly in x^*, i.e., $f^{(n)}(x) = x^*$, we may say that x is in the stable set (consisting of only a finite number of points) of the unstable fixed point x^*. The point x is immersed in the unstable set of x^*. In other words, the iteration of a small neighbourhood of the unstable fixed point x^* will cover the point x in a finite number of steps. Such a value of x is said to be a homoclinic point of the map. We speak about fixed points, but the arguments may be carried over to unstable periodic points, by merely replacing the map f by the iterate $f^{(n)}$ where n is the period.

If we ask the question: for what parameter value will the critical point $x_c = 0$ be a homoclinic point for an unstable period n orbit of the logistic map? Then the answer may be given in terms of the same set of functions $\{P_n(\mu)\}$ that determines the dark lines and zone boundaries in the bifurcation diagram (see Section 2.1). In fact, according to what has been said in the last paragraph, we must require

$$\begin{aligned} x^* &= f^{(k)}(0), \\ x^* &= f^{(n)}(x^*). \end{aligned} \qquad (2.59)$$

Using the definition (2.2), these equations combine to give

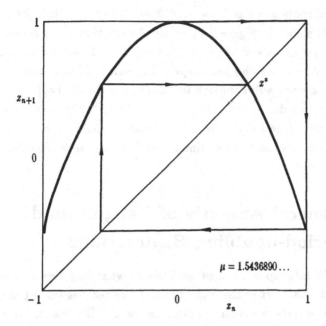

Figure 2.11: A homoclinic orbit at $\mu = 1.5436980$.

$$P_{n+k}(\mu) - P_k(\mu) = 0. \tag{2.60}$$

We have encountered this equation in the particular case of logistic map, see (2.8). The real roots of (2.60) in the interval $(1.401\ldots, 2)$ give the values we are looking for. Usually, the order of this equation may become too high to permit for reliable numerical solution. Nevertheless, if the x^*-μ dependence is known, one may find a short cut. Take, for instance, $k = 3$ and $n = 1$ (a fixed point). From the explicit expression (2.32) for x^*, we may get a simpler equation

$$(2\mu\, P_3(\mu) + 1)^2 = 1 + 4\mu,$$

which yields

$$\mu^4 - 4\mu^3 + 6\mu^2 - 6\mu + 4 = 0.$$

It has two real roots $\mu = 2$ and $\mu = 1.5436980$. The homoclinic situation for the latter value is shown in Figure 2.11, while the solution $\mu = 2$ is clear from our previous discussion: it will be present for any $k > 2$ (see the sequence (2.58)) and $n = 1$. As can be checked graphically, points from a neighbourhood of the homoclinic point will approach the fixed point in the beginning and spiral outwardly afterwards.

We shall introduce a more effective method for the determination of the homoclinic parameters later on by a link to symbolic dynamics (see Section 3.6.7).

2.6 General Analysis of Tangent and Period-doubling Bifurcations

Next, we will take up a detailed analysis of what happens at the two limits (2.36) and (2.37), when the stability condition reaches the ± 1 margin. This analysis will provide instructive examples for the illustration of bifurcation theory, i.e., the mathematical theory of how the number of solutions changes (bifurcates) at a certain value of a varying parameter. The basic tools of this theory are various extensions of the implicit function theorem, only the simplest form of which, formulated in Section 2.2.3, is required. Technically, it is easier to start with the case of tangent bifurcation, which leads naturally to the phenomenon of intermittency.

2.6.1 The Period 3 Window

In order to get some real feeling with regard to tangent bifurcations, let us take a closer look at the most clearly seen window in the bifurcation diagram Fig. 1.2 of the logistic map, namely, the period 3 window. We shall show later on that this window starts precisely at $\mu^* = 7/4$. For the time being, we will draw the function

$$f^3(\mu, x) = f \circ f \circ f(\mu, x)$$

versus x at this parameter value (Fig. 2.12). It touches the bisector tangentially at three points and crosses the latter at one point. The last point corresponds

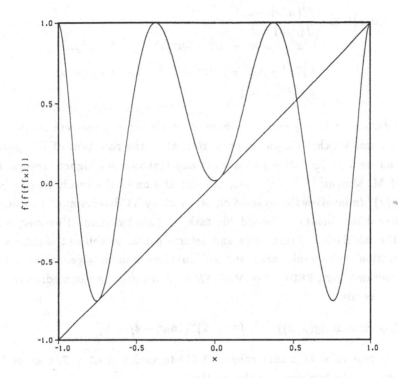

Figure 2.12: The map $f^3(x)$, $f(x) = 1 - 0.75x^2$.

to an unstable fixed point of $f^3(x)$, due to the steep slope, whereas the first three points are marginally stable, because the slope at the point of contact equals exactly +1. If one makes μ slightly larger than 7/4, each of these three contact points will give rise to a pair of fixed points. A little simple arithmetic will show that one fixed point of the pair is stable, while the other one is unstable. Therefore, right at $\mu^* = 7/4$ each pair degenerates into a double root of $f^3(x) = x$. This is the hint which leads to the determination of the exact parameter value of this tangent bifurcation.

In order to simplify the calculation, we note that any root of $f(x) = x$ will also be a root of $f^3(x) = x$, so that it is better to get rid of this trivial root. Hence, for the logistic map, we analyze the quotient

$$g(\mu, x) \equiv \frac{f^3(\mu, x) - x}{f(\mu, x) - x}$$
$$= \mu^6 x^6 - \mu^5 x^5 + (\mu^4 - 3\mu^5)x^4 + (2\mu^4 - \mu^3)x^3$$
$$+ (3\mu^4 - 3\mu^3 + \mu^2)x^2 - (\mu^3 - 2\mu^2 + \mu)x - \mu^3$$
$$+ 2\mu^2 - \mu + 1.$$

The criterion for a polynomial to have multiple zeros is the vanishing of the discriminant, which, in turn, is proportional to the resultant of the polynomial and its first derivative (see almost any textbook on higher algebra, e.g., that of M. Marcus[5]). In our case, the calculation of the resultant of $g(\mu, x)$ and $dg(x)/dx$ involves the evaluation of an 11 by 11 determinant (a so-called Sylvester determinant), a formidable task, if done by hand. However, nowadays, the calculation of resultants and determinants, as well as the subsequent factorization of the results are standard functions of many algebraic manipulation programs (e.g., REDUCE or MACSYMA). Therefore, we immediately write down the result:

$$\text{Discriminant}(g(\mu, x)) \ \propto \ (4\mu - 7)^3 (16\mu^2 - 4\mu + 7)^2 \qquad (2.61)$$

The only real value of μ that causes (2.61) to vanish is $\mu^* = 7/4$ as we have mentioned at the beginning of this section.

A simpler way to get the same result consists of presenting $g(\mu, x)$ as a complete square of a cubic polynomial in x:

$$g(\mu, x) \equiv (A\,x^3 + B\,x^2 + C\,x + D)^2.$$

In order to calculate the four undetermined coeffcients from seven equations, one first looks for the conditions that make these equations compatible. It turns out to be simply $4\mu = 7$.

2.6.2 Tangent Bifurcation

Now we apply the *Implicit Function Theorem* (see Section 2.2.3) to study the general case of tangent bifurcations in one-dimensional mappings, i.e., to elucidate under what conditions n pairs of real solutions of $f^n(x) = x$ will emerge

[5]M. Marcus, *Introduction to Modern Algebra*, Marcel-Dekker, 1978, pp. 230-232.

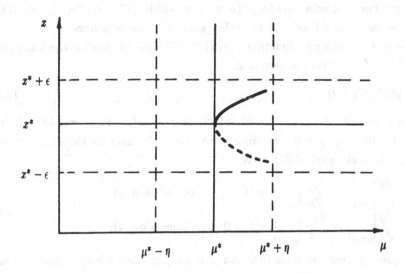

Figure 2.13: The vicinity of a tangent bifurcation point (μ^*, x^*).

"from nothing": no real solutions exist at all on one side of the bifurcation parameter value μ^*, and n pairs of stable and unstable real solutions arise on the other side, giving birth to a period n window. We shall follow closely the presentation of Guckenheimer (1977).

Instead of looking at an n-cycle of $f(x)$, we consider one of the n fixed points of $f^n(x) = x$. The final assertion involves the following four conditions

1. There is a fixed point at μ^*: $f^n(\mu^*, x^*) = x^*$;

2. The stability is marginal with $\frac{\partial}{\partial x} f^n(\mu, x)\big|_{\mu^*, x^*} = +1$;

3. $\frac{\partial}{\partial \mu} f^n(\mu, x)\big|_{\mu^*, x^*} \neq 0$;

4. $\frac{\partial^2}{\partial x^2} f^n(\mu, x)\big|_{\mu^*, x^*} \neq 0$.

If they are satisfied, there exists a small region in the μ-x plane, centered at (μ^*, x^*), say, from $\mu^* - \eta$ to $\mu^* + \eta$ along the μ axis and from $x^* - \epsilon$ to $x^* + \epsilon$ in the x direction (see Fig. 2.13), such that to one side of μ^*, say, $\mu \in (\mu^*, \mu^* + \eta)$, there are two real solutions to $f^n(\mu, x) = x$ within $(x^* - \epsilon, x^* + \epsilon)$, one stable and one unstable, whereas to the other side of μ^* there is none.

We first define a new function $g(\mu, x) \equiv f^n(\mu, x) - x$, so that the fixed point (μ^*, x^*) of f becomes the zero of g:

$$g(\mu^*, x^*) = 0. \tag{2.62}$$

In order to test the existence of solutions $x = x(\mu)$ or $\mu = \mu(x)$, defined implicitly by $g(\mu, x) = 0$ according to the implicit function theorem, we have to know that the partial derivatives

$$\begin{aligned}
\left.\frac{\partial g}{\partial \mu}\right|_{\mu^*,x^*} &= \left.\frac{\partial f^n}{\partial \mu}\right|_{\mu^*,x^*} \neq 0 && \text{(Condition 3)}, \\
\left.\frac{\partial g}{\partial x}\right|_{\mu^*,x^*} &= \left.\frac{\partial f^n}{\partial x}\right|_{\mu^*,x^*} - 1 = 0 && \text{(Condition 2)}.
\end{aligned} \tag{2.63}$$

Consequently, there exists a function $\mu = \mu(x)$ in the vicinity of (μ^*, x^*) such that $g(\mu(x), x) = 0$. Moreover, the derivative

$$\frac{d\mu}{dx} = -\left.\frac{\partial g/\partial x}{\partial g/\partial \mu}\right|_{\mu^*,x^*} = 0. \tag{2.64}$$

Therefore, $\mu(x)$, viewed as a function of x, may have a maximum, or a minimum, in μ at $\mu = \mu^*$, or has a point of inflection at (μ^*, x^*), depending on the second derivative $d^2\mu/dx^2$. In order to calculate the latter, we differentiate twice the implicit relation $g(\mu(x), x) = 0$:

$$\begin{aligned}
&\frac{\partial g}{\partial x} + \frac{\partial g}{\partial \mu}\frac{d\mu}{dx} = 0, \\
&\frac{\partial^2 g}{\partial x^2} + 2\frac{\partial^2 g}{\partial x \partial \mu}\frac{d\mu}{dx} + \frac{\partial^2 g}{\partial \mu^2}\left(\frac{d\mu}{dx}\right)^2 + \frac{\partial g}{\partial \mu}\frac{d^2\mu}{dx^2} = 0.
\end{aligned} \tag{2.65}$$

For these equations, we get at (μ^*, x^*)

$$\left.\frac{d^2\mu}{dx^2}\right|_{\mu^*,x^*} = -\left.\frac{\partial^2 g/\partial x^2}{\partial g/\partial \mu}\right|_{\mu^*,x^*} \neq 0, \tag{2.66}$$

according to Conditions 3 and 4 above. If we take $\partial^2 g/\partial x^2 > 0$, $\partial g/\partial \mu < 0$, or *vice versa*, then $d^2\mu/dx^2 > 0$, and two solutions exist at the $\mu > \mu^*$ side (Fig. 2.13).

Finally, we will study the stability of these solutions. We know from Condition 2 that $\frac{\partial f}{\partial x}^n(\mu^*, x^*) = 1$, and that it is sufficient to expand

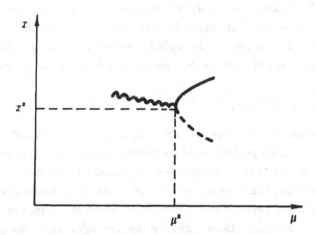

Figure 2.14: A tangent bifurcation (schematic): wavy line — an invisible pair of complex roots, solid line — a stable real solution, dashed line — an unstable real solution.

$$\frac{\partial f}{\partial x}^n (\mu^* + d\mu, x^* + dx)$$

near (μ^*, x^*). However, μ and x are no longer independent arguments due to the $\mu(x)$ relationship

$$\mu(x^* + dx) = \mu^* + \frac{1}{2} \frac{d^2\mu}{dx^2}\bigg|_{\mu^*, x^*} (dx)^2,$$

since $\frac{d\mu}{dx}\big|_{\mu^*, x^*}$ vanishes. Therefore, we have, to the first order in dx,

$$\frac{\partial f}{\partial x}^n (\mu, x) = 1 + \frac{\partial^2 f}{\partial x^2}^n \bigg|_{\mu^*, x^*} dx + \cdots, \tag{2.67}$$

where dx changes sign on x crossing x^*, so there is always one stable and one unstable fixed point in the right half (at our choice of the signs) of the square box in Fig. 2.13. In the left half, there is no real solution at all. At least in the case of the logistic map, it is a pair of complex solutions that become two real ones at (μ^*, x^*). We draw schematically this situation in Fig. 2.14, which will help us to count the number of periodic windows in the next chapter (see

Section 3.5). In Figure 2.14, the wavy line represents a pair of complex roots of the fixed point equation that cannot be seen when iterating with real numbers only. A tangent bifurcation is also called a saddle node bifurcation, because similar situations in two or more dimensions give birth to saddle points.

2.6.3 Intermittency

In the last subsection, we have shown that at a tangent bifurcation a pair of stable and unstable periodic orbits develops "from nothing", since, to the other side of the bifurcation point, they correspond to complex roots of the fixed point equation, and thus are invisible in iterations of real numbers. However, there are other periodic and nonperiodic orbits at parameter values just preceding the bifurcation. Does a nearby tangent bifurcation leave any trace on these orbits? The fact that close to a tangent bifurcation in unimodal maps the orbits are indeed chaotic will become clearer in Chapter 3 after we learn more about symbolic dynamics. For the time being, let us have a qualitative discussion of the characteristic features of the motion. If we call the transition to aperiodic motion via accumulation of period-doubling bifurcations a route to chaos, then we shall recognize here another route to chaos, namely, the intermittent transition, first studied by Pomeau and Manneville (1980) (see also Manneville and Pomeau, 1979).

Again, consider, for example, the period 3 window of the logistic map. Using the notation for composite functions defined in Section 2.2.1, we look at the iterates of the mapping

$$y_{n+1} = F(3, \mu, y_n) \equiv f \circ f \circ f(\mu, y_n). \qquad (2.68)$$

Of course, the iterates $\{y_n\}$ may be obtained from that of $f(\mu, x_n)$ by keeping only the set $\{x_{3n}\}$. A plot of $F(3, \mu, x)$ is given in Fig. 2.15 at the parameter value $\mu = 1.745$, i.e., slightly less than the threshold $\mu_c = 1.75$ for the bifurcation to period 3. (One might have drawn it even closer to 1.75, but the details in the figure would be harder to see.)

We see that there are three narrow "corridors" between the map and the bisector. When iterating the map, if one point falls close to the entrance to one of the corridors, it would take many iterations to get through the passage (see Fig. 2.16). The closer one gets to the bifurcation point, the narrower will

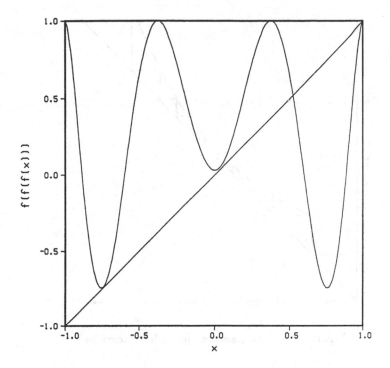

Figure 2.15: The map $f \circ f \circ f$ at $\mu = 1.745$.

be the corridor, and the more iterations would be required. At the beginning, the process resembles the convergence to a fixed point. Since the latter does not at all exist, the iterations sooner or later quit the corridor. Then, after a number of irregular jumps, they approach one or another corridor and the process repeats itself in much the same way.

Now we have a full picture of the iterations. The passage of a narrow corridor looks like marking time at a fictitious fixed point, and corresponds to the "laminar" phase of the motion. The jumps among different corridors give an impression of a "turbulent" phase, interrupting the laminar motion at

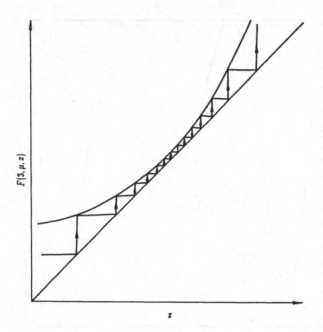

Figure 2.16: The passage through a "corridor".

unpredictable instances. These two phases appear intermittently. At a fixed parameter value, one can only speak about average laminar or turbulent time.

The passage time through a corridor can be estimated easily from the picture described above (Pomeau and Manneville, 1980; Hirsh, Huberman and Scalapino, 1982). Right at the bifurcation point $\mu = \mu_c$, we have three fixed points

$$y_i^* = F(3, \mu_c, y_i^*), \qquad i = 1, 2, 3. \tag{2.69}$$

For (μ, y) close to (μ_c, y^*), where y^* denotes one of the fixed points, we expand $F(3, \mu, y)$ as follows:

$$F(3, \mu, y) = y^* + \frac{\partial F}{\partial y}(y - y^*) + \frac{\partial F}{\partial \mu}(\mu - \mu_c) + \frac{1}{2}\frac{\partial^2 F}{\partial y^2}(y - y^*)^2 + \dots .$$

At $\mu = \mu_c$ we know that $\frac{\partial F}{\partial y} = 1$. Close to any of the corridors in Fig. 2.15 this equation takes the form

$$F(3, \mu, y) = y^* + (y - y^*) + a(\mu_c - \mu) + b(y - y^*)^2 + \cdots, \tag{2.70}$$

where a and b are positive constants, assuming different values at different fixed points. If, incidentally, the coefficient of $(y - y^*)^2$ happens to be zero, then we take the next non-zero term $b(y - y^*)^z$. Therefore, in the vicinity of a corridor like the one shown in Fig. 2.16, we may write the map (2.68) in the form

$$y_{n+1} - y^* = y_n - y^* + a(\mu_c - \mu) + b(y_n - y^*)^z. \tag{2.71}$$

In the corridor, the y_n are close to y^*, so that one can introduce a continuous variable $x = y - y^*$ and consider n to be a discrete time with $\Delta n = 1$. Thus, the map (2.71) may be replaced by a differential equation

$$\frac{dx}{dn} = a(\mu_c - \mu) + bx^z. \tag{2.72}$$

We integrate this equation from the entrance x_1 to the exit x_2. The calculated passage time must be averaged over the distribution of x_1. Under reasonable assumption of a homogeneous distribution of x_1 within certain limits and for $\mu \to \mu_c$, one gets the asymptotic estimate (Hirsch, Huberman and Scalapino, 1982)

$$n \propto [ab(\mu_c - \mu)]^{-(1 - \frac{1}{z})}. \tag{2.73}$$

For $z = 2$, we have the exponent $1/2$. The divergence of n, when μ approaches μ_c, signals a longer and longer laminar, i.e., time-correlated, phase. It can be compared to the divergence of spatial correlation at phase transitions. In the mean-field theory of critical phenomena, the correlation length ξ diverges as

$$\xi(T) \propto |T_c - T|^{-\nu}$$

and $\nu = 1/2$. In this sense, the simple derivation, just described, has led to the same mean-field exponent $1/2$. Obviously, the theory is equally applicable to tangent bifurcations into other periods, i.e., to $F(p, \mu, y)$, $p = 4, 5, \ldots$.

We will generalize the picture of passage through a corridor to a renormalization group equation for intermittency in Section 2.9. A symbolic dynamics description of intermittency will be given in Section 3.6.7.

We mention, in passing, that the above usage of the term "intermittency" is somewhat different from that in the literature on turbulence[6]. In a turbulent

---

[6]See, e.g., D. J. Tritton, *Physical Fluid Mechanics*, Van-Norstand, 1977, p.265.

fluid, there are clear-cut boundaries between the laminar and turbulent regions, but these irregular boundaries move and change with time. If one keeps looking at a given point close to such a boundary, then the recorded velocity will show an intermittent alternation of laminar and turbulent behaviour, as the point enters and leaves the turbulent region. The condition which we call intermittency in this section (and in this book) concerns only the time evolution of a dynamical system, and not at all the spatial distribution.

2.6.4 Period-doubling Bifurcation

Now we are well prepared to analyse the general case of period-doubling bifurcation, i.e., to elucidate the conditions for period-doubling to occur. This analysis provides another instructive example of an application of the implicit function theorem. We will follow again closely the presentation of Guckenheimer (1977), breaking it into steps and making minor modifications, e.g., in the definition of the auxiliary function $k(\mu, x)$ (see Step 3 below) and in the explicit introduction of the Schwarzian derivative. Without loss of generality, we consider a fixed point (μ^*, x^*) of $f^n(\mu, x)$ instead of an n-cycle of $f(\mu, x)$. In order to simplify the writing of formulae, one might have replaced, in what follows, f^n by f and f^{2n} by f^2. Let the following conditions hold true

1. There is a fixed point $f^n(\mu^*, x^*) = x^*$ in the (μ, x) plane.

2. The stability is marginal with $\frac{\partial}{\partial x} f^n(\mu, x) \mid_{\mu^*, x^*} = -1$.

3. The mixed second derivative $\frac{\partial^2}{\partial x \partial \mu} f^n(\mu, x) \mid_{\mu^*, x^*} \neq 0$.

4. The Schwarzian derivative $S(f, x)$, and consequently $S(f^n, x)$ [see Inequality (2.25) in Section 2.2.4], is negative at (μ^*, x^*). Since most mappings of physical interest have negative Schwarzian derivatives on the whole interval of definition, this condition may be dropped for such mappings.

We are going to show that in the vicinity of the fixed point (μ^*, x^*) there is a small region, say, the box

$$\left(\mu^* - \eta < \mu < \mu^* + \eta, \quad x^* - \epsilon < x < x^* + \epsilon\right)$$

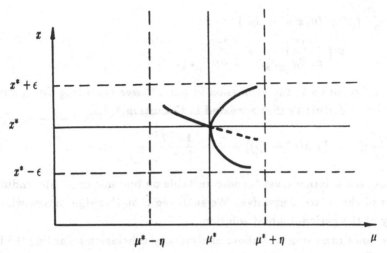

Figure 2.17: Period-doubling bifurcation at (μ^*, x^*) (schematic).

in the x-μ plane, such that to one side of μ^* (depending on the choice of the sign in Condition 3) there is only one stable solution of $f^n(\mu, x) = x$ in the box, which is certainly a trivial stable solution of $f^{2n}(\mu, x) = x$, but in the other half of the box there are three solutions to $f^{2n}(\mu, x) = x$, two non-trivial stable ones and a trivial unstable one (see Fig. 2.17). The form of the solution curves in Fig. 2.17 explains an alternative name for this kind of bifurcation: pitchfork bifurcation. We will divide the discussion into several steps.

Step 1. Let

$$h(\mu, x) = f^n(\mu, x) - x, \tag{2.74}$$

then the fixed point of f^n becomes a zero of h: $h(\mu^*, x^*) = 0$. It follows from Condition 2 that

$$\left.\frac{\partial h}{\partial x}\right|_{\mu^*, x^*} = \left.\frac{\partial f^n}{\partial x}\right|_{\mu^*, x^*} - 1 = -2 \neq 0. \tag{2.75}$$

Therefore, according to the implicit function theorem, there exists a function $x = x(\mu)$ on both sides of μ^* such that $g(\mu(x), x) = 0$ in the vicinity of (μ^*, x^*). Since $x = x(\mu)$ is unique, this solution of the fixed point equation $f^n(\mu, x) = x$ (as well as $f^{2n}(\mu, x) = x$) undergoes nothing but a change of stability. In order to check the stability of this solution, we expand the derivative $\frac{\partial f^n}{\partial x}(\mu, x)$ in the vicinity of (μ^*, x^*), taking into account the $x(\mu)$ dependence. We have

$$\frac{\partial f^n}{\partial x}(\mu^* + d\mu, x(\mu^* + d\mu)) = -1$$
$$+ \left(\frac{\partial^2 f^n}{\partial x \partial \mu} \Big|_{\mu^*, x^*} + \frac{1}{2} \frac{\partial f^n}{\partial \mu} \Big|_{\mu^*, x^*} \frac{\partial^2 f^n}{\partial x^2} \Big|_{\mu^*, x^*} \right) d\mu + \dots .$$

Owing to Condition 2, the expression in parentheses preceding $d\mu$ equals the mixed second derivative that appeared in Condition 3, i.e.,

$$\frac{\partial f^n}{\partial x}(\mu^* + d\mu, x(\mu^* + d\mu)) = -1 - \frac{1}{2} \frac{\partial^2 f^{2n}}{\partial x \partial \mu} \Big|_{\mu^*, x^*} d\mu + \dots . \tag{2.76}$$

Therefore, the solution must become unstable on one side of μ^*, depending on the sign of the mixed derivative. We shall see that this sign agrees with the stability of the period-doubled solution.

We cannot draw from the above analysis any conclusion regarding the fixed point of $f^{2n}(\mu, x) = x$. One has to define another function, in order to perform an analysis similar to what has been done for tangent bifurcation.

Step 2. The existence of a 2-cycle on one side of the bifurcation suggests consideration of the function

$$g(\mu, x) = f^{2n}(\mu, x) - x = f^n(\mu, f^n(\mu, x)) - x. \tag{2.77}$$

The fixed point of f^n remains a fixed point of f^{2n}, thus leading to a zero of g: $g(\mu^*, x^*) = 0$. Now we calculate the derivatives of g:

$$\frac{\partial g}{\partial x} = \frac{\partial f^{2n}}{\partial x} - 1 = f^{n\prime}(f^n(x))f^{n\prime}(x) - 1. \tag{2.78}$$

Taking into account Conditions 1 and 2, one is led to

$$\frac{\partial g}{\partial x} \Big|_{\mu^*, x^*} = (-1)^2 - 1 = 0. \tag{2.79}$$

Next, we have the other partial derivative

$$\frac{\partial g}{\partial \mu} = \frac{\partial f^n(\mu, x)}{\partial \mu} \Big|_{x = f^n(\mu, x)} + \frac{\partial f^n(\mu, x)}{\partial x} \Big|_{x = f^n(\mu, x)} \frac{\partial f^n(\mu, x)}{\partial \mu}. \tag{2.80}$$

At the fixed point, its two terms cancel

$$\frac{\partial g}{\partial \mu} \Big|_{\mu^*, x^*} = 0. \tag{2.81}$$

The vanishing of both partial derivatives $\partial g/\partial x$ and $\partial g/\partial \mu$ at the fixed point shows that g is not the appropriate implicit functional relationship to determine the μ-x dependence. Before looking for a better function, we calculate the higher derivatives of g for later use:

$$\frac{\partial^2 g}{\partial x^2} = \frac{\partial^2 f^{2n}}{\partial x^2} - f^{n\prime\prime}(f^n(x))[f^{n\prime}(x)]^2 + f^{n\prime}(f^n(x))f^{n\prime\prime}(x). \tag{2.82}$$

Taking into account Conditions 1 and 2, one is led again to

$$\left.\frac{\partial^2 g}{\partial x^2}\right|_{\mu^*,x^*} = 0 \tag{2.83}$$

and

$$\begin{aligned}
\frac{\partial^3 g}{\partial x^3} &= \frac{\partial^3 f^{2n}}{\partial x^3} \\
&= f^{n\prime\prime\prime}(f(x))[f^{n\prime}(x)]^3 + 3f^{n\prime\prime}(f(x))f^{n\prime}(x)f^{n\prime\prime}(x) \\
&\quad + f^{n\prime}(f(x))f^{n\prime\prime\prime}(x),
\end{aligned} \tag{2.84}$$

which, at the fixed point, yields

$$\left.\frac{\partial^3 g}{\partial x^3}\right|_{\mu^*,x^*} = -2f^{n\prime\prime\prime}(\mu^*, x^*) - 3[f^{n\prime\prime}(\mu^*, x^*)]^2. \tag{2.85}$$

Recalling the definition of a Schwarzian derivative in Section 2.2.4, we see that the last expression is nothing but the Schwarzian derivative of f^n taken at the fixed point (μ^*, x^*). Thus, taking into account Condition 2, we have

$$\left.\frac{\partial^3 g}{\partial x^3}\right|_{\mu^*,x^*} = 2S(f^n(\mu, x), x)\,|_{\mu^*,x^*}, \tag{2.86}$$

which, in general, is different from zero. The fact that

$$\frac{\partial f}{\partial \mu} = 0, \quad \frac{\partial g}{\partial x} = 0, \quad \frac{\partial^2 g}{\partial x^2} = 0,$$

and only $\partial^3 g/\partial x^3 \neq 0$ at (μ^*, x^*) shows that the implicit relation $g(\mu, x) = 0$ does not provide us with a convenient equation to draw conclusions with regard to the functional dependence of μ on x.

   Step 3. Our difficulty in Step 2 with the function g stems from the fact that any fixed point of $f^n(x) = x$ remains a fixed point of $f^{2n}(x) = x$. It is better to eliminate the trivial fixed point of $f^n(\mu, x) = x$ by defining another function:

$$k(\mu, x) \equiv \frac{g(\mu, x)}{h(\mu, x)} = \frac{f^{2n}(\mu, x) - x}{f^{n}(\mu, x) - x}. \tag{2.87}$$

In order to calculate k and its derivatives at (μ^*, x^*), we have to resolve uncertainties of the type $0/0$ by applying repeatedly L'Hospital's rule. We list only the final results:

$$\begin{aligned}
k(\mu^*, x^*) &= 0, \\
\left.\frac{\partial k}{\partial x}\right|_{\mu^*, x^*} &= 0, \\
\left.\frac{\partial k}{\partial \mu}\right|_{\mu^*, x^*} &= -\frac{1}{2}\left.\frac{\partial^2 f^{2n}}{\partial x \partial \mu}\right|_{\mu^*, x^*} \neq 0, \\
\left.\frac{\partial^2 k}{\partial x^2}\right|_{\mu^*, x^*} &= -\frac{1}{3}S(f^n, x)|_{\mu^*, x^*} > 0.
\end{aligned} \tag{2.88}$$

Therefore, the implicit function theorem guarantees the existence of a function $\mu = \mu(x)$ such that $k(\mu(x), x) = 0$ in the vicinity of (μ^*, x^*). Differentiating $k(\mu(x), x) = 0$ twice, we get

$$\begin{aligned}
\left.\frac{d\mu}{dx}\right|_{\mu^*, x^*} &= 0, \\
\left.\frac{d^2\mu}{dx^2}\right|_{\nu^*, x^*} &= -\frac{2}{3}\left.\frac{S(f^n, x)}{\frac{\partial^2 f^{2n}}{\partial x \partial \mu}}\right|_{\mu^*, x^*}.
\end{aligned} \tag{2.89}$$

We see that $\mu(x)$ exists only on one side of μ^*, depending on the sign of the mixed second derivative in (2.89) and in accordance with (2.76). In other words, it has either a maximum or a minimum in the μ direction, touching the point (μ^*, x^*) (see Fig. 2.17). Therefore, $\mu(x)$, viewed as a function of μ, has two branches, starting from the bifurcation point. This picture looks much like a tangent bifurcation that we have discussed in Section 2.6.2. However, these two cases differ in the stability of the two branches.

Step 4. To determine the stability of the two branches of $\mu(x)$, we expand the first derivative $\partial f^{2n}/\partial x$ near (μ^*, x^*). Due to the first equation (2.88) and the vanishing of $\partial^2 f^{2n}/\partial x^2 |_{\mu^*, x^*}$, which is a direct consequence of Condition 2, first order terms in dx do not appear in the expansion. The final result reads

$$\frac{\partial}{\partial x}f^{2n}(\mu, x)|_{\mu(x^*+dx), x^*+dx} = 1 + \frac{2}{3}S(f^n, x)|_{\mu^*, x^*}(dx)^2 + \dots . \tag{2.90}$$

Therefore, owing to the negativeness of the Schwarzian derivative, the stability criterion holds for small enough dx, independently of its sign, giving rise to a period-doubled regime.

2.7 The Renormalization Group Equation for Period-doubling

Up till now, we have introduced in an empirical way two "universal" exponents related to the period-doubling cascade: the convergence rate δ, the scaling factor α, and a third exponent, the noise scaling factor κ will be introduced in the next section. Their universality has been verified on some one- and two-dimensional dissipative mappings and differential equations, as well as in a few laboratory experiments. In order to justify their universality, we need a mathematical framework which would give, on one hand, a method to calculate these exponents to an arbitrary precision, and, on the other hand, a limitation of the "universality". At this point the renormalization group idea comes to our assistance.

Since what follows involves a little more, although, in essence, very simple mathematics, we will first sketch the relationship of the universal functions and exponents that are to be introduced in this and the subsequent sections. To start with, a nonlinear functional equation

$$\alpha^{-1}g(\frac{x}{\alpha}) = -g \circ g(x) \tag{2.91}$$

will be derived for a universal function $g(x)$ which gives a "fixed point" in the space of all unimodal functions under the renormalization group transformation of period-doublings. The Feigenbaum exponent α figures in (2.91) as a nonlinear "eigenvalue". Then we shall examine the stability of this fixed point function g by adding to it a small perturbation in two ways. At first, we add a deterministic correction $h(x)$:

$$f(x) = g(x) + h(x),$$

and then examine the linear stability with respect to $h(x)$ by expanding it at the fixed point. $h(x)$ will be determined from a linear eigenvalue problem, the largest eigenvalue that exceeds 1 gives the other Feigenbaum exponent δ. If a random noise ξ of strength σ is added, we look at the universal deviation function $D(x)$

$$f(x) = g(x) + \sigma\xi D(x),$$

then the linear eigenvalue problem for $D(x)^2$ leads to the noise scaling factor κ (in fact, its square) as its eigenvalue. We juxtapose these relations in Table 2.1. Since, in a sense, the most significant contribution of physicists to the recent

The Universal Functions	The Universal Exponents
$g(x)$	α — the scaling factor
$h(x)$	δ — the convergence rate
$D(x)$	κ — the noise scaling factor

Table 2.1: The universal functions and exponents

development of the chaotic "fever" consists of the introduction of the renormalization group approach (Feigenbaum, 1978, 1979; Coullet and Tresser, 1978) which has been proved so successful in the modern theory of phase transitions and critical phenomena, we digress for the moment to the latter, confining ourselves to a qualitative exposition.

2.7.1   Renormalization Group Idea in Phase Transition Theory

The two central concepts in the modern theory of phase transitions and critical phenomena are scaling and universality, which can be explained very simply. There are two basic kinds of phase transitions. In the first kind, so-called first order phase transitions, the old and new phases may coexist; they have clear-cut boundaries and there may be latent heat and volume change at the transition. A picture of this kind of transition is the growth of the new phase from a seed embedded in the old phase. In the second kind of transition, the two phases are in "all or none" relation. They cannot coexist, there is no latent heat, volume change, etc. What is present in the old phase, when the system comes very close to the transition temperature T_c (we speak about temperature induced transitions only, there may be geometric analogies as well), is a tendency to change, a growing sphere of influence favouring this tendency. This is called the correlation length ξ which is a singular function of the temperature:

$$\xi(T) \to \infty \quad \text{when} \quad T \to T_c.$$

In order to "visualize" this situation, one imagines a piece of magnet, then T_c may be understood as the Curie temperature (or the critical temperature). The "spheres of influence" are fluctuating patterns of temporally and locally ordered (in all possible directions) spins. These patterns are randomly distributed and nested in each other. There are no clear-cut boundaries and fixed shapes. The only characteristic feature of these fluctuating patterns is that, on an average, their sizes do not exceed the correlation length ξ.

The modern theory of phase transitions and critical phenomena addresses especially to the second kind of transition. Now we are well prepared to grasp the essence of this powerful and beautiful theory. For the construction of a theory, we have three characteristic lengths on hand: the lattice constant a, the "observation" length or the resolution of our "microscope" r, and the correlation length $\xi(T)$. Among these three lengths, only r does not appear to be intrinsic: we have the freedom to change it within a certain range. Suppose we have the inequality

$$a \ll r \ll \xi(T), \tag{2.92}$$

we can increase r, e.g., multiplying it by a power of 2, i.e., we can let $r \to r \times 2^k$, $k = 1, 2, \ldots$, as long as the second strong inequality $\ll$ holds. The most favourable case for doing so occurs, of course, when the system sits exactly at the critical temperature $T = T_c$ and the correlation length diverges:

$$a \ll r \ll \xi(T_c) = \infty, \tag{2.93}$$

for we can perform the rescaling infinitely many times without violating the second inequality. Speaking in terms of the fluctuating patterns, we have now in this picture complete self-similarity. No matter how one adjusts the resolution of the "microscope", one sees almost the same kind of patterns in the field of vision.

Actually, the system is described by a Hamiltonian, or a free energy, or whatsoever function $\mathcal{H}$, depending on the parameters a, T, etc. In practice, the rescaling implies a certain complicated transformation of the function $\mathcal{H}$. We denote this "renormalization" transformation by $\mathcal{R}$. Each time when r is increased, we obtain a new function $\mathcal{H}_k(T)$ from the preceding function:

$$\mathcal{R}\mathcal{H}_{k-1}(T) = \mathcal{H}_k(T), \quad k = 1, 2, \ldots, \tag{2.94}$$

where $\mathcal{H}_0(T)$ is the original function $\mathcal{H}$. If the inequality (2.93) applies more strongly, k can extend to infinity and the a dependence becomes weaker and weaker as the inequality $a \ll r$ holds stronger, implying independence of the theory on details at the lattice constant scale, i.e., universality comes into play. Moreover, there is a good chance that we reach a fixed point $\mathcal{H}^*$ in the space of all possible functions $\mathcal{H}$:

$$\mathcal{H}_k(T_c)|_{k \to \infty} \to \mathcal{H}^*. \tag{2.95}$$

The existence of $\mathcal{H}^*$ must be guaranteed by the underlying physics and the correctness of the model, i.e., the function $\mathcal{H}$ and the renormalization group transformation $\mathcal{R}$. If this is the case, one can linearize the usually highly complicated operator $\mathcal{R}$ in the vicinity of the fixed point $\mathcal{H}^*$. In other words, in the neighbourhood of the fixed point, one can represent the renormalization transformation by a matrix of finite or infinite dimension, depending on the manner of spanning the parameter space for the system. It is remarkable that in most interesting cases the fixed point happens to be a hyperbolic point, i.e., there are expanding as well as contracting eigendirections. Repeated use of the renormalization transformation corresponds to multiplication of the linearized matrices. Clearly, only expanding directions will play an essential role, because their eigenvalues are greater than one. The eigenvalues of the linearized operator along these expanding directions determine various critical exponents which in turn provide the link to experiments. Those parameters which correspond to eigenvalues > 1 are called relevant parameters, while those with eigenvalues < 1 are irrelevant. Another remarkable fact of the critical phenomena theory is that there are only two relevant parameters in typical second order phase transitions.

What we have just mentioned represents a long story on its own. It requires a sound knowledge of theoretical physics to be able to fully appreciate the beauty of the approach. However, period-doubling bifurcation perhaps fur-

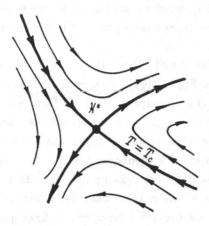

Figure 2.18: The fixed point $\mathcal{H}^*$ and its vicinity.

nishes the simplest realization of the programme, and therefore may serve as a general introduction to renormalization group theory.

If we are not on the critical line $T = T_c$, the renormalization transformation is restricted by the second inequality in (2.92). One can come close to $\mathcal{H}^*$ only up to some finite k; if we keep rescaling beyond this k, we are led away from $\mathcal{H}^*$. In other words, the fixed point $\mathcal{H}^*$ must be a saddle point. This situation is shown schematically in Fig. 2.18. Put simply, the first inequality in (2.92) or (2.93) implies universality, the second inequality scaling invariance. The difference between $\xi(T)$ finite and $\xi(T_c) = \infty$ implies a saddle type fixed point. This is the essence of renormalization group theory.

2.7.2 Derivation of the Feigenbaum Renormalization Group Equation

We start the renormalization group study of period-doubling by looking at the picture of the mappings f, $f^{(2)} = f \circ f \equiv F(2, \mu, x)$, and $f^{(4)} = f^{(2)} \circ f^{(2)} \equiv F(4, \mu, x)$, drawn at a parameter value that is close to the accumulation point μ_∞ for the logistic map (2.3) (see Figs. 2.19 (a), (b) and (c)). We have drawn a few auxiliary lines in order to show how the lower picture can be obtained qualitatively from the upper one by graphically fixing some key points and then linking them appropriately.

We see that in Fig. 2.19(b) the central part (in the dashed-line box) looks much like the entire Fig. 2.19(a), inverted and reduced. The same remark applies to Fig. 2.19(c) with respect to (b). In fact, one can concentrate at the vicinity of one of the fixed points and include the parameter dependence as well. In Figs. 2.20 (a) -(d) we have redrawn the not-well-proportioned figures from the original paper of Feigenbaum (1978). In Figs. 2.20(a) and (b), the functions $f^{(2^{n-1})}$ and $f^{(2^n)}$ are drawn partially, with the origin of coordinates shifted to the center of the graph. They are taken at the same superstable parameter value μ_n for the $f^{(2^n)}$ function. A fixed point of $f^{(2^n)}$ manifests itself as a local period 2 for $f^{(2^{n-1})}$ (the boundary of the hatched square in Fig. 2.20(a)). Now shift the parameter to the next superstable value μ_{n+1} and draw the same function $f^{(2^n)}$ (Fig. 2.20(c)). The story repeats itself. A fixed point in Fig. 2.20(b) now becomes a local 2-cycle (labelled as 1 and 2 in Fig. 2.20(c)). It is sufficient to rescale both the function and the argument to obtain Fig. 2.20(d) which looks much like Fig. 2.20(a). In fact, the coordinates in Fig. 2.20(c) have been reversed in getting (d). This shows up in the negative sign of the scaling factor α_n. It has been verified numerically that, when $n \to \infty$, α_n tends to a universal constant α.

These observations suggest the following transformation:

1. Period-doubling: changing a function F to the composite function $F \circ F$;

2. μ-shifting: changing the parameter from the current superstable value μ_n to the next period-doubled cycle μ_{n+1};

3. Rescaling and inverting: the function $F \to -\alpha F \circ F$ and the argument $x \to -x/\alpha$ (if the function $F(x)$ is symmetric, the $-$ sign may be dropped) with a (as yet unknown) scaling factor α.

We denote this transformation by $\mathcal{T}$:

$$\mathcal{T} F(n, \mu_n, x) = -\alpha F\left(2n, \mu_{n+1}, -\frac{x}{\alpha}\right). \tag{2.96}$$

In order to simplify the derivation, we pursue the superstable periods, i.e., the condition $F' = 0$ is replaced by $F^{(2)'} = 0$ at every application of the operator

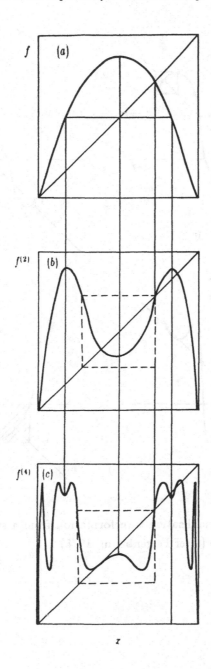

Figure 2.19: Comparison of f, $f^{(2)}$ and $f^{(4)}$.

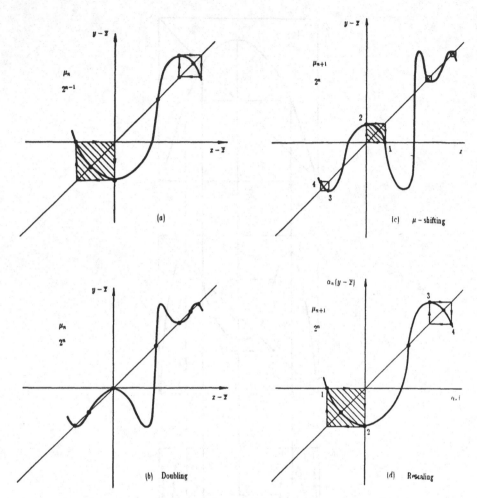

Figure 2.20: The renormalization transformation from a superstable orbit to the period-doubled one (after Feigenbaum, 1978).

$\mathcal{T}$. Taking the original map $f(\mu_n, x)$ at some fixed μ_n as the initial function, we have explicitly

$$
\begin{aligned}
\mathcal{T}f(\mu_n, x) &= -\alpha f(\mu_{n+1}, f(\mu_{n+1}, -\frac{x}{\alpha})) \\
&= -\alpha F(2, \mu_{n+1}, -\frac{x}{\alpha}), \\
\mathcal{T}^2 f(\mu_n, x) &= (-\alpha)^2 F(2^2, \mu_{n+2}, \frac{x}{(-\alpha)^2}),
\end{aligned}
\tag{2.97}
$$

...   .

Feigenbaum has given some plausible arguments to conjecture the existence of a universal limit $g(x)$, i.e., a limit which does not depend on the starting function $f(x)$, apart from a certain trivial rescaling of coordinates:

$$
\lim_{k \to \infty} (-\alpha)^k F(2^k, \mu_{n+k}, \frac{x}{(-\alpha)^k}) = g(x).
\tag{2.98}
$$

This conjecture has been proved later (Lanford, 1982; Eckmann and Wittwer, 1987). From the definition of the composite function F [see (2.14) in Section 2.2.1], it follows that

$$
F(2^k, \mu_{n+k}, \frac{x}{(-\alpha)^k}) = F(2^{k-1}, \mu_{n+k}, F(2^{k-1}, \mu_{n+k}, \frac{x}{(-\alpha)^k}));
\tag{2.99}
$$

taking the limit $k \to \infty$ on both sides of (2.99) yields the renormalization group equation of Feigenbaum:

$$
g(x) = -\alpha g(g(-\frac{x}{\alpha})).
\tag{2.100}
$$

The normalization of $f(\mu, 0) = 1$ and the superstable condition (2.35) imply for (2.100) the boundary conditions

$$
\begin{aligned}
g(0) &= 1, \\
g'(0) &= 0.
\end{aligned}
\tag{2.101}
$$

It is worth mentioning that the conditions (2.101) alone do not determine uniquely the solution of (2.100). At least the behaviour of $g(x)$ near $x = 0$, i.e., near its maximum, should be given. For example, the series solution

$$
g(x) = 1 + Ax^z + Bx^{2z} + Cx^{3z} + \ldots + Nx^{nz} + \ldots
\tag{2.102}
$$

will lead to different solutions to (2.100) for different choices of z. We shall return to the problem of boundary conditions for this type of equations in the next chapter in connection with the scaling properties of so-called period-n-tupling sequences, a straightforward generalization of period-doubling. No analytical solution is known for (2.100) with the boundary conditions (2.101). However, if the second condition in (2.101) is replaced by $g'(0) = 1$, an analytical solution does exist. We shall discuss this case when dealing with the theory of intermittency (see Section 2.6.3).

It should be emphasized that by introducing the fixed point function $g(x)$ through the renormalization transformations we have "lifted" the original fixed point problem for a unimodal function f to a fixed point problem in the space of all possible unimodal functions. This comparison may be seen more clearly from the following Table 2.2: Equation (2.100) may be viewed as a nonlinear

f – function acting on number z	$\mathcal{T}$ – operator acting on function f
$f(x_n) \;\to\; x_{n+1}$	$\mathcal{T} f \;\to\; f^{(2)}$
$f^{(n)}(x)\,\|_{n\to\infty}\to$ fixed point x^*	$\mathcal{T}^{(n)}f\,\|_{n\to\infty}\to$ fixed function $g(x)$
Independent on initial choice of x_0	Independent on initial choice of f

Table 2.2: From fixed point x^* to fixed function $g(x)$.

eigenvalue problem with the eigenvalue α. Both the universal function $g(x)$ and α must be obtained by solving (2.100). The fixed function $g(x)$ is a saddle type fixed point in the functional space of unimodal functions, similar to the situation in phase transition theory that we have briefly reviewed in Section 2.7.1. Therefore, one must be careful in developing any systematic approximation to its solution.

However, if an algebraic manipulation language such as REDUCE or MAC-SYMA is available on hand, one can construct a systematic approximation to the solution of (2.100) by first truncating the expansion (2.102) of $g(x)$ at a given order n with undetermined coefficients A, $B, C, \dots$, N and then substituting it into both sides of (2.100). By equating coefficients of the same powers of x, one gets a system of algebraic equations for α, A, B, C, ..., which is then solved step by step numerically. The first two equations for α and A can be solved by hand. In fact, in this approximation, we have $\alpha = 1 + \sqrt{3} = 2.73205\ldots$. Next, these values of α, A and $B = 0$ are used as initial values to solve the next system of equations for α, A, and B, etc. This method appears to converge quickly and is readily programmed. In order to demonstrate its efficiency, we give in Table 2.3 the numerical results for the Feigenbaum period-doubling case, where the last row referred to as "exact" was taken from Feigenbaum (1978) [see also Lanford (1982) for more terms].

n in Eq. (2.102)	α	A	B	C	D
2	2.73205	-1.366025			
3	2.534030	-1.522426	0.1276133		
4	2.478897	-1.521880	0.0729350	0.0455088	
5	2.50316	-1.52779	0.10533	0.02631	-0.00334
"exact"	2.502907	-1.527633	0.104815	0.0267056	-0.00352

Table 2.3: Convergence of the series solution for period-doubling.

2.7.3 Linearized Renormalization Group Equation and Convergence Rate δ

The Feigenbaum renormalization group equation (2.100) determines a fixed point $g(x)$ in the functional space of unimodal maps. In order to test the stability of this fixed point, let us try to approach $g(x)$ by a sequence of functions $\{g_n(x)\}$, satisfying the set of equations

$$g_{n-1}(x) = -\alpha g_n\left(g_n\left(\frac{x}{\alpha}\right)\right). \tag{2.103}$$

Clearly, if this sequence converges, then $g_n(x) \rightarrow g(x)$ as n goes to infinity, and we shall get a solution of (2.100). The following procedure repeats in spirit the linear stability analysis for a fixed point of a one-dimensional map (see Section 2.3.1), although all the calculations are now being carried out in a functional space.

If we are close to the fixed point function $g(x)$, we can write

$$g_n(x) = g(x) + h_n(x), \tag{2.104}$$

where $h_n(x)$ is a "small" function. Substituting (2.104) into (2.103) and expanding the right-hand side using the chain rule of differentiation, one finds

$$g(x) + h_{n-1}(x) = -\alpha[g(g(\frac{x}{\alpha})) + g'(g(\frac{x}{\alpha}))h_n(\frac{x}{\alpha}) + h_n(g(\frac{x}{\alpha})) + \cdots],$$

We have neglected here high order terms in h_n, retaining only the linear ones. Removing the first terms on both sides in accordance with the fixed point equation (2.100), and assuming

$$h_n(x) = \frac{h(x)}{\delta^n}, \tag{2.105}$$

where $h(x)$ is an as yet unknown function of the order of 1 and δ must be a constant greater than 1 to ensure the smallness of the $h_n(x)$, we arrive at a linear eigenvalue problem

$$\mathcal{L}h(x) = \delta h(x) \tag{2.106}$$

with δ being the eigenvalue and $\mathcal{L}$ the newly defined linear operator

$$\mathcal{L}h(x) = -\alpha[h(g(\frac{x}{\alpha})) + g'(g(\frac{x}{\alpha}))h(\frac{x}{\alpha})]. \tag{2.107}$$

Note that the definition of the linear operator $\mathcal{L}$ depends on α and the function $g(x)$, which should have been known by solution of (2.100), e.g., by the method mentioned in the last subsection. Equation (2.106) can be solved by a similar procedure; its numerical solution leads to a universal function $h(x)$ (up to normalization) and the Feigenbaum convergence rate $\delta = 4.66920\ldots$, which is indeed greater than one.

Everything looks quite nice so far. We have a nonlinear functional equation (2.100) the "eigenvalue" of which yields the scaling factor α. Its linearization has led to an eigenvalue problem which determines the convergence rate δ. Equation (2.106), however, is a linear eigenvalue problem in a functional space, so that there might be infinitely many eigenvalues. In fact, Feigenbaum (1979) has constructed another set of eigenfunctions

$$\psi_\rho(x) = [g(x)]^\rho - x^\rho g'(x). \tag{2.108}$$

The first term is the ρ-th power of $g(x)$. Therefore, it is reasonable to take $\rho \geq 1$. Applying the linear operator $\mathcal{L}$ to ψ_ρ, we get

$$\mathcal{L}\psi_{\rho(x)} = -\alpha\{[g(g(\frac{x}{\alpha}))]^{\rho} - (\frac{x}{\alpha})^{\rho}\frac{d}{dx}g(g(\frac{x}{\alpha}))\}.$$

Replacing the nested $g(g(x/\alpha))$, according to the renormalization group equation (2.100), one arrives at

$$\mathcal{L}\psi_{\rho}(x) = (-\alpha)^{1-\rho}\psi_{\rho}(x). \tag{2.109}$$

Indeed, the ψ_{ρ} are eigenfunctions with eigenvalues $(-\alpha)^{1-\rho}$. With the known values of α and $\rho > 1$, all these eigenvalues are less than 1, and therefore irrelevant. Are there other eigenfunctions? Feigenbaum (1979) conjectured that the functions $h(x)$ and ψ_{ρ} comprise the complete set of eigenfunctions of (2.100) with δ being the only relevant eigenvalue. Numerical studies seem to support this conjecture. There have been several computer-assisted proofs of the Feigenbaum conjecture on the hyperbolic nature of the fixed point, see, e.g., Lanford (1982), Eckmann and Wittwer (1987).

2.7.4 A Generalized Renormalization Group Equation

The renormalization group equation (2.100) has been obtained at one single value of the parameter, namely, the accumulation point μ_{∞}, by realizing a limiting process through the superstable parameter values. The exponent α describes the scaling property in the phase space, i.e., along the x direction. In contrast, the linearized equation (2.106) has been obtained by looking for a systematic approach towards $g(x)$, and the eigenvalue δ happens to be the exponent that describes the scaling property along the parameter axis. This result suggests the possibility of scaling μ and x simultaneously along the x and control parameter axes. Indeed, Young and coworkers (Liu, Lo and Young, 1984; Young, 1985; Liu and Young, 1987) have succeeded in deriving a generalized renormalization group equation

$$\psi(\frac{\lambda}{\delta}, \psi(\frac{\lambda}{\delta}, -\frac{y}{\alpha})) = -\frac{1}{\alpha}\psi(\lambda, y), \tag{2.110}$$

where λ is the deviation of the control parameter μ from the accumulation point μ_{∞} and y is the deviation of x from x_{∞} that makes $f(\mu_{\infty}, x)$ a maximum:

$$\lambda = \mu - \mu_{\infty}, \quad y = x - x_{\infty}.$$

If one expands ψ in terms of λ:

$$\psi(\lambda, y) = g(y) + \lambda h(y) + O(\lambda^2), \tag{2.111}$$

then $g(y)$ and $h(y)$ satisfy the Feigenbaum equation (2.100) and the linearized equation (2.106), respectively. Skipping the derivations, we note only that (2.110) holds *around* μ_∞, while the Feigenbaum equation (2.100) holds right *at* μ_∞. The difference originates from the fact that the scaling property underlying (2.100) pertains to a sequence of isolated parameter values, say, the superstable μ_n, whereas that underlying (2.110) pertains to a sequence of μ ranges.

2.8   Scaling Properties Related to External Noise

External noise is an inevitable factor in modelling and understanding any phenomenon in nature. Experiments are subject to ambient fluctuations and instrumental noise. Numerical experiments can hardly be rid of round-off errors, if performed for a long enough time. All secondary factors, that have not been taken into account in a mathematical model, may be thought of as some noise affecting the chosen variables. In a sense, the requirement of stability for one or another solution of a dynamical system is equivalent to its immunity to implicit perturbations. In order to obtain a complete understanding of a model, one also has to analyze the consequence of noise, when it is included explicitly.

Moreover, in the theory of chaotic phenomena, external noise plays a more positive role than just being a notorious, but unavoidable villain. This suggestion originates again from a comparison with the theory of phase transitions and critical phenomena. A phase transition takes place when a certain order parameter becomes different from zero, signaling the onset of a new order. In a ferromagnet, it is the macroscopic magnetization M that plays the role of the order parameter. It can appear either spontaneously at the Curie temperature T_c without the presence of any external magnetic field, or it can be induced by an external field H at some value $T \neq T_c$. A complete phase diagram of a magnet must be drawn at least in the $T \sim H$ plane. In general, the magnetic field is replaced by a "conjugate field" coupled to the order parameter. It so happened that the most successful scaling theory of critical phenomena was developed in terms of temperature and the conjugate field. These two quantities are the relevant parameters in the renormalization group analysis of phase transitions.

In the renormalization group theory of period-doubling, that we have developed in the last section, the control parameter μ clearly figures as the analogue of temperature. Do we have a conjugate field in this context? We have emphasized repeatedly that chaos is not simply disorder. In Chapter 6, we shall learn how to distinguish quantitatively chaos from random noise. There exists a kind of global order in the organization of chaos in the parameter space, as well as in the characteristics of an entire (infinitely long!) chaotic trajectory. Nevertheless, if one looks at a finite segment of an orbit, in the vicinity of a particular transition, chaos manifests itself much like a kind of disorder. In this sense, external noise may well play the role of a conjugate field, inducing disorder in the system. This has been the motivation for the scaling theory of external noise that we are going to describe (Crutchfield and Huberman, 1980; Shraiman, Wayne and Martin, 1981; Crutchfield, Nauenberg and Rudnick, 1981).

In order to include external noise, a random source term is inserted into the nonlinear mapping (2.1), which is thus transformed into a discrete Langevin equation

$$x_{n+1} = f(\mu, x_n) + \sigma \xi_n, \tag{2.112}$$

where ξ_n are random numbers obeying a certain statistical distribution with the constraints

$$\overline{\xi_n} = 0, \qquad \overline{\xi_n \xi_m} = \delta_{nm}. \tag{2.113}$$

Now let us introduce a new universal exponent κ, related to the system's response to external noise. We shall proceed in a hand-waving manner by extending the approach discussed in the last section. We know that the closer one is to $x = 0$, the better a unimodal function $f(x)$ is represented by the universal function $g(x)$. The same holds for its iterate $f^{(2)}$. We juxtapose what has just been said in the following two lines

$$\begin{aligned}
f(x) &\to g(x), \\
f^{(2)}(x) &\to g^{(2)}(x) = \alpha^{-1} g(\alpha x),
\end{aligned} \tag{2.114}$$

the equality in the last line being a direct consequence of the renormalization group equation (2.100). Now suppose that small noise near $x = 0$ will
modify (2.114) into

$$f(x) + \xi \to g(x) + \xi D(x),$$
$$(f + \xi)^{(2)}(x) \to (g + \xi D)^{(2)}(x), \tag{2.115}$$

where $D(x)$ is a new scaling function, the *dispersion*. We require that the
right-hand side of the second line satisfies the same type of equation as that in
the second line of (2.114), namely,

$$(g + \xi D)^{(2)}(x) = \alpha^{-1}[g(\alpha x) + \xi \kappa D(\alpha x)], \tag{2.116}$$

where we have introduced a new noise scaling factor κ, because for no apparent
reason one factor α would be enough to scale the two functions $g(x)$ and $D(x)$.
Writing down the left-hand side of (2.116) explicitly and expanding it to the
first order in ξ, we have

$$\begin{aligned}(g + \xi D)^{(2)}(x) &= g(g(x) + \xi_1 D(x)) + \xi_2 D(g(x) + \xi_1 D(x)) \\ &= \alpha^{-1}g(\alpha x) + \xi_1 g'(g(x))D(x) + \xi_2 D(g(x)) \\ &\quad + \dots, \end{aligned} \tag{2.117}$$

where we have used (2.100) and taken into account the fact that, in general,
at two iterations we encounter two different random numbers ξ_1 and ξ_2. We
cannot equate the last expression to the right-hand side of (2.116) as it stands.
However, from the statistical property (2.113), we know that the squares of the
coefficients at the random numbers must satisfy the relation

$$[g'(g(x))D(x)]^2 + [D(g(x))]^2 = \alpha^{-2}[\kappa D(\alpha x)]^2. \tag{2.118}$$

Recalling the same linear operator $\mathcal{L}$ of (2.107), now acting on $D(x)^2$,

$$\mathcal{L}D(x)^2 = \alpha^2\{D(g(x))^2 + [g'(g(x))D(x)]^2\}, \tag{2.119}$$

we rewrite (2.118) as a linear (in $D(x)^2$) eigenvalue equation

$$\mathcal{L}D(x)^2 = \kappa^2 D(x)^2. \tag{2.120}$$

It is interesting to note that this equation has precisely the same structure
as the linear eigenvalue equation (2.106) for δ and $h(x)$. They can be solved
numerically, using one and the same program.

Employing a polynomial interpolation for $D(x)$, one finds (Crutchfield, Nau-
renberg and Rudnick, 1981):

$\kappa = 6.61903\ldots$

To conclude this section, we summarize the role of noise as follows. Firstly, external noise smears the details in the high order bifurcation structure, including both period-doubling cascades and inverse sequences of chaotic bands. Roughly speaking, in order to see one more bifurcation, the noise level must be reduced by a factor of κ. This explains the meaning of the noise scaling exponent κ. Secondly, the presence of noise makes the bifurcation diagram fuzzy and lowers the accumulation point μ_∞ slightly, as should be expected from the role of noise as a disordering field.

2.9 Renormalization Group Equation for Intermittency

The scaling properties of the intermittent transition from chaos to periodic motion, that takes place at tangent bifurcations (see Section 2.6), can be described by a renormalization group equation. There is a widespread statement in the literature that the renormalization group equation for intermittency is the same as the Feigenbaum equation (2.91) or (2.100) for period-doubling with a mere change of boundary conditions. We shall see that this statement is not entirely precise. In a sense, the renormalization group equation for intermittency bears more resemblance to the scaling theory of phase transitions and a detour to the latter will ease the discussion afterwards.

2.9.1 The Scaling Theory and Generalized Homogeneous Functions

The scaling theory of phase transitions, developed by L. Kadanoff and others in the late 1960s, was a precursor of the renormalization group approach of K. Wilson. While mastery of the renormalization group technique requires a sound training in mathematics and even in quantum field theory, the scaling theory is easy and transparent. For this reason, it remains a source of inspiration for the development of new theories. In one of its formulations, the scaling theory makes use of the notion of generalized homogeneous functions.

We first recall the notion of a homogeneous function. A function $f(x)$ is said to be homogeneous of order m if it acquires a multiplier l^m under the scaling transformation $x \to lx$, i.e.,

$$f(lx) = l^m f(x). \tag{2.121}$$

Differentiating both sides of this equation and then letting $l = 1$, one finds the Euler equation

$$x \frac{df(x)}{dx} = m f(x), \tag{2.122}$$

which determines the function $f(x) = C x^m$ apart from a constant C. The definition (2.121) generalizes to functions of two or more variables:

$$f(lx, ly, \ldots) = l^m f(x, y, \ldots), \tag{2.123}$$

when the Euler equation reads

$$x \frac{\partial f}{\partial x} + y \frac{\partial f}{\partial y} + \ldots = m f. \tag{2.124}$$

If on the left-hand side of (2.123) it is required to scale the variables by different powers of l, in order to get the factor l^m on the right-hand side, f is called a generalized homogeneous function. Interchanging the two sides, we write it as

$$f(x, y, \ldots) = l^{-m} f(l^\rho x, l^\nu, \ldots). \tag{2.125}$$

Generalized homogeneous functions have many nice properties, e.g., their derivatives, integrals and Fourier transform are also generalized homogeneous functions of different orders [7].

In order to see how these generalized homogeneous functions appear in the scaling theory of phase transitions, let us take a piece of magnet at a temperature T close to the Curie point T_c when there may be a magnetic field H (the conjugate field, mentioned in Section 2.8). Usually one introduces two normalized dimensionless parameters $t = |T - T_c|/T_c$ and h. Now recall the picture of fluctuating spin patterns, described in Section 2.7.1. One can imagine the magnet to be divided into small blocks of linear size l. If we assume that each block contains one averaged spin and adjust the interaction between

[7]See, e.g., A. Hankey, and H. E. Stanley (1972), *Phys. Rev.*, **B6**, 3515.

them in such a way as to cast it back into the form of the original interaction, we would have the same kind of fluctuating pattern that we have had with the original spins at a different temperature t_l and magnetic field h_l. Now we have two ways to estimate, say, the free energy per spin: either in terms of the original spins, when we get a function $F(t, h)$, or in terms of the averaged spin-blocks, when we get the same function $F(t_l, h_l)$. The unknown functional form of F must be the same, due to the similarity of the two patterns, but different arguments would lead to different numerical values. However, we do know the ratio of these two values, because the number of our original spins in a block is proportional to the volume of the block l^d, where d is the spatial dimension of the magnet. Therefore, we have

$$F(t_l, h_l) = l^d F(t, h).$$

Not knowing the dependence of t_l and h_l on l, we make the simplest assumption that

$$t_l = l^\rho t, \qquad h_l = l^\nu h, \tag{2.126}$$

which may be verified only by comparison with experiments or by a more fundamental theory. Therefore, the unknown function F must be a generalized homogeneous function

$$F(t, h) = l^{-d} F(l^\rho t, l^\nu h). \tag{2.127}$$

The relation (2.127) alone yields many concrete quantitative results. For example, since the temperature T has not been fixed, we can relate it to the choice of block size l in such a way as to keep $l^\rho t = 1$, i.e., let $l = t^{-1/\rho}$. From (2.127) we find

$$F(t, h) = t^{d/\rho} F(1, \frac{h}{t^{\nu/\rho}}). \tag{2.128}$$

Assuming that $F(1, 0)$ is not singular, as is usually the case, we get the singularity of $F(t, 0)$, when t goes to zero, i.e., when $T \to T_c$ in the zero magnetic field. Since all physical quantities of interest are derivatives of F with respect to t and h, we can relate their singularities to only the two unknown exponents ρ and ν (and the dimension d) that figure in (2.127). If, furthermore, we succeed in calculating ρ and ν from some model, then we would have solved the central problem of phase transition theory, i.e., calculate the critical exponents that characterize the singularities. We shall see how this programme is carried out in the context of intermittency in the next section. For the time being, we make only the anticipating remark that the idea just discussed appears to be crucial for an understanding of the "singularity point of view" in studying high order information dimensions which in turn leads to the thermodynamic formalism for describing strange sets (see Chapter 6).

2.9.2 Scaling Theory for Intermittency

Now let us return to the intuitive explanation of intermittency as a passage along narrow corridors with intervening random jumps between corridors. Look at the magnification of a corridor in Fig. 2.16. Shifting the origin of coordinates to the center, i.e., the location of the would-be fixed point y^*, one can write the map in the form

$$x_{n+1} = x_n + b|x_n|^z + \epsilon. \tag{2.129}$$

This is nothing but (2.71) with $y_n - y^*$ replaced by x_n and $a(\mu_c - \mu)$ denoted by ϵ. It applies to the vicinity of any tangent bifurcation to period k orbits, and is not restricted to $F(3, \mu, x)$ only. Tangent bifurcation takes place when $\epsilon = 0$. For infinitesimal $\epsilon > 0$, a corridor like the one shown in Fig. 2.16 forms near each of the k would-be fixed points of $F(k, \mu, x)$. It is sufficient to concentrate on one of these corridors. In general, one might have included in (2.129) a random source $\sigma\xi$ in order to simulate external noise (cf. (2.112)).

Suppose an analysis like that performed in Section 2.6.3 has given us the passage time n through the corridor as a function of ϵ and σ, the only two relevant parameters in the theory, i.e., a formula similar to (2.73) with the dependence on σ taken into account. We do not as yet know the exact form of $n(\epsilon, \sigma)$, but this does not prevent us from developing a scaling theory in the following manner.

To start with, we consider first the usual case $z = 2$. Suppose that there is a way to present two successive iterations $x \rightarrow x' \rightarrow x''$ of the map (2.129) as a single iteration $x \rightarrow x''$ of the same map with redefined parameters ϵ' and σ'. Of course, this can be done only as long as x remains small and high powers of x may be neglected. If we count the number of steps needed to pass through the corridor using the new map, the functional form of $n(\epsilon', \sigma')$ would not change, but the numerical value would be just half of the previous number:

$$n(\epsilon', \sigma') = \frac{1}{2} n(\epsilon, \sigma),$$

because we have taken two old steps as one new step. We could have grouped l old steps into one new step and got

$$n(\epsilon, \sigma) = l\, n(\epsilon', \sigma'). \qquad (2.130)$$

For the sake of brevity, let us call this regrouping process an l-transformation. Not knowing the explicit dependence of ϵ' and σ' on ϵ and σ, we make the scaling assumptions

$$\epsilon' = l^{\rho}\, \epsilon, \qquad \sigma' = l^{\nu}\, \sigma. \qquad (2.131)$$

Thus (2.130) acquires the form of a generalized homogeneous function

$$n(\epsilon, \sigma) = l\, n(l^{\rho} \epsilon, l^{\nu} \sigma). \qquad (2.132)$$

As we have the freedom to shrink the corridor in a convenient manner, we keep

$$l^{\rho}\, \epsilon = 1,$$

when changing ϵ, i.e.,

$$l = \epsilon^{-1/\rho}.$$

Inserting the last expression for l into (2.132), we find the scaling form for n:

$$n(\epsilon, \sigma) = \epsilon^{-\frac{1}{\rho}} \Phi(\sigma / \epsilon^{\frac{\nu}{\rho}}). \qquad (2.133)$$

The new (unknown) function $\Phi(x)$ stands for $n(1, x)$. If $n(1, 0)$ is not singular, then (2.133) determines the singularity of $n(\epsilon, 0)$ when ϵ goes to zero, i.e., how the number of steps diverges with vanishing ϵ in the absence of external noise. The two exponents $1/\rho$ and ν/ρ will be calculated by solving and linearizing the corresponding renormalization group equation (see Section 2.9.4). Looking ahead, we note that the two leading eigenvalues of the linearized renormalization group equation at its fixed point in the direction of the two relevant parameters, λ_{ϵ} and λ_{σ}, describe the stretching rate in (2.124), i.e.,

$$\lambda_{\epsilon} = l^{\rho}, \qquad \lambda_{\sigma} = l^{\nu}.$$

Taking the logarithm of these relations, the two exponents will be given by

$$\frac{1}{\rho} = \frac{\log l}{\log \lambda_\epsilon}, \qquad \frac{\nu}{\rho} = \frac{\log \lambda_\sigma}{\log \lambda_\epsilon}. \tag{2.134}$$

The number l, reflecting how many old steps have been grouped into a single new step, must be cancelled by a similar l-dependence of the eigenvalues and drop out from the final expression for the physical exponents. We shall see how this happens in the following sections.

2.9.3   Renormalization Group Equation and Its Exact Solution

What will happen if one executes the l-transformation indefinitely? The map $f(x)$ [see (2.129)] may reach a universal function $g(x)$ that will no longer change under the l-transformation. Postulating the existence of such a function, it must satisfy the fixed point condition in the functional space of all functions, similar to (2.129):

$$\alpha^{-1} g(\alpha x) = g^{(l)}(x), \tag{2.135}$$

where α is an as yet undetermined scaling factor. Equation (2.135) must be supplemented by boundary conditions which follow naturally from an inspection of Fig. 2.16. Firstly, we normalize $g(x)$ at the origin in the same manner as $f(0)$ does when $\epsilon \to 0$. Secondly, we are dealing with tangent bifurcations, so that the slope must be 1 at the origin. Therefore, the boundary conditions for (2.135) become

$$\begin{aligned} g(0) &= 0, \\ g'(0) &= 1. \end{aligned} \tag{2.136}$$

In the past,, people have only worked on the case $l = 2$ (Hirsch, Nauenberg and Scalapino, 1982; Hu and Rudnick, 1982a), so that (2.135) looked exactly like the Feigenbaum renormalization group equation (2.91), or (2.100) for period-doubling with a mere change of boundary conditions from (2.101) to (2.136). However, the similarity is illusive, as one could work with any l and put the renormalization group equation for intermittency in the form of that for period-l-tupling bifurcations (see Section 3.4.3). In the Feigenbaum case, the number 2 (or l) is essential to the physics and all exponents do depend on it implicitly. In the theory of intermittency, l is merely a rescaling constant and the exponents must not depend on it.

In a sense, the renormalization group equation for intermittency is not so profound as that for period-doubling. Moreover, there exists no sound mathematical theory about the existence of its solution, etc. However, as an advantage over its period-doubling counterpart, a simple analytical solution is known for (2.135) (Hirsch, Nauenberg and Scalapino, 1982a) which we will consider next.

Let us apply functional composition to the linear fraction

$$g(x) = x/(1 - a\,x), \qquad (2.137)$$

where a is an arbitrary constant. We have

$$g \circ g(x) = x/(1 - 2a\,x),$$
$$g \circ g \circ g(x) = x/(1 - 3a\,x),$$
$$\cdots \quad \cdots;$$

and, in general,

$$g^{(l)}(x) = \frac{x}{1 - l\,a\,x}. \qquad (2.138)$$

It is clear now that by letting $\alpha = l$, the function (2.137) does satisfy the renormalization equation (2.135). The boundary conditions (2.136) have been satisfied by the choice of $g(x)$ at the outset.

The foregoing discussion applies to the particular case $z = 2$. For general z, however, the analysis goes much the same way. In order to find a more general solution of (2.135), we write (2.137) in the form (Hu and Rudnick, 1982a)

$$g(x) = \frac{1}{G(x) - a}, \qquad (2.139)$$

with $G(x) = 1/x$. Now forget for the time being this specific form of G and try to apply functional compositions. We get

$$g^{(l)}(x) = \frac{1}{G(x) - l\,a}. \qquad (2.140)$$

We see that the iteration of $g(x)$ corresponds to a shift transformation of $G(x)$:

$$x' = g(x),$$
$$G(x') = G(x) - a. \tag{2.141}$$

Let us look for a more general link between $g(x)$ and $G(x)$, that would generalize (2.139) and preserve (2.141). Define

$$g(x) = \Phi(G(x) - a),$$
$$G(x) = \Phi^{-1}(x), \tag{2.142}$$

with a function Φ which for the moment is undetermined. We have

$$x' = g(x),$$
$$G(x') = \Phi^{-1}(x') = \Phi^{-1}(g(x)) = G(x) - a,$$
$$g(x') = g^{(2)}(x) = \Phi(G(x') - a),$$
$$\cdots \quad \cdots$$
$$x^{(l)} = g^{(l)}(x),$$
$$G(x^{(l)}) = G(x) - l\,a,$$
$$g^{(l)}(x) = \Phi(G(x) - l\,a).$$

Therefore, in order to satisfy the renormalization group equation

$$g(\alpha x) = \alpha g^{(l)}(x),$$

we must require

$$\Phi(G(\alpha x) - a) = \alpha \Phi(G(x) - l\,a). \tag{2.143}$$

If $G(x)$ is a homogeneous function

$$G(x) = l\,G(\alpha x), \tag{2.144}$$

then (2.143) implies that

$$\Phi(G(\alpha x) - a) = \alpha \Phi(l[G(\alpha x) - a]), \tag{2.145}$$

i.e., $\Phi(x)$ is also a homogeneous function. Let it be a homogeneous function of order $-m$:

$$\Phi(l\,x) = l^{-m}\,\Phi(x), \tag{2.146}$$

then the renormalization group equation (2.135) holds for

$$\alpha = l^m. \tag{2.147}$$

We know from (2.122) that the condition (2.146) determines the function $\Phi(x)$ up to a constant factor, which, by the way, may be absorbed into the arbitrary constant a. Therefore, a general solution to (2.135) would be

$$g(x) = \frac{1}{\left(\frac{1}{x^{1/m}} - a\right)^m} - \frac{x}{(1 - a\,x^{1/x})^m}. \tag{2.148}$$

We see that the boundary conditions (2.136) hold for a positive choice of m. However, the conditions (2.136) alone do not fix the solution uniquely, just as it has happened with the period-doubling case (see what has been said following (2.102)). In order to determine the value of m, the local behaviour of $g(x)$ near $x = 0$ is required to match that of the map (2.129) which was the starting point of the entire discussion.

Expanding (2.148) at the origin, we get

$$g(x) = x + m\,a\,x^{\frac{1}{m}+1} + \frac{m(m+1)}{2}a^2 x^{\frac{2}{m}+1} + \dots. \tag{2.149}$$

A comparison with (2.129) yields

$$m = \frac{1}{z - 1}. \tag{2.150}$$

The requirement of m being positive implies that $z > 1$. We will write the final form of the renormalization group equation and its solution in the form

$$
\begin{aligned}
g(\alpha x) &= \alpha g^{(l)}(x), \qquad \alpha = l^{1/(z-1)}, \\
g(x) &= x\big(1 - a\,x^{z-1}\big)^{-1/(z-1)}.
\end{aligned}
\tag{2.151}
$$

2.9.4 Linearized Renormalization Group Equations and the Exponents

Since we have an analytical expression for the fixed point function (2.151) at our disposal, all the coefficient functions in the linearized renormalization group equation may be calculated explicitly. In performing the linearization, one can consider either a deterministic deviation $h(x)$ or a stochastic disturbance $\xi D(x)$, as we have done in Sections 2.7.3 (Eq. (2.104)) and 2.8 (Eq. (2.115)), respectively. The two cases are largely parallel. We will list only the final results (Hu and Rudnick, 1982a):

1. Deterministic deviation:

$$h(x) = \frac{1 - [1 - (z-1)ax^{z-1}]^{(2z-1)/(z-1)}}{(2z-1)ax^{z-1}[1 - (z-1)ax^{z-1}]^{z/(z-1)}},$$

$$\lambda_\epsilon = l^{z/(z-1)}.$$

2. Stochastic deviation:

$$D(x)^2 = \frac{1 - [1 - (z-1)ax^{z-1}]^{(3z-1)/(z-1)}}{(3z-1)ax^{z-1}[1 - (z-1)ax^{z-1}]^{2z/(z-1)}},$$

$$\lambda_\sigma = l^{\frac{1}{2}(z+1)/(z-1)}.$$

In these formulae, a is the arbitrary constant in (2.137). Indeed, the parameter l drops out from (2.134), leading to the following exponents in the scaling relation (2.133):

$$\rho = \frac{z}{z-1}, \quad \nu = \frac{z+1}{2(z-1)}. \tag{2.152}$$

We see that when $z = 2$, the choice $l = 2$ also leads to $\alpha = \rho = 2$, resulting in a quite degenerate case that hinders the general structure of the equations. Conceptually, it is better to work with an arbitrary value of l.

Chapter 3

Elementary Symbolic Dynamics

In this chapter, we study the problem of labelling and ordering all the periodic windows in one-dimensional mappings. This can be accomplished by the method of symbolic dynamics, an algebraic approach which originated from the abstract topological theory of dynamical systems, developed in the 1930's[1] and made concrete, when applied to one-dimensional mappings, since the 1970's. Although symbolic dynamics provides the most rigorous way to define chaotic motion in dynamical systems, its abstract formulation (see, e.g., Alekseev and Yakobson, 1981) still prevents a non-mathematically minded scientist from appreciating the power of this beautiful theory. An easier way to overcome this difficulty is just to learn the game by playing it with simple examples.

Our emphasis will be on symbolic dynamics of two letters, associated with unimodal mappings. After a general discussion of the idea of symbolic dynamics, we shall show the usefulness of symbolic description of periodic orbits by "lifting" a word to an equation, in order to determine the superstable parameter without actually constructing the symbolic dynamics. Then comes the

---

[1]See, e.g., M. Morse, and G. A. Hedlund, *Am. J. Math.* **60** (1938) 815, reprinted in *Collected Papers of M. Morse*, vol. 2, World Scientific, 1986. According to historical references, given in the paper of Procaccia, Thomae and Tresser (1987), the use of symbolic sequences dates back to 1851.

main part of this Chapter, namely, the symbolic dynamics of two letters. We
shall treat it twice: first in the conventional way, as formulated by Metropolis,
Stein and Stein (1973), and Derrida, Gervois and Pomeau (1978), concentrating on the ordering of superstable periodic orbits along the parameter axis;
then, in an easier and more natural way, due to Zheng (Zheng, 1988a and b;
Hao and Zheng, 1988), extending the formalism to all symbolic sequences, both
in the phase space and along the parameter axis. We believe this circulation
of notions may help the reader gain a deeper feeling of the methods.

Section 3.4 on scaling properties of period-n-tupling sequences should have
been included in Chapter 2, but the nomenclature of periods, using symbolic
dynamics, furnishes a good context for us to insert the material into the present
chapter. Some repetition of the discussion of Chapter 2 serves the same purpose of deepening the understanding of the renormalization group idea. We
touch briefly on symbolic dynamics of three and more letters in Section 3.7,
and conclude this chapter with a discussion of the many facets of the enumeration problem related to the number of periodic windows in one-dimensional
mappings.

3.1 Introduction

What is symbolic dynamics to a practitioner in the physical sciences? It is just
a kind of coarse-grained description of an evolutionary process which can be
explained in the following manner.

The most complete description of the discrete-time evolution of a map like

$$x_{n+1} = f(\mu, x_n), \qquad x \in I, \tag{3.1}$$

would require the knowledge of the whole set $\{x_i, \ i = 0, 1, \ldots\}$ for all possible
initial choices of x_0. However, one can develop a coarse-grained description by
ignoring the actual numbers in the set, but retaining the essential feature of
the evolution. In order to do so, we divide the "phase space", i.e., the interval
I, into several regions and label each region by a letter, say, R, L, M ....
Replacing each number x_i by the label of the interval into which it falls, every
set $\{x_i\}$ will become a sequence of letters. It is clear that different sets $\{x_i\}$
may correspond to one and the same symbolic sequence. However, instead

of being a drawback, this is rather an advantage, because it opens up the possibility of introducing a classification scheme for the numerical sequences. All numerical sequences, corresponding to one and the same symbolic sequence, may be taken to be equivalent. Although one may lose some numerical details of the iterations, such essential features of the evolution as periodicity will be preserved.

What has just been said holds for an arbitrary partition of the interval I (as mathematicians usually do). However, if we divide the interval more thought-fully, in accordance with the underlying "physics", sometimes a beautiful set of rules may be defined for the construction and ordering of the symbolic sequences. In fact, we shall divide the interval into segments, according to the monotonic branches of the map. These segments are separated by critical points of the map (and by the end points of I, if those do not belong to the critical ones). Each segment needs a letter to label, and critical points may require additional symbols in the scheme. In most cases, one has to deal with a finite set of symbols.

Using a finite number of symbols, e.g., two letters R and L, one can form infinitely many symbolic sequences, among which periodic ones play a specific role. This is due to the fact that in numerical studies on computers, and in laboratory experiments, periodic orbits of not very long periods are the only kind of motion that one can recognize with confidence. As we shall see, the systematics of periodic windows alone tells us much about the location and nature of the chaotic regions, and most of the knowledge acquired from mappings can be carried over to higher dimensional systems. Therefore, we shall study in the first place the symbolic dynamics of periodic orbits.

By symbolic dynamics of periodic orbits we understand the following:

1. The rules to generate all admissible words made up of these letters, and, in particular, the rule to generate a word of the shortest period in between two given words (called also a median word, for short) and the rule to generate all admissible words of a given period. This is necessary due to the obvious fact that by far not every combination of the letters makes a word that corresponds to an existing orbit in the mapping.

2. The rule to introduce an order for all admissible words and the corre-

spondence of this ordering to the value of the parameter μ.

3. The rule to check the admissibility of an arbitrarily given word.

4. The rule to calculate the number of distinct admissible words for a given period.

Before discussing these rules in subsequent sections, we would like to make two remarks. Firstly, it is not always possible to construct all these rules for an arbitrary map. However, even in this case, the use of symbolic description may be quite helpful in an analysis of a map, as we shall see in the next section. Secondly, the usefulness of symbolic dynamics is not restricted to periodic orbits. In fact, it furnishes the only mathematically rigorous way to define chaotic motion, and it may be used to classify chaotic orbits and to evaluate their topological entropy. We shall refer to these problems in Section 3.6.7 and Chapter 6.

3.2   The Location of Superstable Orbits

In order to demonstrate the usefulness of a symbolic description of periodic orbits without construction of specific rules, we will study in this section the problem of how to determine the parameter value for a given type of periodic orbit. We recall from Section 2.3 that an orbit is superstable, if the product of the first derivatives taken at the periodic points equals zero, or, equivalently, if the orbit contains at least one of the critical points of the map. (We only consider continuous mappings in this section.)

There are at least three methods to calculate the superstable parameter value. The first method was used by M. J. Feigenbaum (1978). It is a straight-forward calculation from the definition. Just write down the iterations for the required period n and the superstability condition

$$x_2 = f(\mu, x_1),$$
$$x_3 = f(\mu, x_2),$$
$$\ldots$$
$$x_1 = f(\mu, x_n),$$
$$\prod_{i=1}^{n} \mid f'(\mu, x_i) \mid = 0.$$

This is a system of $n + 1$ equations for $n + 1$ unknowns: x_i, $i = 1$ to n, and μ. However, it works well only for $n = 1, 2, 3$ (in the case of the unimodal mappings). The problem is when n gets larger, there may exist many different solutions to this system, and that, one needs very good initial values to separate them numerically, what may become even impossible for not very large n, say, $n = 11$. In addition, this method, if successful, gives also the periodic points x_i which are of little use.

A better method is based on the solution of one equation for μ only, namely,

$$P_n(\mu) = x_c, \tag{3.2}$$

where $P_n(\mu)$ is the composite function defined in Section 2.1 [see (2.2)]. It is obtained from (3.1) by noting that a superstable orbit must contain the critical point x_c; thus the last equation of (3.1) holds automatically and for $x_1 = x_c$ the n iterations lead precisely to P_n. Equation (3.2), however, suffers from the same defect, i.e., it will have multiple solutions when n gets big enough. For example, the logistic map has 93 different $n = 11$ solutions, all populated in the tiny interval $\mu = 1.543689 \cdots$ to 2. No existing numerical algorithm would be able to separate them to high precision, e.g., the subroutine in the NAG library is better than that in IMSL, but it still does not work properly, not to mention the problem of estimating initial values for μ (see, however, Section 3.4.3). Therefore, we need a method to approach each root of (3.2) separately with as high a precision as we desire. At this point, the third method comes to our help.

3.2.1　Word-Lifting Technique

Take a superstable orbit of period n and write it down explicitly as an n-cycle starting from and ending at the critical point x_c:

$$\underbrace{f \circ f \circ f \cdots \circ f}_{n\ times}(x_c) = x_c. \tag{3.3}$$

Now let us shift all but one of the functions f from the left-hand side to its right-hand side. However, since the inverse function f^{-1} is multi-valued, we must attach a subscript to each term to indicate which single-valued branch has been used in taking the inverse:

$$f(\mu, x_c) = f_{\alpha_1}^{-1} \circ f_{\alpha_2}^{-1} \circ \ldots \circ f_{\alpha_{n-1}}^{-1}(\mu, x_c). \tag{3.4}$$

Each of the subscripts α_i is equal to one of the letters which we have been using to denote the monotonic branches of the map. Putting together all the subscripts in the order in which they appear in (3.4), we get a word W made of $n - 1$ letters

$$W = \alpha_1 \alpha_2 \ldots \alpha_{n-1}. \tag{3.5}$$

Note that an n-cycle is described by a word made of $n - 1$ letters, if we do not count the first or last letter corresponding to x_c. Let us introduce a new notation by naming each single-valued branch of the inverse function by its subscript. This is a simple matter of changing notation, i.e., defining

$$\alpha_i(y) \equiv f_{\alpha_i}^{-1}(y). \tag{3.6}$$

Using this new notation, we can "lift" the word W to a composite function $W(y)$, depending on the parameter μ:

$$W(y) = \alpha_1 \circ \alpha_2 \circ \ldots \circ \alpha_{n-1}(y). \tag{3.7}$$

Now (3.4) assumes the very simple form

$$f(\mu, x_c) = W(x_c). \tag{3.8}$$

Since f, f^{-1}, and x_c are all known, Equation (3.8) appears to be an equation which determines the superstable parameter value μ for the n-cycle described by the word W. We shall call this method the "word-lifting" technique.

We shall extend the word-lifting technique to determine the precise parameter value of a certain type of chaotic orbits in Section 3.6.7.

3.2.2   Example 1: The Logistic Map

For the logistic map $x_c = 0$, $f(\mu, 0) = 1$, Equation (3.8) looks even simpler:

$$W(0) = 1.$$

We shall see in Section 3.3.2 that there exists one period 2 orbit R, one period 3 word RL and three period 5 orbits RLR^2, RL^2R and RL^3. They lead to the following equations, respectively:

$$R(0) = 1,$$
$$R \circ L(0) = 1,$$
$$R \circ L \circ R \circ R(0) = 1, \tag{3.9}$$
$$R \circ L \circ L \circ R(0) = 1,$$
$$R \circ L \circ L \circ L(0) = 1,$$

where

$$R(y) = +\sqrt{(1-y)/\mu},$$
$$L(y) = -\sqrt{(1-y)/\mu}.$$

Written out explicitly, with both sides multiplied by μ, the last three equations (3.9) become

$$RLR^2 : \quad \mu = \sqrt{\mu + \sqrt{\mu - \sqrt{\mu - \sqrt{\mu}}}},$$

$$RL^2R : \quad \mu = \sqrt{\mu + \sqrt{\mu + \sqrt{\mu - \sqrt{\mu}}}}, \tag{3.10}$$

$$RL^3 : \quad \mu = \sqrt{\mu + \sqrt{\mu + \sqrt{\mu + \sqrt{\mu}}}}.$$

If one tries to get rid of the square roots in (3.2) by squaring them several times, all these three equations would lead to one and the same polynomial equation $P_5(\mu) = 0$, i.e., (3.2) for the logistic case, and all three roots would mix up again. A better way to solve Equations (3.10) is to convert them into iterations. Take, for example, the first equation (3.10) and write it in the form

$$\mu_{n+1} = \sqrt{\mu_n + \sqrt{\mu_n - \sqrt{\mu_n - \sqrt{\mu_n}}}},$$

The iteration converges very quickly and we obtain a good estimate for the initial value μ_0: any number between 1.4 and 2 will do. We get in this way

$$\mu_{RLR^2} = 1.625413725,$$
$$\mu_{RL^2R} = 1.860782522,$$
$$\mu_{RL^3} = 1.985424253,$$

where we have attached the word as a subscript to μ. It is very easy to read off the rule for writing down these iteration equations directly from the words, and to program the procedure to solve them. The question we are facing now is how to find all the admissible words. Before turning to this question in the following sections, we discuss another example, where the symbolic dynamics has not yet been constructed, but where the symbolic description does assist greatly.

3.2.3   Example 2: The Sine-square Map

The sine-square map

$$x_{n+1} = A \sin^2(x_n - B) \tag{3.11}$$

(see Fig. 1.10 in Section 1.6) has four monotonic branches. Therefore, we require four letters R, L, M, N to denote these branches. Taking into account the convention for the principal values of the arccosine function, the inverse functions can be written down explicitly:

$$
\begin{aligned}
N(y) &\equiv f_N^{-1} = B + \tfrac{1}{2} \arccos(1 - 2y/A), \\
L(y) &\equiv f_L^{-1} = N(y) - \pi = B - \pi + \tfrac{1}{2} \arccos(1 - 2y/A), \\
M(y) &\equiv f_M^{-1} = 2B - N(y) = B - \tfrac{1}{2} \arccos(1 - 2y/A), \\
R(y) &\equiv f_R^{-1} = M(y) + \pi = B + \pi - \tfrac{1}{2} \arccos(1 - 2y/A).
\end{aligned}
\tag{3.12}
$$

The map (3.11) has four critical points

$$
\begin{aligned}
x_{c1} &= B - \pi/2, \quad x_{c3} = B + \pi/2 \quad \text{(peaks)}, \\
x_{c2} &= B, \qquad\qquad x_{c4} = B + \pi \quad \text{(valleys)}.
\end{aligned}
$$

Therefore, we have to attach a subscript i $(i = 1, 2, 3, 4)$ to the word W, i.e., to its last letter, to indicate which critical point it contains. Equation (3.8) now reads

$$f(A, B, x_{ci}) = W(x_{ci}), \qquad i = 1, 2, 3, 4. \tag{3.13}$$

Since there are two parameters A and B in the map, it is not so easy to work out, in general, the symbolic dynamics of these four letters. However, by inspecting the drawings at different parameter values, one can find the following period 2 orbits: $N_1, R_1, M_1, M_2, L_2, R_3, L_4, M_4, N_4, R_4$, and the following period 3 orbits: $ML_1, NL_1, RL_1, RM_1, RN_1, NM_1, MN_2, LN_2, MR_2, LR_2, LM_2,$

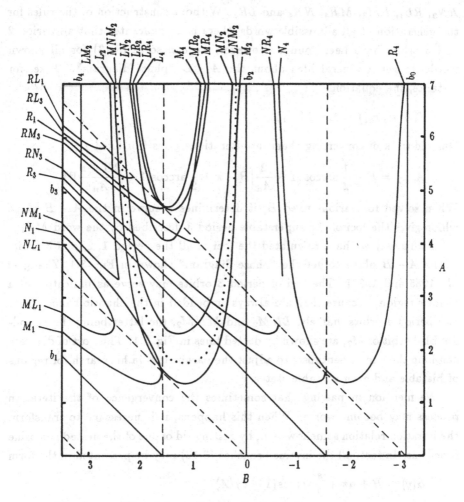

Figure 3.1: The "phase diagram" of the sine-square map.

RN_3, RL_3, RM_3, MR_4, NR_4 and LR_4. Without construction of the rules for the generation of all admissible words, there is no guarantee that all period 2 and 3 orbits have been found. However, we can solve (3.13) for all known words, to get a general idea about the $A - B$ "phase diagram". Take, for instance, the equation

$$f(A, B, x_{c1}) = M \circ L(x_{c1}).$$

Our old trick of converting them into iterations gives (for fixed B)

$$A_{n+1} = B - \frac{1}{2} \arccos\{1 - \frac{2}{A_n}[B - \pi + \frac{1}{2}\arccos(1 - \frac{2B - \pi}{A_n})]\}.$$

When solved for various fixed B, it determines a curve in the $A - B$ plane which gives the loci of the superstable period 3 described by the word ML_1.

In this way we have calculated the loci of all the period 1, 2, and 3 orbits in the $A - B$ plane to get the "phase diagram" shown in Fig. 3.1 (Zhang *et al.*, 1985 and 1987). The loci of period 3 orbits may serve as indicators of a chaotic region, because they are always embedded in the chaotic bands. Only two period 4 orbits, namely, LNM_2 and MNM_2, the latter being the period-doubled orbit of M_2, are shown by dotted lines in Fig. 3.1. This sort of diagram can help the experimentalists to adjust the parameters in his search for regions of bistable and even tristable states.

We mention in passing that sometimes the convergence of the iteration process may become worse. When this happens, it is necessary to transform the iteration relation somehow, e.g., by getting rid of one of the nested arccosine functions. In fact, all the inverse branches (3.12) may be presented in the form

$$\alpha(y) = B + n\pi + \frac{s}{2}\arccos(1 - 2y/A)$$

with $n = 0$ or ± 1, $s = \pm 1$. Next, separating the leftmost letter α in a word, i.e., letting $W = \alpha\overline{W}$, we have

$$\frac{A}{2}[1 - \cos(2f(x_c) - 2B)] = \overline{W}(x_c) = f^2(x_c).$$

We see that in the expression of $\alpha(y)$ n and s drop out, indicating that all αW with different α actually lead to the same function in the $A - B$ plane. This discussion can be repeated to yield the result that all words of the same length lead to one and the same function, if one gets rid of the arccosines. These

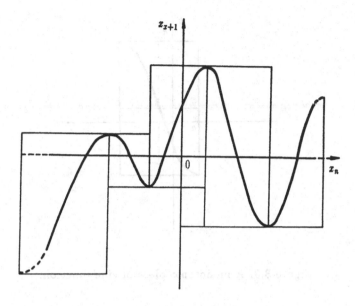

Figure 3.2: Construction of a complicated map by matching trigonomical functions.

conclusions explain why the curve for M_2 goes smoothly into that for L_2, etc., as seen in Fig. 3.1.

3.2.4 Generalizations

Our method, outlined in this section, may be applied to any one-dimensional map, provided the inverse branches $f_{\alpha_i}^{-1}$ are known explicitly. Moreover, even when the analytical form of a complicated continuous map is unavailable, one can still construct a C^1 multimodal map to imitate it by matching appropriately scaled pieces of sines or cosines. Take, for example, the map shown in Fig. 3.2. By enclosing each monotonic branch in a box and replacing the enclosed curve by a sine function

$$y = \frac{c+d}{2} + \frac{d-c}{2}\sin\left(\frac{2x-a-b}{b-a}\right)$$

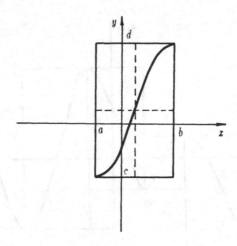

Figure 3.3: A monotonic piece of sine function.

(compare Fig. 3.3), one can get another map. There is a good hope that for most purposes this new map would mimic the old one. We have in mind the *structural*, not the *metric* properties. (We have mentioned the structural and metric universality at the end of Section 1.3.) The latter depends on the behaviour of the mapping function at critical points and a piecewise matching may lead to different renormalization group equations (cf., e.g., Arneodo, Coullet and Tresser, 1979). Since now all the inverse branches are explicitly known, the calculation of the "phase diagram" follows the same course which we pursued for the sine-square map and involves only elementary mathematics.

3.3 Symbolic Dynamics of Two Letters

We have already mentioned that symbolic dynamics may be understood as a kind of coarse-grained description of the time evolution of a dynamical system. The symbolic dynamics of two letters for unimodal mappings has been constructed, before and concurrently with Feigenbaum's discovery of the scaling properties of these mappings, by Metropolis, Stein and Stein (1973), and by Derrida, Gervois and Pomeau (1978, 1979). Since there are a few, somewhat boring, definitions and theorems in the construction, it is better to adopt at the beginning a pragmatic point of view, i.e., to first formulate the related rules without going into detailed proofs or derivations, and then to show how they work in practice. Having mastered the practical rules, some readers may desire to learn more mathematical details from the subsequent sections and from the original literature. In fact, we follow in this section the "traditional" formulation of symbolic dynamics of two letters, concentrating on superstable orbits only. In Section 3.6, we shall present a more natural and convenient form of the theory, extending it to all possible orbits.

3.3.1   The Ordering of Words

We will start with a number of useful definitions.

Admissible Word. A word W is said to be admissible, if it does correspond to a stable periodic orbit of the unimodal map. A word made up of $n-1$ letters corresponds to period n and is simply said to be of length n. We shall discuss later on how to verify the admissibility of a word..

The Parity of a Word. A word made of a certain number of letters R and L is said to be odd (even), if it contains an odd (even) number of R's. Thus the word RL is odd, while RLR is even. Anticipating somewhat, we point out that that the parity of a word is determined by the number of descending segments of the map involved in the orbit represented by the word. This is the lead for generalization of the notion of parity of words to symbolic dynamics of more letters. For example, both the letters R and M in the sine-square map (3.11) (see Fig. 1.10) represent descending branches, therefore, we shall count the total number of R's and M's in a word to determine its parity.

The Order of Two Words. Two words W_1 and W_2 can be ordered in the following way. Compare these two words, letter by letter, from left to right, and denote the largest common part by W^*, i.e., let

$$W_1 = W^* \sigma_1 \sigma_2 \ldots,$$
$$W_2 = W^* \tau_1 \tau_2 \ldots,$$

where $\sigma_1 \neq \tau_1$. Furthermore, the critical point x_c will play a special role and we denote it by the letter C, meaning Critical (or Central in the case of the logistic map). Now an order is defined on the basis of the natural order on the interval I:

$$L < C < R. \tag{3.14}$$

If W^* is even, then the order of W_1 and W_2 is determined by the order of σ_1 and τ_1 in the sense of (3.14), i.e., $W_1 > W_2$ if $\sigma_1 > \tau_1$, and *vice versa*. If W^* is odd, then the order of W_1 and W_2 is the anti-order of σ_1 and τ_1, i.e., $W_1 > W_2$, if $\sigma_1 < \tau_1$ in the sense of (3.14).

Example 1. A blank word (denoted by b if necessary) corresponds to a period 1, i.e., a fixed point, and may be thought as consisting of a single letter C. We have

$$b < R < RLR < RLR^3 LR < RLR^3 LRLRLR^3 LR < \cdots \tag{3.15}$$

We shall see that this sequence determines the direction of bifurcations for the main period-doubling cascade of Feigenbaum.

Example 2. It is easy to verify that RL^{n-2} is the last word among all admissible words of length n.

Example 3. The chain of inequalities

$$R < RL < RL^2 < \cdots < RL^n < RL^{n+1} < \cdots \tag{3.16}$$

determines the most clearly seen periodic windows in many mappings and ordinary differential equations. When n gets large and other windows in between RL^n and RL^{n+1} cannot be resolved, the sequence (3.16) has a close relation to the so-called period-adding sequence, studied by Kaneko (1982, 1983a).

Example 4. All finite and infinite symbolic sequences made up of letters R and L are ordered in between L^∞ and RL^∞, i.e.,

$$L^\infty < \text{any symbolic sequence} < RL^\infty. \tag{3.17}$$

This remark is more relevant to the understanding of chaotic orbits (see Section 3.6 below).

For those unimodal mappings, whose height (maximum) varies monotonically with the parameter μ, e.g., the logistic map, the order we have just defined is identical to the order of appearance of the corresponding periods along the parameter μ axis. If $W_1 > W_2$, then the parameter values for these periods must satisfy $\mu_{W_1} > \mu_{W_2}$. This is no longer true even for unimodal maps, when there is a non-monotonic dependence on the parameter. We shall return to this point at the end of the next section.

3.3.2 Finding the Median Word Between Two Given Words

Next, we formulate the rule for the generation of the median word describing the shortest periodic orbit in between two given admissible words. Historically, this has been accomplished by using the notion of *harmonic* and *antiharmonic* of a word (Metropolis, Stein and Stein, 1973).

Harmonic of a Word. Given a word W, one can construct another word $H(W)$, called the harmonic of W, according to the following rule:

$$H(W) = W\sigma W, \quad \text{where} \quad \sigma = \begin{cases} R & \text{for } W \text{ even,} \\ L & \text{for } W \text{ odd.} \end{cases}$$

Thus, $H(R) = RLR$, $H(RLR) = RLR^3LR$, etc. If W contains $n-1$ letters, then there are $2n-1$ letters in $H(W)$. If W is an admissible word, then $H(W)$ is also admissible and corresponds to the period-doubled cycle of W.

Antiharmonic of a Word. The antiharmonic $A(W)$ of a word W is constructed according to a rule, opposite to that for the harmonic, i.e.,

$$A(W) = W\sigma W, \quad \text{where} \quad \sigma = \begin{cases} R & \text{for } W \text{ odd,} \\ L & \text{for } W \text{ even.} \end{cases}$$

The antiharmonic of a word is an artificial construct, needed only during the intermediate stage of a discussion. It is not an admissible word, and does not correspond to a real orbit of the map. In Section 3.6.5, we will describe a new method of generating the median words without invoking the notion of harmonics and antiharmonics.

High Order Harmonics and Antiharmonics. The definition of harmonics and antiharmonics can be applied repeatedly to generate high order harmonics and antiharmonics, e.g.,

$$H^2(R) = H(H(R)) = H(RLR) = RLR^3LR$$

The Feigenbaum main period-doubling sequence (3.15) may be written in the form

$$b < H(b) < H^2(b) < H^3(b) < \cdots,$$

(A blank word b is considered to be of even parity.) Now we can search the shortest admissible word included in between two given words W_1 and W_2. Suppose that $W_1 < W_2$ and W_1 is of length n_1, then there may exist another admissible word W, corresponding to the shortest period located in between W_1 and W_2, i.e.,

$$W_1 < W < W_2$$

which can be constructed in the following way. (We say "may exist", because one cannot exclude the possibility that W_1 and W_2 are adjacent words and the constructed W coincides with one of them.)

We first construct $H(W_1)$ and $A(W_2)$, and then denote their leading common part of length n^* by W^*, i.e., we let

$$H(W_1) = W^*\sigma_1\sigma_2\ldots,$$
$$A(W_2) = W^*\tau_1\tau_2\ldots.$$

Two cases must be distinguished:

1. $n^* > 2n_1$, then the word of the shortest period is given by the harmonic of W_1, i.e., $H(W_1)$.

2. $n^* < 2n_1$, then the word is given by W^* itself.

In this process, we may encounter cases where the above rules cannot be readily applied. Then one must go to higher order harmonics or antiharmonics or both instead of only using first order ones.

We consider a few examples.

Example 1. Given $W_1 = RLR^2$ and $W_2 = RL$, we have $H(W_1) = RLR^2LRLR^2$ and $A(W_2) = RLR^2L$. Since formally $W^* = A(W_2)$, but an antiharmonic cannot correspond to any admissible period, we have to construct the second order antiharmonic

$$A^2(W_2) = A(A(W_2)) = RLR^2LR^2LR^2L.$$

Now $W^* = RLR^2LR$. This is a period 7, in fact, it is the shortest word in between period 5 (RLR^2) and period 3 (RL).

Example 2. Given the blank word b and the last period 7 word RL^6, we can construct all the words corresponding to periods $n \leq 7$. The results are listed in Table 3.1 with a few remarks added. The Appendix of the paper by Metropolis, Stein and Stein (1973) contains a list of all admissible word for periods $n \leq 11$. In the next section, we will give a program to generate all admissible words of periods less than or equal to a given number. In Table 3.1, mP stands for period m, mI for chaotic band of "period m" or m-piece chaotic Islands, and PDB for Period-Doubling Bifurcation sequence.

It is remarkable that the order of appearance of the words in Table 3.1 is in monotonic correspondence to the increasing parameter value of the logistic map. Metropolis, Stein and Stein (1973) examined these words for four different maps and ended up with the same order, whence they called this ordering of words a U-sequence, i.e., a Universal sequence. It is known in the literature also as the MSS sequence. In other words, one can compile a dictionary of these words and there exists a correspondence between the position of words in the dictionary and the place of their occurrence on the parameter axis.

However, the universality of the MSS sequence holds only for those unimodal mappings whose height depends on the parameter in a monotonic manner. In general, the ordering of words, as shown in Table 3.1, should be understood in a local sense, since one can redefine the logistic map by letting the parameter μ depend on a new parameter λ in a non-monotonic way:

$$x_{n+1} = g(\lambda, x_n) \equiv 1 - \mu(\lambda)\, x_n^2. \tag{3.18}$$

For a functional dependence of $\mu(\lambda)$, sketched in Fig. 3.4, the ordering of words in terms of λ will no longer be universal: those words which correspond to parameter values

$$\mu_2 < \mu < \mu_1$$

will appear three times, twice in normal order and once in reversed order. In other words, the book binding of the MSS dictionary is not always of a good quality: some parts of it may be bound repeatedly with the local order of pages and words preserved or reversed. In fact, this kind of folding occurs

Word	Period	Remarks
R	2	The only $2P$, beginning of 2×2^n PDB
RLR	4	The harmonic of R, next member in the above PDB
RLR^3	6	$3P$ embedded in $2I$, beginning of $2 \times 3 \times 2^n$ PDB
RLR^4	7	The 1st $7P$, beginning of the first 7×2^n PDB
RLR^2	5	$5P$ embedded in $1I$, beginning of the 1st 5×2^n PDB
RLR^2LR	7	The 2nd $7P$, beginning of the 2nd secondary 7×2^n PDB
RL	3	The only $3P$, beginning of 3×2^n PDB
RL^2RL	6	The harmonic of RL, next member in the 3×2^n PDB
RL^2RLR	7	The 3rd $7P$, beginning of the 3rd secondary 7×2^n PDB
RL^2R	5	The 2nd $5P$ embedded in $1I$, beginning of 5×2^n PDB
RL^2R^3	7	The 4th $7P$, beginning of the 4th secondary 7×2^n PDB
RL^2R^2	6	The 1st primitive $6P$, beginning of a 6×2^n PDB
RL^2R^2L	7	The 5th $7P$, beginning of the 5th secondary 7×2^n PDB
RL^2	4	The only primitive $4P$, beginning of 4×2^n PDB
RL^3RL	7	The 6th $7P$, beginning of the 6th secondary 7×2^n PDB
RL^3R	6	The 2nd primitive $6P$, beginning of a 6×2^n PDB
RL^3R^2	7	The 7th $7P$, beginning of the 7th secondary 7×2^n PDB
RL^3	5	The last $5P$, beginning of the 3rd 5×2^n PDB
RL^4R	7	The 8th $7P$, beginning of the 8th secondary 7×2^n PDB
RL^4	6	The last $6P$, beginning of the last 6×2^n PDB
RL^5	7	The 9th and last $7P$, beginning of the last 7×2^n PDB

Table 3.1: Admissible words for periods ≤ 7 (for abbreviations see the text).

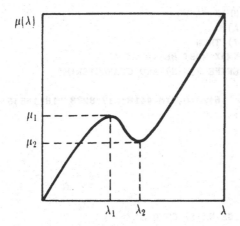

Figure 3.4: Nonmonotonic depenence on a new parameter (schematic).

frequently in the parameter spaces of differential equations when the parameter dependence becomes more intricate (see, e.g., Fig. 5.20 in Chapter 5).

3.3.3   A Program to Generate Median Words

We give now a FORTRAN program that generates all the admissible words of length less than or equal to a given number NMAX and included in between two given words W1 < W2. It implements the method described in the two previous sections in a straightforward way. No attempt has been made to optimize the program, which, however, works well.

```
      PROGRAM MSSTBL
C
C THIS IS A PROGRAM TO GENERATE ALL ADMISSIBLE WORDS OF
C LENGTH <= NMAX IN BETWEEN TWO GIVEN WORDS W1 < W2.
C
      CHARACTER*42 WA,WB,HWA,AWB,WP,HAR,AHAR,PSTAR
      CHARACTER*20 W(8500)
      INTEGER KK(100)
      N=8500
      OPEN(UNIT=4,FILE='MSS.TBL',STATUS='NEW')
      WRITE(*,*) 'INPUT NMAX,W1 AND W2, USING CAPITAL R,L.'
```

```
          READ(*,500) NMAX,WA,WB
500       FORMAT(I2/A/A)
          IF(NMAX.GT.17) THEN
          WRITE(*,*) 'NMAX MUST BE <= 17,'
          STOP 'OR REDEFINE N,W(N) AND CHARACTER*M'
          ENDIF
C NMAX AND W(N) SIZE: 15:2370, 16,4418; 17:8273; 18:15555
C 19:29352; 20:55567; ...
          IA=0
          IB=N
          W(IB)=WB
1         IA=IA+1
          W(IA)=WA
          HWA=HAR(WA)
          IF(LLEN(HWA).GE.NMAX) GOTO 3
          WA=HWA
          GOTO 1
3         WB=W(IB)
          AWB=AHAR(WB)
4         WP=PSTAR(HWA,AWB)
          IF(WP.EQ.HWA) THEN
              HWA=HAR(HWA)
              GOTO 4
          ENDIF
          IF(WP.EQ.AWB) THEN
              AWB=AHAR(AWB)
              GOTO 4
          ENDIF
          KSTAR=LLEN(WP)
          K1=LLEN(WA)
          IF(KSTAR.GE.2*K1) THEN
              IF(LLEN(HWA).LT.NMAX) THEN
                  WA=HWA
                  GOTO 1
              ELSE
                  IF(LLEN(WP).LT.NMAX) THEN
                      WA=WP
                      GOTO 1
                  ELSE
                      GOTO 5
                  ENDIF
              ENDIF
          ELSE
              IF(LLEN(WP).LT.NMAX) THEN
```

```
                 IB=IB-1
                 W(IB)=WP
                 GOTO 3
              ELSE
                 GOTO 5
              ENDIF
              ENDIF
5       WA=W(IB)
        IB=IB+1
        IF(IB.GE.N+1) GOTO 6
        GOTO 1
6       IA=IA+1
        W(IA)=WA
        DO 7 I=1,NMAX
7       KK(I)=0
        DO 8 I=1,IA
        WP=W(I)
        L=LLEN(WP)
        KK(L)=KK(L)+1
8       WRITE(4,501) I,L+1,KK(L),WP
501     FORMAT(1X,3I8,6X,A)
        STOP 'END OF JOB'
        END
C
C HARMONIC OF P.
C
        CHARACTER*42 FUNCTION HAR(P)
        CHARACTER C
        CHARACTER*42 P
        L=LLEN(P)
        C='L'
        IF(IPARTY(P).EQ.1) C='R'
        HAR=P(L:L)//C//P(L:L)
        RETURN
        END
C
C ANTIHARMONIC OF P.
C
        CHARACTER*42 FUNCTION AHAR(P)
        CHARACTER C
        CHARACTER*42 P
        L=LLEN(P)
        C='R'
        IF(IPARTY(P).EQ.1) C='L'
```

```
          AHAR=P(L:L)//C//P(L:L)
          RETURN
          END
C
C R-PARITY OF P.
C
          FUNCTION IPARTY(P)
          CHARACTER*42 P
          IPARTY=1
          DO 10 I=1,LLEN(P)
10        IF(P(I:I).EQ.'R') IPARTY=-IPARTY
          RETURN
          END
C
C COMMON LEADING PART OF TWO WORDS
C
          CHARACTER*42 FUNCTION PSTAR(P,Q)
          CHARACTER*42 P,Q
          DO 20 I=1,42
20        PSTAR(I:I)=' '
          DO 30 I=1,MIN(LLEN(P),LLEN(Q))
          IF(P(I:I).NE.Q(I:I)) GOTO 40
30        PSTAR(I:I)=P(I:I)
40        RETURN
          END
C
C THE LENGTH OF P.
C
          FUNCTION LLEN(P)
          CHARACTER*42 P
          LLEN=0
          DO 50 I=1,42
          IF(P(I:I).NE.'R'.AND.P(I:I).NE.'L') GOTO 60
50        LLEN=LLEN+1
60        RETURN
          END
```

3.3.4   The ∗-composition and Fine Structure of Power Spectra

Derrida, Gervois and Pomeau (1978) introduced a very useful operation on words, namely, the ∗-composition. The harmonics and antiharmonics defined in the above subsection happen to be particular cases of this composition rule. The *-composition of two words P and $Q = \sigma_1\sigma_2\ldots\sigma_{k-1}$ is defined as

$$P * Q = P\overline{\sigma}_1 P\overline{\sigma}_2 P \ldots P\overline{\sigma}_{k-1} P, \tag{3.19}$$

where $\overline{\sigma}_i = \sigma_i$, if P is even, and $\overline{\sigma}_i$ equals the conjugate of σ_i, i.e., with R and L interchanged, if P is odd. Therefore, a harmonic (antiharmonic) is nothing but the *-composition with $Q = R$ (L):

$$H(W) = W * R,$$
$$A(W) = W * L.$$

Using *-composition, the main period-doubling sequence in (3.15) acquires the more concise form

$$b \ < \ R \ < \ R*R \ < \ R*R*R \ < \ R*R*R*R \ < \ \ldots \ , \tag{3.20}$$

If both P and Q are admissible, then $P * Q$ will be admissible as well. It is readily verified that *-composition is associative, but non-commutative. The existence of *-composition is the mathematical manifestation of the self-similar structure of the bifurcation diagram of the unimodal mappings.

If a word cannot be decomposed using * operation, it is called a primitive word. The primitive components of a non-primitive word, say, P, Q and S in $W = P * Q * S$, describe periodic motion at different scales, P corresponding to the largest, while S to the smallest scale. Reflected in the power spectrum[2], these different scales give rise to fine structures of different order. Take, for example, the two period 6 orbits $W_1 = R * RL = RLR^3$ and $W_2 = RL * R = RL^2RL$, the former being the period 3 window embedded in the two-band chaotic zone and the latter representing the period-doubled regime of the period 3 in the one-band region. Their power spectra are shown schematically in Fig. 3.5. The fine structure of power spectra turns out to be quite useful in experiments aiming to tell the whereabouts in parameter space.

As a last example, let us find the symbolic sequences describing the band-merging points in the bifurcation diagram. All the words in the two-band region may be decomposed into $R * W$, where W is a period in the one-band region. The rightmost boundary of the one-band region, corresponding to the

[2]For the definition of power spectrum, see Section 5.6.

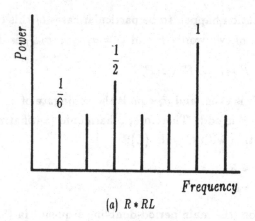

(a) $R * RL$

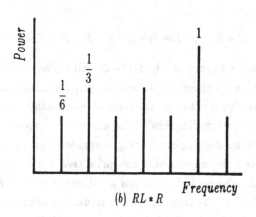

(b) $RL * R$

Figure 3.5: The fine structure of power spectra.

word RL^∞, may be taken as the limit $n \to \infty$ of RL^n. Therefore, in order to find the point where two chaotic bands merge into one, we look at the limit

$$\lim_{n\to\infty} R * RL^n = \lim_{n\to\infty} RLR^{2n+1} = RLR^\infty = RL(RR)^\infty. \qquad (3.21)$$

The last version in (3.21) will appear to be more convenient later on (see Section 3.6.7).

Similarly, for the $2^2 \to 2^1$ band-merging point, we start from

$$R * R * RL^n = RLR^3 LR(LRLR)^n,$$

and arrive at $RLRR(RLRL)^\infty$. The next, $2^3 \rightarrow 2^2$ merging point would be $RLR^3 LRL(RLRRLRR)^\infty$, etc. The general rule will become clear after a study of the periodic window theorem in Section 3.6.4.

3.4 Scaling Property of the Period-n-tupling Sequences

We have seen in the last section that the main period-doubling sequence of Feigenbaum acquires a very simple symbolic representation using *-composition rule, namely,

$$R^{*n} = \underbrace{R * R * \cdots * R}_{n \ times}, \qquad n = 0, 1, 2, \ldots ,$$

with $R^{*0} \equiv b$, $R^{*1} \equiv R$ [see (3.20)]. In general, every primitive word W gives birth to a period-doubling sequence $W * R^{*n}$, $n = 0, 1, 2, \ldots$, the simplest one being $W = RL$, corresponding to the period-doubling sequence seen in the period 3 window starting at $\mu = 1.75$ for the logistic map (Fig. 1.2). We have indicated a number of such sequences in Table 3.1.

Period-doubling sequences, however, are not the only ones that can be selected from the infinitely many periodic windows embedded in the chaotic regime. In fact, there are infinitely many ways to select other sequences having one or another kind of scaling property. In particular, every primitive word W leads to a sequence W^{*n}, $n = 1, 2, \ldots$. For example, $W = RL$ leads to a period-tripling sequence with period 3^n, $W = RL^2$ to a period-quadrupling sequence, and the three different period 5 words $W = RLR^2$, RL^2R, and RL^3 to three different period 5^n sequences. These are the period-n-tupling sequence we will study in this section (Zeng, Hao, Wang and Chen, 1984; cf. Chang and McCown, 1985).

3.4.1 The Convergence Rate δ_W

Unlike period-doubling sequences, periodic orbits in period-n-tupling sequences are not adjacent on the parameter axis, but they do enjoy scaling properties, characterized by various exponents: the scaling factor α_W, the convergence rate δ_W, the noise exponent κ_W, etc., all depend on the particular primitive word W. These scaling properties can be checked numerically or studied almost analytically, starting from a generalized renormalization group equation (see Section 3.4.4).

We first calculate the superstable parameter values, using the method described in Section 3.2, i.e., by lifting a word to an equation. Since these periods populate more and more densely towards the $\mu = 2$ end (in the logistic map), even double precision arithmetic may happen to be insufficient to distinguish between neighbouring periods. In Table 3.2, we list the results of calculations using quadruple precision (128 bits per word) on an IBM computer.

We note that, in order to use the iteration formula (3.10) in the general case $z \neq 2$, one has to rescale the $R(y)$ and $L(y)$ functions by the definition

$$\tilde{R}(y) = \mu R(y/\mu) \quad \tilde{L}(y) = \mu L(y/\mu) \ .$$

The parameter values listed in Table 3.2 permit to make estimate of the convergence rate; cf. Table 3.3. The limiting parameter value μ_∞ can also be estimated from Table 3.2. For many practical purposes, the last μ_n shown for each sequence may be adopted for μ_∞. If necessary, one can improve this value by linear interpolation

$$\mu_\infty = \frac{\delta}{\delta - 1}\mu_n - \frac{1}{\delta - 1}\mu_{n-1},$$

with δ taken from Table 3.3. These rough estimates for μ_∞ will be used to estimate the α in the next section.

3.4.2   The Renormalization Group Equations

In order to generalize the Feigenbaum renormalization group equation (2.100), i.e.,

$$g(x) = \alpha g(g(\frac{x}{\alpha})), \tag{3.22}$$

(we have absorbed here the minus sign into the constant α) to the period-n-tupling case, it is better to carry its α dependence to the left-hand side, rewriting it in the form

Sequence	n	Period	$z = 2$	$z = 4$
	1	3	1.7548776662466927601	1.8566748838545028749
	2	9	1.7858656464106735307	1.9086948554177016753
$(RL)^{*n}$	3	27	1.7864298580557610603	1.9093280558741658584
	4	81	1.7864400673590786787	1.9093353842572771913
	5	243	1.7864402521570437541	1.9093354697862036017
	1	4	1.9407998065294847522	1.9819644358540345969
	2	16	1.9427024138565849815	1.9855018451716632914
$(RL^2)^{*n}$	3	64	1.9427043527781808803	1.9855046581439668025
	4	256	1.9427043547534536107	1.9855046603481163514
	5	1024	1.9427043547554659200	1.9855046603498450337
	1	5	1.6254137251233037474	1.7257015664643988397
	2	25	1.6319011677161277913	1.7432904624239859055
$(RLR^2)^{*n}$	3	125	1.6319265547550026433	1.7433508076287708905
	4	625	1.6319266540998245644	1.7433510143345554613
	5	3125	1.6319266544885808146	1.7433510150425949792
	1	5	1.8607825222048548712	1.9420753830002419457
	2	25	1.8622229026519960391	1.9458559215825065416
$(RL^2R)^{*n}$	3	125	1.8622240217485018214	1.9458585817437407417
	4	625	1.8622240226179871817	1.9458585836193512608
	5	3125	1.8622240226186627 3309	1.9458585836206738236
	1	5	1.9854242530542053106	1.9977462432204825039
	2	25	1.9855395232518538783	1.9979740106043500921
$(RL^3)^{*n}$	3	125	1.9855395300600108962	1.9979740214653725113
	4	625	1.9855395300604130165	1.9979740214658899074
	5	3125	1.9855395300604130403	1.9979740214658899321

Table 3.2: Superstable parameters for period-n-tupling sequences.

Sequence	n	Period	$z = 6$	$z = 8$
	1	3	1.8986537126286988928	1.9215993196339830063
	2	9	1.9484544724364238925	1.9665038018590607680
$(RL)^{*n}$	3	27	1.9488631633769899069	1.9667749565343987154
	4	81	1.9488662451610181563	1.9667764263358868050
	5	243	1.9488662689376093020	1.9667764346235884588
	1	4	1.9819644358540345969	1.9914570155119417993
	2	16	1.9855018451716632914	1.9942041292551912581
$(RL^2)^{*n}$	3	64	1.9855046581439668025	1.9942054169741862964
	4	256	1.9855046603481163514	1.9942054175518741326
	5	1024	1.9855046603498450337	1.9942054175521347459
	1	5	1.7767449868088250059	1.8091255734485670547
	2	25	1.7958758361715660892	1.8276439402472996440
$(RLR^2)^{*n}$	3	125	1.7959199425362660922	1.8276748189941343331
	4	625	1.7959200446660657791	1.8276748712929236889
	5	3125	1.7959200449026512594	1.8276748713816601944
	1	5	1.9676397329650952331	1.9791041712847113288
	2	25	1.9709712592487771335	1.9817785479618168510
$(RL^2R)^{*n}$	3	125	1.9709726145514439150	1.9817792366628464760
	4	625	1.9709726151082085726	1.9817792368440701576
	5	3125	1.9709726151084374322	1.9817792368441179298
	1	5	1.9992876616816040558	1.9996900199133528145
	2	25	1.9994324377862196520	1.9997794103770296427
$(RLR^2)^{*n}$	3	125	1.9994324411784617403	1.9997794115493810270
	4	625	1.9994324411785402030	1.9997794115493958713
	5	3125	1.9994324411785402049	1.9997794115493958715

Table 3.2: Superstable parameters for period-n-tupling sequences (continued).

W	$z = 2$	$z = 4$	$z = 6$	$z = 8$
R	4.669	7.285	9.298	10.9
RL	55.25	85.81	130.3	178.9
RL^2	981.6	1275	2220	3542
RLR^2	255.5	291.9	431.9	590.4
RL^2R	1287	1418	2434	3800
RL^3	16929	20990	43190	78500

Table 3.3: Convergence rate δ_W for period-n-tupling sequences.

$$\alpha^{-1}g(\alpha x) = g^{(2)}(x), \tag{3.23}$$

which readily generalizes to the period-n-tupling case

$$\alpha^{-1}g(\alpha x) = g^{(n)}(x). \tag{3.24}$$

In order to fix the solution, it is insufficient to impose the normalization condition $g(0) = 1$ and the superstability condition $g'(0) = 0$. Even in the period-doubling case, one has to indicate the behaviour of $g(x)$ near its maximum, i.e., indicate the power z in the expansion

$$g(x) = g(0) + ax^z + bx^{2z} + \ldots . \tag{3.25}$$

The well-known Feigenbaum constants $\delta = 4.6692\ldots$ and $\alpha = 2.5029\ldots$ have been obtained for the particular case $W = R$ and $z = 2$. In the general period-n-tupling case, Equation (3.24) may have several solutions, corresponding to different primitive words of the same length n. For instance, when $n = 5$, it has three different solutions, corresponding to the primitive words RLR^2, RL^2R and RL^3, respectively. In order to pick up the desired solution, one has to carefully choose the initial value for α. We shall consider this aspect in the next section.

Meanwhile, we continue to discuss the solution of (3.24). Since this equation determines a saddle type fixed point $g(x)$ in the functional space of unimodal functions, just as we have learnt from the period-doubling case in Section 2.7.2, precautions must be taken in applying any systematic approximation to it. The trouble is that an initially convergent approximation scheme will sooner or later get away from the true solution, unless one sticks to the "critical line" (in the language of phase transition theory), or, in other words, keeps staying at μ_∞ from the very beginning — a numerically unrealistic requirement.

First of all, we can still use the method of truncated series, as we did with the Feigenbaum renormalization group equation (2.100) in Section 2.7.2 with the help of an algebraic manipulation language such as REDUCE or MACSYMA. For the period-tripling sequence, we get $\alpha = 9.0497$ and 9.2764, by retaining in (2.102) three and four terms, respectively. This is to be compared with the "exact" value 9.2773 computed in the following section. Retaining only two terms in (2.102) we have $\alpha = 0.03$, -46.21 and 161.9 for RLR^2, RLR^2 and RL^3, respectively. However, in order to obtain these approximate values, we need a rough estimate to solve the abovementioned system of algebraic equations separately for each given word.

We shall discuss the direct numerical solution of the renormalization group equation in Section 3.4.4. In concluding this subsection, we derive a few formulae from the renormalization group equation (3.24). Replacing x by $g^{(n)}(x)$ repeatedly in the equation, we find

$$g^{(jn)}(x) = \alpha^{-1} g^{(j)}(\alpha x), \quad j = 1, 2, \ldots, n-1.$$

When $j = n$, we have

$$g^{(n^2)}(x) = \alpha^2 g(\alpha^2 x).$$

Now, repeated substitution $x \to g^{(n^2)}$ leads to

$$g^{(jn^2)}(x) = \alpha^{-2} g^{(j)}(\alpha^2 x).$$

Combined use of the two substitutions yields the relation

$$g^{(jn^k)}(x) = \alpha^{-k} g^{(j)}(\alpha^k x).$$

These relations as well as their derivatives are useful in analyzing scaling properties of the limiting sets of period-n-tupling sequences.

3.4.3   Estimating the Scaling Factors α_W

We can use the polynomials $P_n(\mu)$, defined in Section 2.1, and the μ_∞ values, estimated in Section 3.4.1, to estimate the scaling factor α for various period-n-tupling sequences. For this purpose we replace the yet unknown universal function $g(x)$ in the renormalization group equation (3.24) by $f(\mu_\infty, x)$ and set $x = 0$. Thus, we are led to

$$\alpha = \frac{1}{g^{(n)}(0)} \approx \frac{1}{P_n(\mu_\infty)}. \qquad (3.26)$$

The α values obtained in this way are only approximate, but can be used as initial values in the search for more precise solutions to the renormalization group equation (3.24). We list these values of α in Table 3.4. Estimates from numerical experiments are given in parentheses in the same table.

W	$z=2$	$z=4$	$z=6$	$z=8$
R	2.50(2.50)	1.68(1.69)	1.46(1.47)	1.36(1.36)
RL	9.53(9.09)	3.27(3.15)	2.37(2.28)	2.00(1.95)
RL^2	39.7(40.0)	6.56(6.18)	3.90(3.66)	2.98(2.80)
RLR^2	20.1(20.2)	4.34(4.30)	2.82(2.79)	2.29(2.24)
RLR^2	-46.0(-46.1)	-6.51(-6.4)	-3.82(-3.73)	-2.91(-2.83)
RL^3	161.0(161.3)	13.11(12.3)	6.43(6.0)	4.42(4.10)

Table 3.4: Estimated scaling factor α_W for period-n-tupling sequences.

3.4.4 Numerical Solution of the Renormalization Group Equations

As has been shown by Feigenbaum (1978) in the period-doubling case, the fixed point described by (3.24) corresponds to a saddle point in functional space. Therefore, if one tries to approach the universal scaling function $g(x)$ by an iteration

$$g_{k+1}(x) = \alpha_k g_k^{(n)}\left(\frac{x}{\alpha_k}\right), \qquad (3.27)$$

g_k will first approach the scaling function g, but then, owing to the existence of the eigenvalue $\delta > 1$ of the linearized operator, it will eventually go away from g as $g_k = g - \delta^k h$, unless one has adhered to the parameter μ_∞ with high precision. This situation resembles the renormalization group analysis of continuous phase transitions, where the rescaling transformation can be carried out an infinite number of times only when one settles down on the critical surface $T = T_c$ from the outset, otherwise the "relevant" variables corresponding to eigenvalues greater than 1 will sooner or later lead the transformation away from the saddle point.

In order to solve the renormalization group equation numerically, Feigenbaum (1978) has mentioned an idea, based on iteration and coordinate rescaling, but he did not give any elaboration or application of the idea. In fact, he adopted a direct parametrization for $g(x)$. Inspired by this idea, an iteration scheme has been devised to solve (3.24) for $n \geq 3$ (Zeng, Hao, Wang and Chen, 1984). The essentials of this scheme will now be described.

For unimodal maps, normalized at $f(0) = 1$, one can take the rescaling factor for the coordinate as a control parameter. For an initial function, in order to start the iteration (3.27), we use some $f(\lambda_0 x)$ with λ_0 chosen as well as possible:

$$g_0(x) = f(\lambda_0 x). \tag{3.28}$$

Because λ_0 cannot be exactly equal to $\lambda_\infty = \sqrt{\mu_\infty}$, we rescale x during the iteration, choosing λ_k in such a way that at $x = 1$ the equality

$$\alpha_k(\lambda_k)g_k^{(n)}\left(\lambda_k, \frac{1}{\alpha_k(\lambda_k)}\right) = g_k(\lambda_k) \tag{3.29}$$

holds, where

$$g_k^{(n)}(\lambda, x) \equiv \underbrace{g_k(\lambda g_k(\lambda g_k \ldots \lambda g_k(\lambda x) \ldots))}_{n \ times}$$

and

$$\alpha_k(\lambda) = \frac{1}{g_k^{(n)}(\lambda, 0)}.$$

Once λ_k has been determined, let

$$g_{k+1}(x) = \alpha_k(\lambda_k)g_k^{(n)}\left(\lambda_k, \frac{x}{\alpha_k(\lambda_k)}\right) \tag{3.30}$$

be the result of the $(k+1)$th iteration. In practice, $g(x)$ has been approximated, using the Lagrange interpolation formula

$$g(x) = \sum_{i=0}^{N} g(x_i) \frac{\prod_{j \neq i}^{N}(x^2 - x_j{}^2)}{\prod_{j \neq i}^{N}(x_i{}^2 - x_j{}^2)} \tag{3.31}$$

with $g(0) = 1$. The iterate g_{k+1} is represented by $g_{k+1}(x_i)$ at N points x_i.

Numerical calculations show that the iteration converges quickly; the difference between two successive iterations becomes less than 10^{-D} where D is the number of digits in the computer, e.g., $D = 10$. The error of the approximation can be estimated by the order of magnitude of the coefficient of $x^{2(N+1)}$. Because $\delta\alpha = \alpha^2 \delta(g^{(n)}(0))$, the precision of α and the corresponding $g(x)$ drops for larger α. Skipping the computational technicalities, we summarize the calculated scaling factors in Table 3.5. The first number in Table 3.5

W	RL	RL^2	RL^2R	RLR^2	RL^3
α	9.2773	38.8189	20.128	-45.804	160.0

Table 3.5: Scaling factors α_W for period-n-tupling sequences.

may be compared with $\alpha = 9.2774$ in Derrida, Gervois and Pomeau (1979), and $\alpha = 9.28$ in Hu and Satija (1983).

All the results derived for the scaling factor κ and dispersion function D, related to the external noise in the case of a period-doubling cascade, can be carried over to period-n-tupling sequences (Wang and Chen, 1986). We skip the detailed calculation and only list the results in Table 3.6.

W	RL	RL^2	RLR^2	RL^2R	RL^3
κ	89.522	1558.7	431.91	2182.6	26458

Table 3.6: The noise scaling factor κ_W for period-n-tupling sequences.

3.5 Admissibility of Words and the Tent Map

In Section 3.3.1, we have called a word admissible, if it does correspond to a stable orbit, observable in iterations of the unimodal map. This is not quite correct, as actually the admissibility of a word is related to the property of a symbolic sequence to be maximal (see Section 3.6.3). In practice, there are few criteria to check the admissibility (or rather the inadmissibility) of a word. First of all, speaking about superstable orbits only, an admissible word must start with the two letters RL (except for the blank word and the single-letter word R corresponding to periods 1 and 2). This is clear from an inspection of the figure of a unimodal map: if one starts from the critical point x_c, the next iterate leads to the rightmost point, which in turn leads the next iterate to the leftmost point of the map. In fact, if we write down explicitly the understood letter C, then, in any admissible word

$$CRL\ldots\ldots ,$$

we only know the relative position of the first three points: they are central, rightmost and leftmost. All other letters correspond to points included in between the two extremes. Conversely, if a one-dimensional map consists of a well instead of a hump, then an admissible word would start with $CLR\cdots$. We shall make use of this simple remark in Chapter 5 when assigning letters to the numerically observed periodic orbits in differential equations.

In order to formulate a working criterion for checking the admissibility of words, we need some more mathematical structure. In what follows, we shall temporarily break our convention of dealing with smooth maps only, to discuss a piecewise linear map, namely, the tent map. We shall see repeatedly throughout this book that piecewise linear maps have the merit of allowing considerable analytical treatment. In this manner, we shall obtain a criterion for the admissibility of a word (Derrida, Gervois and Pomeau, 1978). A simpler and more general form of admissibility conditions, applicable to maps with more than one critical point, will be given at the end of Section 3.7.4.

3.5.1 The Tent Map

We consider the piecewise linear map

$$x_{n+1} = T(\lambda, x_n) \equiv \begin{cases} \lambda\, x_n, & \text{for } 0 < x_n < 1, \\ \lambda\,(2 - x_n), & \text{for } 1 < x_n < 2. \end{cases} \qquad (3.32)$$

The function $T(\lambda, x)$ maps the interval $(0, 2)$ into itself, when the parameter λ lies between 1 and 2; T is often called the tent map, from its shape (see

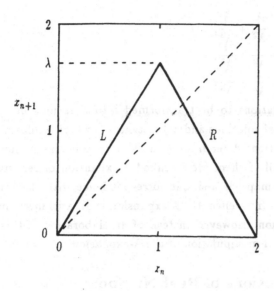

Figure 3.6: The tent map.

Fig. 3.6). Since the stability condition $T'(\lambda, x) < 1$ never holds, the tent map does not have any stable periods, but still we can use the method of Section 3.2 to determine the parameter value for a periodic orbit containing the "critical value" $x_c = 1$. (In fact, the derivation of (3.8) did not require stability of the period.) Now, the inverse branches are given by

$$
\begin{aligned}
R(y) &= 2 - y/\lambda, \\
L(y) &= y/\lambda.
\end{aligned}
\tag{3.33}
$$

Taking again the three different period 5 orbits and writing down the analogue of (3.10), using the functions (3.33); we have

$$
\begin{aligned}
RLR^2: \quad \lambda &= 2 - \frac{2}{\lambda^2} + \frac{2}{\lambda^3} - \frac{1}{\lambda^4}, \\
RL^2R: \quad \lambda &= 2 - \frac{2}{\lambda^3} + \frac{1}{\lambda^4}, \\
RL^3: \quad \lambda &= 2 - \frac{1}{\lambda^4}.
\end{aligned}
\tag{3.34}
$$

Since the degenerated value $\lambda = 1$ provides a solution to each of the equations $T(\lambda, 1) = W(1)$ for any word W, we cancel a factor $\lambda - 1$ in (3.34) and rewrite them in the form

$$\lambda = 1 + \frac{1}{\lambda} - \frac{1}{\lambda^2} + \frac{1}{\lambda^3},$$
$$\lambda = 1 + \frac{1}{\lambda} + \frac{1}{\lambda^2} - \frac{1}{\lambda^3}, \tag{3.35}$$
$$\lambda = 1 + \frac{1}{\lambda} + \frac{1}{\lambda^2} + \frac{1}{\lambda^3}.$$

These are the equations to be transformed into iterations in λ_n. We skip the actual numerical solutions and call attention to their uniform appearance. Indeed, these equations differ from each other only in the $\pm$ signs of some of the coefficients. All of them are so-called λ-expansion of real numbers. We note that the tent map has a simple correspondence with the shift map (see, e.g., Section 6.4.1). In essence, the λ-expansion is an arithmetic manifestation of the shift operation. However, instead of an elaboration of this remark, we turn to the practical manipulation of the λ-expansion.

3.5.2 λ-expansions of Real Numbers

Given two real numbers $1 < x < 2$ and $1 < \lambda < 2$, it is required to express x as a sum of inverse powers of λ. Obviously, the first two terms are

$$1 + \frac{1}{\lambda};$$

if this sum exceeds x, we subtract the next power λ^{-2}, otherwise we add it. This process either repeats *ad infinitum* or stops when the sum happens to be precisely equal to x. This is the definition of the λ-expansion of x

$$x = \sum_{i=0}^{\infty} \frac{a_i}{\lambda^i}, \tag{3.36}$$

where $a_0 = a_1 = 1$, $a_i = \pm 1$ or 0. If some a_k happens to be zero, then $a_i = 0$ for all $i > k$, i.e., the sum becomes finite. In particular, one may take x to be λ and get an auto-expansion of λ in terms of its own inverse powers:

$$\lambda = \sum_{i=0}^{\infty} \frac{a_i}{\lambda^i} \tag{3.37}$$

It is clear that not every set $\{a_i = +1, -1, 0\}$ will correspond to the auto-expansion of some $1 < \lambda < 2$. It so happens that the condition for auto-expansion to hold has a close relation to the admissiblity of words in the symbolic dynamics of two letters (Derrida, Gervois and Pomeau, 1978).

Skipping the detailed proof, we state only the condition for the $\{a_i\}$ to be the coefficients of the auto-expansion of a certain λ: the sequence $(a_0, a_1, \ldots)$ must satisfy all the inequalities

$$\pm (a_n, a_{n+1}, \ldots) < (a_0, a_1, \ldots), \quad \forall\, n > 1. \tag{3.38}$$

When left-shifting a finite sequence, it is complemented by 0's at the right end. The comparison in the above inequalities are made on a term-by-term basis from the left to the right. The $<$ relation holds for two sequences of numbers

$$(c_1, c_2, \ldots, c_{k-1}, c_k, \ldots) < (d_1, d_2, \ldots, d_{k-1}, d_k, \ldots)$$

as far as $c_i = d_i$ for $i = 1, 2, \ldots, k-1$, and $c_k < d_k$ in the sense of $-1 < 0 < 1$. In other words, the first appearance of an unequal relation determines the direction of the inequality between the two sequences.

Furthermore, the coefficients for an admissible expansion may be represented as products of ± 1's:

$$a_i = \beta_1 \beta_2 \ldots \beta_i, \qquad i = 1, 2, \ldots, \tag{3.39}$$

($a_0 = -\beta_0 = 1$, by definition). These β_i are in one-to-one correspondence to the letters in the admissible words: $-1 \to R$ and $+1 \to L$.

If the expansion (3.37) is finite, i.e.,

$$\lambda = \sum_{i=0}^{n-2} \frac{a_i}{\lambda^i}, \tag{3.40}$$

then it corresponds to an admissible word made of $n - 1$ letters and describes an orbit of period n.

Now we can formulate the procedure to check the admissibility of any given word:

1. Assign $+1$ to the first R.

2. Assign $+1$ or -1 letter by letter, referring to the parity of the preceding pattern. If the parity is odd, then R corresponds to -1 and L to $+1$, if even, then R corresponds to $+1$ and L to -1. These are the coefficients a_i.

3. Check the inequalities (3.38). If neither of them fails, the word is admissible.

Take, for example, the word RLR^3L, with the assignment

$$
\begin{array}{cccccc}
R & L & R & R & R & L \\
+1 & +1 & -1 & +1 & -1 & -1
\end{array}
$$

Among the inequalities, there is at least one that fails, namely, we have

$$(1, 1, 0, 0, 0, 0) \nless (1, 1, -1, 1, -1, -1)$$

Therefore, the word RLR^3L is inadmissible. In fact, all words of the form

$$RL^m \cdots RL^n \cdots$$

with $n > m$ are inadmissible (see Section 3.8.3 for further discussion).

The following FORTRAN program checks the admissibility of any symbolic sequence made up of letters R and L. Again, the program has not been optimized and is only used for demonstration.

```
        PROGRAM ADMISS
C THIS IS PROGRAM CHECKING THE ADMISSIBILITY OF A WORD
C MADE OF THE TWO LETTERS R AND L
        CHARACTER C(500)
        INTEGER A(500)
        OPEN(UNIT=4,FILE='OUTFILE',STATUS='NEW')
1       WRITE(*,*) 'INPUT # OF LETTERS IN THE WORD N=PERIOD-1'
        READ(*,*) N
        IF(N.EQ.0) STOP 'END OF JOB'
        WRITE(*,*) 'TYPE THE WORD TO BE CHECKED'
        READ(*,500) (C(K),K=1,N)
        WRITE(4,501) (C(K),K=1,N)
        IF(C(1).NE.'R') GOTO 40
500     FORMAT(500A1)
501     FORMAT(1X,'The word',500A1)
        P=-1
        A(1)=1
        IF(N.EQ.1) GOTO 31
        DO 10 M=2,N
        IF(C(M).EQ.'R') THEN
            A(M)=P
            P=-P
```

```
              ELSEIF(C(M).EQ.'L') THEN
                  A(M)=-P
              ENDIF
10            CONTINUE
              DO 30 M=1,N-1
              DO 20 K=1,N
              IF(A(M+K)-A(K)) 22,20,40
20            CONTINUE
22            DO 25 K=1,N
              IF(-A(M+K)-A(K)) 30,25,40
25            CONTINUE
30            CONTINUE
31            WRITE(4,*) '                    IS ADMISSIBLE.'
              GOTO 1
40            WRITE(4,*) '                    IS INADMISSIBLE.'
              GOTO 1
              END
```

3.5.3 Method to Generate All Admissible Words of Given Length

The results of the last section may be made to yield a rule for the generation of all admissible words of a given length. We will demonstrate this by the example of period 6 sequences. Firstly, the coefficients β_i are given all possible values ± 1. Then the coefficients a_i are calculated from (3.39). The 8 possible combinations are listed in Table 3.7. After checking the inequalities (3.38), only 5 of these 8 combinations remain valid, as has been indicated in the last column of Table 3.7. The three admissible words RL^2R, RLR^2 and RL^3 are generated in the same way. Obviously, one can easily program this procedure to generate all admissible words of a given length.

A question arises naturally in connection with the above scheme. How many admissible words of a given length exist? It is also the number of different values of λ that possess a finite auto-expansion in terms of their own inverse powers. This is yet another facet of the enumeration problem, related to the number of periodic windows in one-dimensional mappings. We shall discuss it at length in the next section.

We emphasize once more that no stable periods whatsoever exist in the tent map; however, it does have unstable periods corresponding to all primitive

β_0	β_1	β_2	β_3	β_4	a_0	a_1	a_2	a_3	a_4	Remark
-1	1	1	1	1	1	1	1	1	1	RL^4
-1	1	1	1	-1	1	1	1	1	-1	RL^3R
-1	1	1	-1	1	1	1	1	-1	-1	RL^2RL
-1	1	1	-1	-1	1	1	1	-1	1	RL^2R^2
-1	1	-1	1	1	1	1	-1	-1	-1	inadmis.
-1	1	-1	1	-1	1	1	-1	-1	1	inadmis.
-1	1	-1	-1	1	1	1	-1	1	1	inadmis.
-1	1	-1	-1	-1	1	1	-1	1	-1	RLR^3

Table 3.7: Possible combinations of $\{a_i\}$ and their admissibility.

admissible words in the symbolic dynamics of two letters. One can imagine the stability ranges to have been contracted to zero. Anyway, the tent map is useful for the generation of words required in the study of other unimodal maps. This assertion holds for more complicated piecewise linear maps and their counterparts with more than one critical point.

3.6   Symbolic Dynamics of Two Letters Revisited

In essence, the rules in the symbolic dynamics of two letters are quite simple. We consider the justification and some generalizations of these rules in this section, using again only elementary mathematics. In fact, we shall put most of the results of Section 3.3 into a more general and natural context. The use of symbolic description of orbits has a long history. Many important results were contained in Metropolis, Stein and Stein (1973), Derrida, Gervois and Pomeau (1978), Collet and Eckmann (B1980), with slight differences in some definitions. The presentation of this section is largely due to the work of Zheng (Zheng, 1988a and b; Hao and Zheng, 1988).

3.6.1 Monotonicity of Functions and Ordering of Words

The ordering of words in the symbolic dynamics of a finite number of letters is based on the natural order

$$L < C < R \qquad (3.41)$$

of the interval and on the monotonicity of the map branches that are represented by the letters. In order to explain this statement in more detail, let us recall a few simple consequences of the monotonicity of a function:

Property 1. If a function f is monotonic increasing on an interval, then for points $x_1 < x_2$, belonging to the interval, we have the same order $f(x_1) < f(x_2)$. If f is monotonic decreasing, then $x_1 < x_2$ implies the reverse order $f(x_1) > f(x_2)$.

Property 2. If f is monotonic increasing (decreasing), so is the inverse function f^{-1}, since the chain rule of differentiation applied to $f(f^{-1}(x)) = x$ leads to the same sign for the derivatives $df(x)/dx$ and $df^{-1}(y)/dy$.

Property 3. If two functions f and g are both monotonic increasing or decreasing, then the composition $f \circ g$ is monotonic increasing. If one of these two is monotonic increasing, whereas the other one is monotonic decreasing, then $f \circ g$ will be monotonic decreasing. Therefore, we can represent what has been said by a multiplication table with $+1$ assigned to monotonic increasing and -1 to monotonic decreasing functions. In other words, the parity of a composite function is the product of parities of its components. This observation generalizes naturally to compositions of several monotonic functions.

In listing the above properties, we have treated each monotonic branch of a map as a separate function. Therefore, they work equally well for the construction of the symbolic dynamics of the logistic map, the cubic map, or the sine-square map, etc. For the time being, we will confine ourselves to unimodal mappings.

We introduce the order of symbolic sequences by first considering numerical sequences. Suppose we have started the iteration of a map $f(x)$ from two different initial points x_1 and x_2 and obtained two numerical sequences

$$\Sigma_1 : \quad x_1, \ x_2 = f(x_1), \ x_3 = f(x_2), \ldots,$$
$$\Sigma_2 : \quad x_1', \ x_2' = f(x_2'), \ x_3' = f(x_2'), \ldots.$$

We *define* their order by the order of the leading elements, i.e., let $\Sigma_1 > \Sigma_2$, if $x_1 > x_1'$, and *vice versa*. The two sequences coincide if $x_1 = x_1'$, and there is no need to order them.

This ordering carries over to symbolic sequences on the basis of their natural order (3.41). Namely, if we have two sequences

$$\begin{aligned} \Sigma_1 &= \sigma_1, \sigma_2, \sigma_3, \ldots, \\ \Sigma_2 &= \tau_1, \tau_2, \tau_3, \ldots, \end{aligned} \tag{3.42}$$

then we say $\Sigma_1 > \Sigma_2$, if $\sigma_1 > \tau_1$ in the sense of (3.41). However, σ_1 and τ_1 may happen to be the same letter, but still the two sequences may be different. In this case, one has to compare the succeeding letters. Now parity comes into play. Inequality $\sigma_2 > \tau_2$ implies $\Sigma_1 > \Sigma_2$ only when the first numbers, i.e., σ_1 and τ_1, are located on the L side, so that they iterate to σ_2 and τ_2, using the monotonically increasing, even parity branch f_L. Otherwise, $\sigma_2 > \tau_2$ leads to the opposite relation $\Sigma_1 < \Sigma_2$. In general, we compare two sequences

$$\begin{aligned} W_1 &= W^* \sigma \ldots, \\ W_2 &= W^* \tau \ldots, \end{aligned} \tag{3.43}$$

where Σ^* denotes their common leading string and σ differs from τ. It is clear now that the order of the two sequences is determined by the order of σ and τ in the sense of (3.41) only when the common part Σ^* is of even parity. When Σ^* has odd parity, the order of the two sequences is determined by the anti-order of σ and τ. We have added "$\ldots$" in (3.43), because the order of words is determined by the first different pair of letters, counting from the left to the right, and it does not matter what comes next.

For a given map, each point x in the interval may serve as a seed that starts an iteration to yield a symbolic sequence. This sequence is sometimes called the itinerary of x. We shall see that mong all the itineraries, the itinerary of the critical point plays a special role (see the end of Section 3.7.4). The order of any two different sequences is essentially given by the location of the corresponding seeds.

3.6.2 Auxiliary Notation

In order to facilitate the discussion in the following sections, we introduce additional auxiliary notation. Firstly, we define a symbol-valued function

$$s(\epsilon) = \begin{cases} R & \epsilon = -1, \\ C & \epsilon = 0, \\ L & \epsilon = 1, \end{cases} \tag{3.44}$$

and its inverse

$$\epsilon(s) = \begin{cases} -1 & s = R, \\ 0 & s = C, \\ 1 & s = L. \end{cases} \tag{3.45}$$

Sometimes we will write the leading part of a symbolic sequence

$$\Sigma = s_0 s_1 \cdots s_{k-1} s_k \cdots s_n \cdots$$

as a partial sequence, using the subscripted notation

$$\Sigma_k = s_0 s_1 \cdots s_{k-1}$$

As a rule, Greek capital letters represent whole or partial sequences, lower case Roman or Greek letters represent single letters.

Next, we define a parity function $P(\Sigma_n)$ for the sequence Σ_n

$$P(\Sigma_n) = - \prod_{i=0}^{n-1} \epsilon(s_i). \tag{3.46}$$

Note that a $-$ sign has been introduced in the above definition. It makes an even word (in the sense of Section 3.3.1) to have parity -1. With this convention it is possible to "order" parity functions in connection with the ordering of sequences. In particular, we have

$$P(R) = 1, \quad P(C) = 0, \quad P(L) = -1.$$

In order to be consistent, one must assign parity -1 to a blank sequence.

If one splits a symbolic sequence into two, then the parity function decomposes into factors with an additional minus sign

$$P(\Lambda\Sigma) = -P(\Lambda)P(\Sigma). \tag{3.47}$$

Finally, note the identity

$$P(s(\epsilon)) = -\epsilon \tag{3.48}$$

which follows from the definitions.

3.6.3   Maximal Sequences and Admissibility of Words

In this section, we shall extend our discussion to all possible orbits of the unimodal map, and not restrict ourselves to superstable periodic orbits only. A period n orbit will be described by a word of n letters. Superstable periods will be distinguished by the presence of the letter C. It is convenient to write all symbolic sequences, whether periodic or not, as infinite sequences. For example, the period 3 word RL becomes $(RLC)^\infty$. Being an infinite succession, an ambiguity arises in the way of writing the sequence: the same sequence might be written as $(LCR)^\infty$ or $(CRL)^\infty$. We will follow the convention of always writing the *maximal sequence*.

In order to define the maximal sequence, we introduce the *shift operator* S acting on a sequence Σ:

$$
\begin{aligned}
\Sigma &= s_0 s_1 s_2 \ldots s_k s_{k+1} \ldots, \\
S(\Sigma) &= s_1 s_2 \ldots s_k s_{k+1} \ldots, \\
&\quad \ldots \qquad \ldots \\
S^k(\Sigma) &= s_k s_{k+1} \ldots \ldots.
\end{aligned}
\tag{3.49}
$$

A sequence Σ is called maximal, if it precedes all its shifts in the sense explained in Sections 3.3.1 or 3.6.1, i.e.,

$$
\Sigma \geq S^k(\Sigma) \qquad \forall\, k. \tag{3.50}
$$

Clearly, among the three ways of presentation of the period 3 sequence only $(RLC)^\infty$ is maximal. Similarly, $(RRL)^\infty$ is not, but $(RLR)^\infty$ is maximal. The importance of maximal sequences arises from the fact that they have the privilege of becoming real orbits in a mapping, i.e., they correspond to admissible words, when they are superstable. This can be seen by looking at Fig. 3.7, where a few "inverse paths" of the critical point x_c are shown. The two solid circles drawn on the map are the two possible locations of $f^{-1}(x_c)$, i.e., starting from any of these points, one arrives at x_c in one iteration. Similarly, the four squares indicate various $f^{-1} \circ f^{-1}(x_c)$, i.e., one arrives at x_c in two steps, starting from any of these squares. At the particular height, i.e., for the given parameter value of the drawing, only 6 points (triangles) correspond to $f^{-1} \circ f^{-1} \circ f^{-1}(x_c)$. Now look at the dangling arrow in the upper right-hand corner. One can close it, to achieve a periodic orbit by slightly adjusting the

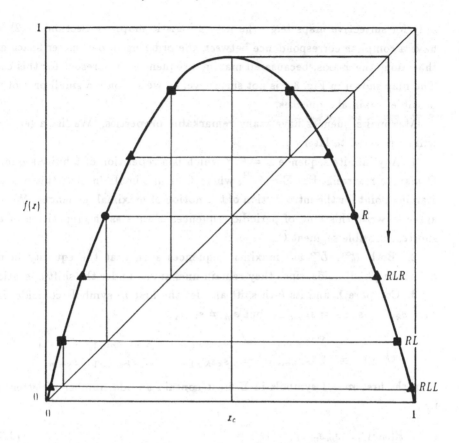

Figure 3.7: The maximal sequence among inverse paths.

height of the map. In doing so, only the rightmost point of some inverse paths may have a chance to be connected to the arrow. Since the symbolic sequences are ordered essentially according to the coordinate of the rightmost point, it is clear that only maximal sequences may become admissible.

It can be seen from the same Fig. 3.7 one realizes that the order of appearance of the words, when the parameter is varied, coincides with the order of the maximal sequences, as defined by their rightmost points on the x-axis, provided the height of the map changes monotonically with the parameter. However, only for surjective maps (e.g., the $\mu = 2$ logistic map, see Section 2.5.2), we have a complete correspondence between the ordering in parameter space and that along the x-axis, because all maximal sequences are present for this case. The map shown in Fig. 3.7 is not surjective, so we see only a small part of the possible maximal sequences.

Maximal sequences have many remarkable properties. We list a few that will be referred to later on.

1. Any infinite sequence $\Sigma = \Theta^\infty$ which is a repetition of a finite sequence Θ may be rearranged as $\Sigma = \overline{\Theta}^\infty$, where $\overline{\Theta}$ is maximal. In fact, this was our starting point for the introduction of the notion of maximal sequences. We will agree to write this kind of periodic sequences always as the repetition of the shortest possible segment $\overline{\Theta}$.

2. Both R^∞, L^∞ are maximal sequences such that the equality in the definition (3.50) holds, since they remain unchanged under the shift operation.

3. Compare Σ and its k-th shift and let the first m symbols coincide, i.e., $s_0 = s_k, \ldots s_{m-1} = s_{k+m-1}$, but $s_m \neq s_{k+m}$:

$$\Sigma \equiv \Sigma^* s_m \ldots = s_0 s_1 \ldots \ldots \ldots s_{m-1} \ s_m \ldots,$$
$$S^k(\Sigma) \equiv \Sigma^* s_{k+m-1} = s_k s_{k+1} \ldots \ldots s_{k+m-1} \ s_{k+m};$$

then the first $m + 1$ symbols in Σ must contain an odd number of letters R, i.e.,

$$P(s_0 s_1 \ldots s_m s_{m+1}) = 1. \tag{3.51}$$

4. If a symbolic sequence starts with RL^n, then any succeeding pattern of the form RL^m with $m > n$ will violate its maximality. From this observation, there follows a method for making an infinite sequence Σ^∞ maximal: shift Σ cyclically until a pattern RL^n with the largest n appears as the leading one.

3.6.4   The Periodic Window Theorem

Analytically, the existence of a periodic window around each superstable parameter value follows from continuity considerations in the vicinity of the point, where the superstable condition (2.35) holds. Moreover, using the language of symbolic dynamics, we are able to write down the symbolic sequences representing the whole window. The width, i.e., the end points of the window, depend, however, on the mapping explicitly (see Section 2.7.2). Consequently, it cannot be determined from symbolic dynamics alone. In this section, we treat merely the symbolic dynamics part of the problem, namely, we are going to prove the following

Periodic Window Theorem (Zheng, 1988a). If $(\Sigma C)^\infty$ is a maximal sequence, where $\Sigma = s_0 s_1 \ldots s_n$, $s_i = R$ or L, $i = 0, 1, \ldots, n$, then $(\Sigma t)^\infty$, where $t = R$ or L, are both maximal sequences.

The proof of this theorem employs a few tricks that are typical for manipulations with symbolic sequences; for this reason, we will present it in full.

The given condition of $(\Sigma C)^\infty$ being a maximal sequence implies that

$$s_0 s_1 \ldots C \geq s_k s_{k+1} \ldots s_n C \ldots, \quad k = 0, 1, \ldots, n. \tag{3.52}$$

We intend to prove that

$$s_0 s_1 \ldots t \geq s_k \ldots s_n t s_0 \ldots, \quad k = 1, 2, \ldots, n. \tag{3.53}$$

In (3.50) and (3.53), $s_i, i = 0, 1, \ldots, n$, and t are either R or L. For the sake of clarity, we put the left-hand side of (3.53) on top of its right-hand side

$$s_0 s_1 \ldots \quad s_{n-k} \quad s_{n-k+1} \quad s_{n-k+2} \cdots \quad s_n t \quad s_0 \ldots,$$
$$s_k s_{k+1} \ldots \quad s_n \quad t \quad s_0 \quad \ldots \quad s_{k-2} s_{k-1} \quad s_k \ldots,$$

and consider four cases, step by step.

Case a.

$$s_0 s_1 \ldots s_{n-k} \neq s_k s_{k+1} \ldots s_n.$$

It is a direct consequence of Condition (3.50) that Relation (3.53) holds.

Case b.

$$s_0 s_1 \ldots s_{n-k} = s_k s_{k+1} \ldots s_n,$$

but

$$s_{n-k+1} \neq t.$$

In this case, it follows from Condition (3.50) and (3.51) that

$$P(s_0 s_1 \ldots s_{n-k} s_{n-k+1}) = 1. \tag{3.54}$$

Therefore, we have

$$P(s_0 s_1 \ldots s_{n-k})P(s_{n-k+1}) = -1.$$

There are two alternatives:

$$t = R > s_{n-k+1} = LP(s_0 s_1 \ldots s_{n-k}) = +1$$

or

$$t = L < s_{n-k+1} = RP(s_0 s_1 \ldots s_{n-k}) = -1.$$

Either of them leads to Relation (3.53), according to the ordering rule reformulated in Section 3.6.1.

Case c.

$$s_0 s_1 \ldots s_{n-k+1} = s_k s_{k+1} \ldots s_n t = \rho. \tag{3.55}$$

(A temporary symbol ρ will be used in the study of Case d.) The order of the two sides of (3.53) is determined by the succeeding sequences

$$s_{n-k+2} \ldots s_n t \ldots \neq s_0 \ldots s_{k-2} s_{k-1} \ldots. \tag{3.56}$$

Condition (3.52) applied to (3.56) now yields

$$s_{n-k+2} \ldots s_n t \ldots < s_0 \ldots s_{k-2} s_{k-1} \ldots,$$

and (3.54) still holds true. Therefore, this case comes under Criterion (3.51), and (3.53) is true.

Case d. In addition to the equality (3.55), we have instead of (3.56) a second equality

$$s_{n-k+2} \ldots s_n t = s_0 \ldots s_{k-2} s_{k-1} = \lambda. \tag{3.57}$$

Therefore, the sequence $s_0 s_1 \ldots s_n t$ may be written either as $\rho\lambda$ or as $\lambda\rho$. The longer one of λ and ρ, say λ, must contain ρ and be further decomposed into $\lambda = \lambda'\rho = \rho\lambda'$. This process continues until the two subsegments equal each other or become multiples of a common pattern σ which may be a single letter or a pattern that cannot be represented as repetition of a smaller subsegment. Therefore, we have

$$(\Sigma t)^\infty = \sigma^\infty,$$

and we know that this sequence may be suitably shifted to acquire the form of a maximal sequence. In fact, this is just the case when the equality sign in (3.53) holds true. Thus we have proved the theorem.

The periodic window theorem tells us that, whenever there is a superstable orbit described by the symbolic sequence ΣC, one can change the letter C to R and L to get orbits in the window. These sequences are ordered in the following way:

$$(\Sigma s(-P(\Sigma)))^\infty \; < \; \Sigma C \; < \; (\Sigma s(P(\Sigma)))^\infty. \tag{3.58}$$

We shall call $(\Sigma s(-P(\Sigma)))^\infty$ the *lower sequence*, and $(\Sigma s(P(\Sigma)))^\infty$ the *upper sequence* of Σ. It is easy to see that the parity of the fundamental string of the lower sequence, i.e., of its unit of repetition, is always -1, whereas that for the upper sequence is $+1$. They will be used in the next section to construct the admissible sequence between two given sequences.

Example: the Period-Doubling Cascade. The superstable fixed point C extends to a window (L, C, R), the leftmost orbit may be written as L^∞, which, as we know, is the smallest among all possible sequences. The subsequent period 2, 4, and 8 superstable orbits extend to

$$(RR, RC, RL),$$
$$(RLRL, RLRC, RLRR),$$
$$(RLRRRLRR, RLRRRLRC, RLRRRLRL),$$

respectively. We see that the rightmost symbol in a window coincides with the leftmost symbol in the next window, if written in their "infinite" form $((RR)^\infty$, $(RLRL)^\infty$, and $(RLRRRLRR)^\infty$ in the above windows), leaving no room for other words to get in between. This remark will become clearer in the next section.

3.6.5 The Median Sequence In Between Two Given Sequences

We give another method to construct a median word, i.e., a symbolic sequence included between two given sequences, based on the periodic window theorem. This natural approach does not invoke such artificial notion as antiharmonic.

Suppose we are given two superstable orbits described by the words $\Sigma_1 C <$ $\Sigma_2 C$. Construct the lower sequence of Σ_2 and the upper sequence of Σ_1 according to (3.58). Then find their common leading part Σ_*, i.e.,

$$
\begin{aligned}
(\Sigma_2 s(-P(\Sigma_2)))^\infty &= \Sigma^* \sigma_1 \ldots, \\
(\Sigma_1 s(P(\Sigma_1)))^\infty &= \Sigma_* \tau_1 \ldots.
\end{aligned}
\tag{3.59}
$$

The common leading part Σ^* provides the median sequence we are looking for. It is easy to verify that $(\Sigma^* t)^\infty$, where $t = R$, L, or, C, are maximal sequences. If the lower sequence of Σ_2 and the upper sequence of Σ_1 coincide, then Σ_1 and Σ_2 are adjacent sequences, and no median sequence can be inserted in between.

Now we return to the examples that we have discussed in Section 3.3.2.

Example 1. $W_1 = RLR^2 C < W_2 = RLC$. Construct the two window patterns

$$
\begin{aligned}
(RLR^3)^\infty &< RLR^2 C < (RLR^2 L)^\infty, \\
(RLR)^\infty &< RLC < (RL^2)^\infty.
\end{aligned}
\tag{3.60}
$$

The common pattern of the upper sequence of W_1 and the lower sequence of W_2 yields the period 7 pattern $RLR^2 LR$.

Example 2. $W_1 = C$, $W_2 = RL^6 C$. We have

$$
\begin{aligned}
L^\infty &< C < R^\infty, \\
(RL^6 R)^\infty &< RL^6 C < (RL^7)^\infty,
\end{aligned}
\tag{3.61}
$$

the common part being R. Now construct the window patterns of RC

$$
R^\infty < RC < (RL)^\infty.
$$

The lower sequence coincides with the upper one of C, indicating that C and R are adjacent periods. The common part in the other direction gives RL. In this way we recover Table 3.1.

These two examples demonstrate the advantage of the procedure: no artificial constructs such as antiharmonics appear in the intermediate stage, all the sequences built at one step may be useful at the next step. Furthermore, the sole idea is based on continuity consideration and is readily generalized to maps with more than one critical points (see the end of Section 3.7.4).

3.6.6   The Generalized Composition Rule

Next, we consider a far-reaching generalization of the ∗-composition rule, described in Section 3.3.4. Equipped with this generalized composition rule, we shall obtain a unifying view on period-doubling, period-n-tupling, intermittency and chaos in one-dimensional mappings. Furthermore, it will provide a bridge from symbolic dynamics of interval mappings to the Farey sequence description of circle maps.

We first recall the ∗-composition rule, introduced in Section 3.3.4, from a slightly different point of view. Given two words Θ and $\Sigma = \sigma_0 \sigma_2 \cdots \sigma_k C$ (we add the terminating C explicitly), the ∗-composition was defined as [see (3.19)]

$$\Theta * \Sigma = \Theta \bar{\sigma}_1 \Theta \bar{\sigma}_2 \Theta \cdots \Theta \bar{\sigma}_k \Theta C,$$

where the rule for $\bar{\sigma}_i$ may be expressed by using the auxiliary function $s(\epsilon)$ and parity function P introduced in Section 3.6.2:

$$R \implies \Theta * R = \Theta s(P(\Theta)),$$
$$C \implies \Theta * C = \Theta C,$$
$$L \implies \Theta * L = \Theta s(-P(\Theta)).$$

In fact, these three cases can be combined into the single case

$$\sigma \implies \bar{\sigma} = \Theta s(P(\sigma) P(\Theta)). \tag{3.62}$$

A remarkable property of the above substitution rule is that it preserves the parity. Indeed, using (3.47) and the identity (3.48), we have

$$
\begin{aligned}
P(\Theta s(P(\sigma) P(\Theta))) &= -P(\Theta) P(s(P(\sigma(P(\Theta))))) \\
&= [-P(\Theta)]^2 P(\sigma) = P(\sigma).
\end{aligned}
$$

This result provides a hint for the generalization of the ∗-composition rule. We will search more general parity-preserving substitutions of R and L by strings of letters ρ and λ which would transform a maximal sequence into a maximal one.

In this book we only formulate the generalized composition rule, referring the interested reader to the original paper (Zheng, 1988b) for mathematical proofs. In addition, we note that in its present form this rule is applicable only to unimodal mappings. In order to demonstrate its power, a series of examples will be discussed in the next section. Thus, we turn to

The Generalized Composition Rule. Let

$$\Sigma = s_0 s_1 \ldots s_n \ldots$$

be a finite or infinite maximal symbolic sequence, where $s_i = R$ or L for $i = 0, 1 \ldots$. Then the substitutions

$$R \to \rho,$$
$$L \to \lambda. \tag{3.63}$$

transform Σ into another maximal sequence, provided the new symbolic sequences ρ and λ satisfy the following 5 conditions:

1. $P(\lambda) = -1$, $P(\rho) = 1$.

2. $\rho > \lambda$.

3. ρ with its last letter replaced by C, denoted hereafter as $\rho|_C$, is maximal.

4. $\rho\lambda|_C$ is maximal.

5. $\rho\lambda^\infty$ is maximal.

We note that the first two conditions preserve the parity, while Conditions 3 through 5 guarantee the maximality of the transformed sequence. It is easy to verify that the old *-composition rule, i.e., $\rho = R$ and $\lambda = L$, satisfies these conditions. Notice also that if ρ and λ satisfy the above conditions, so do the new strings $\rho' = \rho\lambda^n$ and $\lambda' = \lambda^m$ with n, m being finite non-negative integers.

3.6.7   Applications of the Generalized Composition Rule

A good guide to the choice of ρ and λ in particular cases is provided by the periodic window theorem studied in Section 3.6.4, since any finite maximal sequence ΣC will give rise to ΣR and ΣL that are of opposite parity. A few examples may help us reach a better understanding.

The Period-doubling Cascade

The simplest choice starts with a blank (in the notation of Section 3.3) or a single letter C (in our present notation). We already know that it extends to the window of the fixed point (L, C, R) (see the example at the end of Section 3.6.4).

These R and L are, of course, trivial choices for ρ and λ. Then comes the letter R, corresponding to the period 2 window (RR, RC, RL). Therefore, we have

$$R \to \rho = RL, \qquad\qquad L \to \lambda = RR. \qquad\qquad (3.64)$$

One might have added a third substitution $C \to RC$ for the letter C. It is easy to verify that these ρ and λ satisfy all the conditions listed in Section 3.6.6. Therefore, we can define the following sequence by successively substituting (3.64):

$$\begin{aligned} RC \prec RLRC \prec \quad & RLRRRLRC \prec \\ & RLRRRLRLRLRRRLRC \prec \cdots \end{aligned} \qquad (3.65)$$

This is just the period-doubling sequence

$$R \prec R*R \prec R*R*R \prec R*R*R*R \prec \cdots$$

[see (3.20)].

Now take the next member $RLRC$ in (3.65). It leads to the window $RLRL$ — $RLRC$ — $RLRR$, and we can define

$$\rho_1 = RLRR = \rho\lambda, \qquad\qquad \lambda_1 = RLRL = \rho\rho, \qquad\qquad (3.66)$$

where ρ and λ were given in (3.64). The next period starts from

$$\lambda_2 = \rho_1\rho_1 = RLRRRLRR$$

and extends to

$$\rho_2 = \rho_1\lambda_1 = RLRRRLRL,$$

etc. In general, the words of the ith period are related to the previous ones as $\lambda_i = \rho_{i-1}\rho_{i-1}$ and $\rho_i = \rho_{i-1}\lambda_{i-1}$.

Repeated use of any of these ρ_i, λ_i pairs, as a new substitution rule, will pick out a subsequence from the main sequence (3.65). For instance, the substitutions $R \to \rho_1$, $L \to \lambda_1$ and $C \to \rho_1|_C$, select the period-quardrupling members from the doubling sequence (3.65).

In general, if there is a window $(\rho, \rho|_C, \lambda)$, then its period-doubled regime will be given by

$$(\rho\rho, \rho\lambda|_C, \rho\lambda).$$

This observation holds not only for the main period-doubling sequence, but also for any period-doubling sequence developed from a primitive word. However, one must be clearly aware of the fact that the above symbolic description tells nothing about the stability range of the window which depends on the particular nonlinear function defining the map.

Period-n-tupling Sequences

Take, for instance, one of the three period 5 words, say, $\Sigma = RL^2R$. We have the substitutions

$$
\begin{aligned}
R &\to \rho = RL^2RL, \\
L &\to \lambda = RL^2RR, \\
C &\to RL^2RC,
\end{aligned}
\qquad (3.67)
$$

and

$$
RL^2RC \prec RL^2RLR\ L^2RR\ RL^2RR\ RL^2RL\ RL^2RC \prec \cdots \qquad (3.68)
$$

This is just the 5^k sequence, generated from RL^2R, i.e.,

$$
RL^2R \prec (RL^2R) * (RL^2R) \prec \cdots.
$$

These examples show clearly that the ρ, λ substitutions include the $*$-composition. Moreover, they give a lead to the form of the renormalization group equations. For example, the substitution (3.64) implies a rescaling of the form

$$
\begin{aligned}
g_R &\to g_L \circ g_R, \\
g_L &\to g_R \circ g_R,
\end{aligned}
\qquad (3.69)
$$

while the substitution (3.67) suggests

$$
\begin{aligned}
g_R &\to g_L \circ g_R \circ g_L \circ g_L \circ g_R, \\
g_L &\to g_R \circ g_R \circ g_L \circ g_L \circ g_R.
\end{aligned}
\qquad (3.70)
$$

Note the reverse order of functions on the right-hand sides of (3.69) and (3.70). If we had used the inverse functions, they would appear in the same order as the letters in the corresponding word. For symmetric functions g, there is no need to distinguish g_R from g_L. Apart from a scaling factor, the above relations lead to the renormalization group equations (2.100) or (3.24).

Fully Developed Chaos

The nature of the chaotic orbit described by RL^∞ at $\mu = 2$ for the logistic map has been well studied since the pioneering work of Ulam and von Neumann (1947). We have devoted Section 2.5 to its discussion. The most prominent feature consists of the existence of a continuous, invariant density for the distribution of the $\{x_i\}$ when the number of points goes to infinity (we have calculated it in Section 2.5.3). Therefore, it may serve as a prototype for other chaotic orbits. This assertion may be made precise by means of the generalized composition rule: whenever we are able to bring an infinite sequence into the form $\rho\lambda^\infty$ by defining symbolic strings ρ and λ, satisfying the conditions listed in Section 3.6.6, we say that the motion is as chaotic as the RL^∞ orbit. The orbit may exhibit minor details at the scale of ρ and λ (recall what we have said on the fine structure of power spectra in Section 3.3.4), but coarse-grained over the scale of ρ and λ, it looks just like the RL^∞ orbit. Moreover, close to these orbits there should exist a non-zero measure on the parameter axis for aperiodic orbits just as in the vicinity of RL^∞ (this is a statement which is still awaiting a rigorous mathematical proof).

It is worth mentioning that the RL^∞ orbit itself in the surjective logistic map exhibits nothing chaotic. In fact, it yields a regular sequence of numbers

$$0, 1, -1, -1, -1, -1, \cdots$$

We say the map is chaotic, because all possible sequences formed of letters R and L, included between L^∞ and RL^∞ (recall Example 4 in Section 3.3.1), must be present now. There are as many as real numbers. Whenever there is a $R \to \rho$, $L \to \lambda$ substitution, the interval between λ^∞ and $\rho\lambda^\infty$ may be put into correspondence with that between L^∞ and RL^∞, and all possible coarse-grained orbits must be present.

Crises and Homoclinic Orbits

The "microscopic" structure of chaotic bands does not depend continuously on the parameter, whereas their general shape varies smoothly with the parameter, except for certain values where the size of the band changes abruptly. This change takes place when an unstable object in the phase space touches the attractor; this phenomenon has been called *crisis* of attractors by Grebogi, Ott and Yorke (1982 a and b). In one-dimensional mappings, the $\rho\lambda^\infty$ sequences also determine the crisis points, as λ^∞ describes the unstable orbit that collides with the chaotic attractor.

Furthermore, a $\rho\lambda^\infty$ sequence describes a homoclinic orbit. As we have learnt in Section 2.5.5, homoclinic orbits in one-dimensional mappings are not so fully fledged as in higher dimensional systems. The stable set degenerates into a finite number of points, represented by the symbolic string ρ. The sequence λ^∞ always represents an unstable orbit, and the ρ string describes the finite number of steps that leads to the unstable orbit. For example, the homoclinic orbit, shown in Fig. 2.11, is just $RL(RR)^\infty$. The unstable fixed point R^∞ is written as $(RR)^\infty$, in accordance with the parity of λ.

Band-Merging Points

We know from Section 3.3.4 that the merging of two chaotic bands into one takes place at $RL(RR)^\infty$ [see (3.21)]. We know from Section 3.6.7 that the period 2 window (RR, RC, RL) suggests the definitions $\rho = RL$ and $\lambda = RR$. We have λ_∞ at the point where the period 2 starts and $\rho\lambda_\infty$ at the point where the 2-band region ends, i.e., the two bands merge into one. Since these ρ and λ satisfy all the 5 conditions of the generalized composition rule, the corresponding segment of the bifurcation diagram bears a full similarity to the whole parameter range from L^∞ to RL^∞. This is another manifestation of the internal similarity of the bifurcation diagram. A similar assertion holds for the range from the start of the period 4 orbit $((RLRL)^\infty)$ to the end of the 4-band region $(RLRR(RLRL)^\infty$, (see Section 3.3.4); just let $\rho = RLRR$ and $\lambda = RLRL$. This is nothing but the substitution (3.66) defined in Section 3.6.7.

Moreover, we can calculate the exact parameter values for these band-merging points by slightly extending the word-lifting technique of Section 3.2.1. The $2 \rightarrow 1$ merging point $RL(RR)^\infty$ lifts to an infinitely nested square-root equation which reduces to two finite equations on introducing one more variable

$$\begin{aligned}
\mu &= \sqrt{\mu + \sqrt{\mu - \nu}}, \\
\nu &= \sqrt{\mu - \nu}.
\end{aligned} \tag{3.71}$$

The $4 \rightarrow 2$ merging point $RLRR(RLRL)^\infty$ leads to

$$\mu = \sqrt{\mu + \sqrt{\mu - \sqrt{\mu - \sqrt{\mu - \nu}}}},$$
$$\nu = \sqrt{\mu + \sqrt{\mu - \sqrt{\mu + \sqrt{\mu - \nu}}}}. \qquad (3.72)$$

The situation resembles what we have encountered in calculating the superstable parameter values (see Section 3.2.2). By eliminating ν from, say, (3.71) one ends up with a polynomial equation, and then has to single out the required root. Our old trick of transforming these equations into iterations again comes to our aid. For instance, Equations (3.72) yield a quick converging iteration scheme

$$\mu_{n+1} = \sqrt{\mu_n + \sqrt{\mu_n - \sqrt{\mu_n - \sqrt{\mu_n - \nu_n}}}},$$
$$\nu_{n+1} = \sqrt{\mu_n + \sqrt{\mu_n - \sqrt{\mu_n + \sqrt{\mu_n - \nu_n}}}}, \qquad (3.73)$$

which leads to $\mu = 1.4303576\ldots$ and $\nu = 1.3248379\ldots$ from any reasonable initial values, say, $\mu_0 = 2.0$ and $\nu_0 = 1.95$.

In fact, any infinite word of the form $\rho\lambda^\infty$ leads to a pair of iteration relations like (3.73), the first corresponding to the ρ and the second to the λ part. The only exception is RL^∞ which reduces to the simple equation

$$\mu = \sqrt{\mu + \mu},$$

yielding $\mu = 2$ (apart from the trivial solution $\mu = 0$). Actually, there is nothing peculiar in RL^∞. It may be considered as the zeroth band-merging point beyond which the chaotic band ceases to exist.

What has been said applies to all high order band-merging, points seen in the tails of the bifurcation sequences developed from tangent bifurcations. For example, the period 3 window (RLL, RLC, RLR) (see Section 3.6.7 below) has a "mirror" three-band chaotic regime which merges at $RLL(RLR)^\infty$. The iteration scheme generated by this infinite word gives $\mu = 1.79032749\ldots$ which is the exact parameter of the crisis studied first by Grebogi, Ott and Yorke (1982a).

We emphasize that the above-described method is not restricted to the logistic map only. Whenever the inverse branches of a map are known explicitly, all the superstable periodic orbits and all the band-merging points can be located with as high precision as one wishes.

The Intermittency

The description of intermittency provides us with another excellent example demonstrating the power of symbolic dynamics. We start from the period 3 window RL. The periodic window theorem guarantees the existence of the sequences in the following order

$$(RLR)^\infty < RLC < (RLL)^\infty, \tag{3.74}$$

with the middle one being the superstable orbit. Rewrite the lower sequence of the window in the form

$$(RLR)^\infty = R(LRR)^\infty. \tag{3.75}$$

Now let $\rho = R$, $\lambda = LRR$; they satisfy all the conditions for the generalized composition rule. Therefore, there are orbits described by $(RLR)^\infty = \rho\lambda^\infty$ embedded in the lower part of the window. According to what we have just said, they are indeed chaotic orbits. Being overwhelmed by the stable period 3 orbits, these unstable orbits remain invisible in the asymptotic behaviour of numerical iterations.

The lower sequence might as well be transformed into

$$(RLR)^\infty = RL(RRL)^\infty.$$

With $\rho' = RL$, $\lambda' = RRL$, the same may be said with respect to $\rho'\lambda'^\infty$. These unstable chaotic orbits play the role of "strange repellers" and they may show off in transient behaviour (see Chapter 7).

Located before the lower sequence of the window there are many sets of periodic and chaotic orbits that approach the $(RLR)^\infty$ sequence asymptotically. For instance, one may construct a series of superstable periodic orbits

$$\Sigma_k = R(LRR)^k C = RL(RRL)^{k-1} RRC; \tag{3.76}$$

it is readily verified that

$$\Sigma_k^\infty < \Sigma_{k+1}^\infty < (RLR)^\infty. \tag{3.77}$$

Each Σ_k extends to a window of period $(3k + 2)$ orbits. When k gets large enough, these orbits behave much like period 3 orbits.

In order to construct a chaotic orbit with intermittent period 3 behaviour, we take, e.g., the sequences

$$\Lambda_k = R[(LRR)^k RR]^\infty \qquad k = 1, 2, \ldots . \tag{3.78}$$

Clearly,

$$\Lambda_k^\infty < (RLR)^\infty \quad \forall \ k \le 1,$$

Therefore, they are located before the period 3 window. Let $\rho = R$, $\lambda_k = (LRR)^k RR$, both satisfying the conditions for the generalized composition rule; we see that $\Lambda_k = \rho \lambda_k^\infty$ must be in the large chaotic orbits like RL^∞. For sufficiently big k, these orbits spend most of the time travelling around a period 3 pattern. This is precisely what we call intermittency. In order to study the scaling properties of these orbits, one can consider periodic sequences embedded in these orbits. For example, take the periodic orbits

$$\Pi_k = [R(LRR)^k C]^\infty . \tag{3.79}$$

By "embedded in the intermittent orbits" we mean

$$\Lambda_{k-1} < \Pi_k < \Lambda_k \quad \forall \ k \ge 1, \tag{3.80}$$

so that the two sets $\{\Lambda_k\}$ and $\{\Pi_k\}$ approach the same limit as k goes to infinity. The convergence rate of $\{\Lambda_k\}$ is the same as that of $\{\Pi_k\}$, which may be calculated using the word-lifting technique (Section 3.2).

Quasiperiodic Orbits

When speaking about quasiperiodic motion, one usually thinks of a competition of two and more incommensurable frequencies (i.e., the frequency ratios are irrational numbers) and refers to circle mapping as the simplest example. In fact, even in unimodal maps, there are plenty of quasiperiodic orbits. We touch briefly on this problem as the last application of the composition rule.

What happens if one keeps on executing the substitutions $R \to \rho$ and $L \to \lambda$ *ad infinitum*? Take, for example, the period-doubling cascade, discussed at the beginning of this section. The substitution (3.64), applied infinitely many times, leads to an aperiodic orbit. If observed with finite resolution, this orbit may display a periodicity. The higher the resolution, the longer will be the period. In fact, all the limiting sets of the period-n-tupling sequences belong to this kind of aperiodic orbits. They are not chaotic orbits of

the $\rho\lambda^\infty$ type. One could construct composition rules to obtain more complicated sequences of orbits the periods of which would grow, say, like Fibonacci
numbers (Section 4.4.5).

Another angle from which to look at quasiperiodic orbits in unimodal mappings is related to the $P_n(\mu)$ functions, defined in Section 2.1. Being functions
of the parameter μ, some of these functions may intersect or contact tangentially other functions at well defined (by our word-lifting technique) points. It
is easy to show that, if some functions intersect (contact) at a certain point,
then an infinite subset of functions P_n also intersect (contact) at this point.
In fact, intersection points correspond to $\rho\lambda^\infty$ type chaotic orbits, and contact points — to superstable periodic orbits which may extend to a window
of finite width. If intersection points and contact points with their windows,
taken together, do not exhaust the parameter axis, then there will be room
left for quasiperiodic and chaotic orbits. Thus we have returned to the difficult
problem of the measure of aperiodic orbits in one-dimensional mappings (see
the end of Section 2.5.4).

Anyway, symbolic dynamics has provided us with a tool to study the convergence rate and classification of selected types of sequences with a quasiperiodic
limit. This is a problem of current research.

3.7   Symbolic Dynamics of Three and
     More Letters

The rules in symbolic dynamics of unimodal mappings that we have discussed
in Sections 3.3 and 3.6 are essentially based on the monotonicity of the segments
of the nonlinear function (see, especially, Section 3.6.1). Therefore, they are
not restricted to the case of two letters only. The symbolic dynamics rules
extend simply to one-dimensional mappings with more critical points. In this
section, we start from a general discussion of the antisymmetric cubic map,
then use it as an example to analyse the phenomenon of symmetry breaking
and restoration in the bifurcation structure. We conclude this section with a
brief discussion of symbolic dynamics of more letters.

3.7.1 The Antisymmetric Cubic Map

In Section 1.4 we have mentioned that the antisymmetric cubic map

$$x_{n+1} = f(A, x_n) \equiv A\, x_n^3 + (1 - A)\, x_n \tag{3.81}$$

shares the same discrete symmetry with the Lorenz model of three autonomous differential equations, namely, they are both invariant when changing the sign of x (plus y in the Lorenz model). Therefore, we are primarily interested in linking this map to the Lorenz system (see Chapter 5). However, its meaning as a representative of systems with additional discrete symmetry goes beyond the Lorenz model (see, e.g., May, 1979).

We first perform a simple algebraic analysis of the map (3.81) (May, 1979). The antisymmetric function (3.81) maps the interval $(-1, 1)$ into itself when the parameter A varies in $(0, 4)$. However, in the parameter range $(0, 1)$, the map is monotonic and not interesting, so we confine ourselves to the range $(1, 4)$. The shape of the function $F(A, x)$ has been shown in Fig. 1.6. We denote the two critical points by

$$x_c^- = -\sqrt{(A-1)/3A}, \qquad x_c^+ = \sqrt{(A-1)/3A}, \tag{3.82}$$

(sometimes we will denote x_c^+ by C and x_c^- by $\overline{C}$) and label the three monotone branches of the map by the letters R, M and L. Among these three branches, only M is monotonically decreasing, so the parity of a string of letters will be determined by the number of letters M it contains. Occasionally, we will call it the M-parity, as compared to the R-parity in the unimodal case.

Three fixed points exist at all parameter values: $x = \pm 1$, stable for $-1 < A < 0$ and $x = 0$, stable for $0 < A < 2$ (the M branch comes into being when $A > 1$). At $A = 2$, the $x = 0$ fixed point loses stability and period-doubling takes place. Owing to the antisymmetric property of (3.81), there are two kinds of period 2 orbits: symmetric orbits, which may be obtained by iterating twice the inversion relation

$$f(A, x^*) = -x^*, \tag{3.83}$$

and asymmetric orbits which come in pairs:

$$x_1^* \rightarrow -x_2^* \rightarrow x_1^* \quad \text{and} \quad x_2^* \rightarrow -x_1^* \rightarrow x_2^*. \tag{3.84}$$

$(|x_1^*| \neq |x_2^*|$, otherwise they degenerate into the symmetric pair.)

In order to determine the stability range of the symmetric 2-cycle, we note that each element of the pair $(x^*, -x^*)$ may be viewed as a fixed point of an iteration with inversion, i.e.,

$$x_{n+1} = -Ax_n^3 + (1 - A)x_n \ .$$

Its explicit solution yields, besides the trivial unstable fixed point $x^* = 0$,

$$x^* = \sqrt{(A - 2)/A}. \tag{3.85}$$

Its stability is determined by

$$s = \frac{\partial f^{(2)}}{\partial x}\bigg|_{\pm *} = \frac{\partial f}{\partial x}\bigg|_{\mp *}\frac{\partial f}{\partial x}\bigg|_{\pm *} = \left(\frac{\partial f}{\partial x}\bigg|_*\right)^2 = (2A - 5)^2. \tag{3.86}$$

(We will write $|_*$ instead of $|_{x^*}$ hereafter.) Contrary to an orbit, generated by period-doubling, the stability discriminant s can never become negative: $s = 1$ at $A = 2$, as it should be, when due to period-doubling, $s = 0$ corresponds to a symmetric superstable 2-cycle, and again $s = 1$, when it loses stability at $A = 3$. At the last point there occurs a symmetry breaking bifurcation.

In order to calculate the asymmetric 2-cycle, we note that the symmetric 2-cycle still exists when A gets larger than 3, but loses stability. In order to find the asymmetric orbits, we first exclude the trivial zeros of $f^{(2)} - x$, dividing it by the fixed point relation $f - x$ and the inversion point relation $(f + x)/x$, i.e.,

$$\frac{f^{(2)}(x) - x}{(f(x) - x)(f(x) + x)/x} = A\left(A^2x^4 + A(1 - A)x^2 + 1\right). \tag{3.87}$$

Its two zeros

$$x_\pm^* = \sqrt{\frac{A - 1 \pm \sqrt{(A - 3)(A + 1)}}{2A}} \tag{3.88}$$

determine the two 2-cycles as functions of the parameter A, [cf., (3.84)]. Their stability range is

$$3 < A < 1 + \sqrt{5},$$

since

$$\frac{\partial f^{(2)}}{\partial x}\bigg|_{x_1^*} = \frac{\partial f}{\partial x}\bigg|_{-x_2^*}\frac{\partial f}{\partial x}\bigg|_{x_1^*}\frac{\partial f}{\partial x}\bigg|_{-x_1^*}\frac{\partial f}{\partial x}\bigg|_{x_2^*}$$

$$= \frac{\partial f^{(2)}}{\partial x}\bigg|_{x_2^*} = -2A^2 + 4A + 7.$$

The last expression attains -1 at $A = 1 + \sqrt{5}$, where period-doubling into a 4-cycle takes place. Note that both 2-cycles are stable in the above parameter range, each in its own basin. Therefore, only one of them may be observed for a given initial value.

Labelling each point in an orbit by one of the three letters L, M and R according to whether it satisfies $x < x_c^-$, $x_c^- < x < x_c^+$ or $x_c^+ < x$, and denoting x_c^+ and x_c^- by C and $\overline{C}$, respectively, a symmetric superstable orbit starting from $\overline{C}$ must be of the form

$$\overline{C}\sigma_1\sigma_2\ldots\sigma_k C\overline{\sigma}_1\overline{\sigma}_2\ldots\overline{\sigma}_k\overline{C} \equiv \overline{C}\Sigma C\overline{\Sigma}\,\overline{C}, \tag{3.89}$$

where σ_i is L, M, or R, and $\overline{\sigma}_i$ is the conjugate of σ_i, obtained by interchanging L and R, but leaving M unchanged. Similarly, an asymmetric superstable cycle looks like

$$\overline{C}\sigma_1\sigma_2\ldots\sigma_k\overline{C} \equiv \overline{C}P\overline{C}. \tag{3.90}$$

In what follows, we often omit $\overline{C}$, assuming it to be understood, and simply call P a pattern or a word. The superstable parameters can be calculated by solving the lifted equation

$$f(\overline{C}) = P(\overline{C}) \tag{3.91}$$

for an asymmetric orbit, described by the pattern P in (3.90). In the case of a symmetric orbit, one has to solve the equation

$$f(\overline{C}) = \Sigma(C), \tag{3.92}$$

where Σ is the first half-pattern defined in (3.89).

In what follows, we shall require the explicit expressions for the inverse branches of the cubic map (3.81). In order to solve the equation

$$y = Ax^3 - (A - 1)x, \tag{3.93}$$

we rescale the variables, letting $x = \lambda \tilde{x}$ and $y = \lambda \tilde{y}$. By choosing λ to satisfy

$$3A\lambda^2/(A-1) = 4,$$

i.e., $\lambda = 2C$, Equation (3.93) can be expressed in terms of the Chebyshev polynomial

$$3\tilde{y}/(A-1) = 4\tilde{x}^3 - 3\tilde{x} = T_3(\tilde{x}) = \cos(3\arccos(\tilde{x})).$$

Taking into account the multivalueness of the arccosine function, we find

$$\tilde{x} = \cos\left(\frac{1}{3}\arccos\left(\frac{\tilde{y}}{AC^2}\right) + \frac{2k\pi}{3}\right),$$

where $k = 0, \pm 1$.

We shall always rescale x and y by $2C$. With this convention in mind and introducing the notation

$$AC^2 = (A-1)/3 \equiv a,$$

(consequently, $f(\overline{C}) = 2Ca$), we shall name the three inverse functions by their labels R, L, and M, as shown in Fig. 1.6:

$$
\begin{aligned}
R(a,x) &= \cos\left(\frac{1}{3}\arccos(x/a)\right), \\
L(a,x) &= \cos\left(\frac{1}{3}\arccos(x/a) + 2\pi/3\right), \\
M(a,x) &= \cos\left(\frac{1}{3}\arccos(x/a) - 2\pi/3\right).
\end{aligned}
\tag{3.94}
$$

Note the following symmetry property of these functions:

$$
\begin{aligned}
R(a,-x) &= -L(a,x), \\
M(a,-x) &= -M(a,x).
\end{aligned}
\tag{3.95}
$$

Consequently, any composite function $W(a,x)$, made of these functions, will have the property

$$W(a,-x) = -\overline{W}(a,x),$$

where $\overline{W}$ is obtained from W by interchanging R and L, but leaving M unchanged. Sometimes we say that $\overline{W}$ is the mirror image of W. Therefore, dragging a minus sign through a composite function is equivalent to taking its mirror image. The fact that the mirror image orbit $C\overline{W}C\ldots$ has the same parameter is a simple consequence of this symmetry property.

In this notation, the lifted equations (3.91) and (3.92) appear as

$$a = P(-0.5) \quad \text{and} \quad a = \Sigma(0.5), \tag{3.96}$$

where $P(y)$ and $\Sigma(y)$ are composite functions of the functions (3.94). In order to apply these equations, let us consider the symmetric orbits. The only period 2 symmetric superstable orbit is $\Sigma = b$ (a blank letter), leading to the trivial equation

$$a = 0.5,$$

which yields $A = 2.5$. Then follow the period 4 and 6 orbits with $\Sigma = R$, RM, and RR. To solve the corresponding equation, we again transform it into iterations. Take, for example, the period 6 orbit $RMCLM\overline{C}$; the final iteration equation for

$$a = R \circ M(0.5)$$

reads

$$a_{n+1} = \cos(\frac{1}{3} \arccos(\frac{1}{a_n} \cos(\frac{1}{3} \arccos(\frac{0.5}{a_n}) - \frac{2\pi}{3}))),$$

and leads to $A = 3a + 1 = 3.46328\ldots$.

We have collected the superstable parameters for the first symmetric orbits in Table 3.8. The numbering of periods in this table coincides with the corresponding symmetry-broken orbits to be listed in Table 3.10.

i	Period	Word	Parameter
1	2	C	2.5
4	6	$RMCLM$	3.345315558
17	4	RCL	3.830811514
38	6	$RRCLL$	3.981797395

Table 3.8: Superstable symmetric orbits in the antisymmetric cubic map (for the numbering see the text).

3.7.2 The Ordering of Sequences

The symbolic dynamics of three letters has been constructed first by extending the notions of harmonics and antiharmonics (Zeng, Ding and Li, 1985 and

1988; Zeng, 1987), and then recast into a more natural form in line with what we have described in Section 3.6 (Zheng, 1988d). Skipping the details, we only formulate the definitions and practical rules needed in the sequel.

Labelling the three monotonic branches of the map by the letters R, M and L, a numerical orbit

$$x_0, \; x_1 = f(x_0), \ldots, \; x_n = f(x_{n-1}), \ldots,$$

whether finite or infinite, is replaced by a word composed of these letters. The letters C or/and $\overline{C}$ may appear in a word, if the corresponding orbit contains one or both of the critical points. In general, C and $\overline{C}$ are not related. $\overline{C} = -C$ holds only for antisymmetric maps. The monotonic decreasing branch M is assigned odd parity, whereas R and L have even parity. For convenience, C and $\overline{C}$ may be assigned "zero" parity. The parity of a symbolic sequence is determined by the number of letters M it contains.

Referring to the natural order

$$L < \overline{C} < M < C < R, \tag{3.97}$$

the following ordering of symbolic sequences will be introduced. Given two words $W_1 = W^*\sigma \ldots$ and $W_2 = W^*\tau \ldots$, where W^* denotes their common leading pattern and $\sigma \neq \tau$, the order of the two words is determined by the order of σ and τ, if the parity of W^* is even, and by the anti-order of σ and τ, if the parity of W^* is odd, i.e.,

$$\begin{aligned} W_1 &> (<) \, W_2, \quad \text{if} \quad \sigma > (<) \, \tau \quad \text{when } W^* \text{ even,} \\ W_1 &> (<) \, W_2, \quad \text{if} \quad \sigma < (>) \, \tau \quad \text{when } W^* \text{ odd.} \end{aligned} \tag{3.98}$$

Just like unimodal maps, this order is related to the corresponding parameter values of the periods, provided the map changes with the parameter in a monotonic manner. In fact, if A_W denotes the parameter value associated with the sequence W, then it follows from $W_1 < W_2$ that $A_{W_1} < A_{W_2}$, and *vice versa*.

A symbolic sequence Σ is called a *maximal sequence*, if it is larger than or equal to all its shifts $S^k\Sigma$, $k = 1, 2, \ldots$, where S denotes the left-shift operator, i.e.,

$$S\sigma_1\sigma_2\sigma_3 \ldots = \sigma_2\sigma_3 \ldots$$

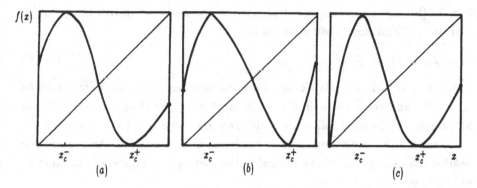

Figure 3.8: Three kinds of "cubic" maps (schematic).

and

$$S^k = \underbrace{S\,S\ldots S}_{k\ \text{times}}.$$

In particular, a periodic sequence

$$(\sigma_1\sigma_2\ldots\sigma_n)^\infty$$

remains the same under cyclic shift of the letters. We will adopt the convention of always writing the maximal sequence.

Starting the iteration from $\overline{C}$, we get a largest (in the sense of the above ordering) sequence or itinerary $I_{\overline{C}}$; it denotes also a point on the interval. Starting from C, one gets a smallest itinerary I_C, represented by another point on the interval. At a fixed parameter A, any initial value that is larger than $I_{\overline{C}}$, or smaller than I_C, will eventually enter the interval $(I_C,\ I_{\overline{C}})$, leaving apart a transient. Therefore, we can confine ourselves to this interval only. Any maximal sequence must satisfy the *admissibility condition*

$$I_C \le S^k\Sigma \le I_{\overline{C}}. \tag{3.99}$$

A sequence satisfying the condition (3.99) will be called an admissible sequence (Zheng, 1988d). For the cubic map the maximality alone does not determine the admissibility; one has to take into account the comparison with the minimal sequence, which is formed by the itinerary of x_c^+. The relation (3.99) holds for general maps described by three letters, as shown schematically in Fig. 3.8(a). Fig. 3.8(b) shows an antisymmetric map, where I_C is the mirror image of $I_{\overline{C}}$. Therefore, Condition (3.99) modifies to

$$\overline{I_{\overline{C}}} \leq S^k\Sigma \leq I_{\overline{C}}. \tag{3.100}$$

Fig. 3.8(c) is a degenerated case where the smallest sequence happens to be $I_C = L^\infty$, and the lower half of Condition (3.99) is satisfied automatically for all possible Σ. In what follows, we will treat only the antisymmetric map.

A superstable orbit must contain the letter C or $\overline{C}$ or both. An *asymmetric orbit* contains only one of the critical points, while a *symmetric orbit* must be of the form

$$\Sigma C\overline{\Sigma C}, \tag{3.101}$$

where $\overline{\Sigma}$ is the mirror image of Σ.

3.7.3 Construction of Median Sequences

The periodic window theorem (Section 3.6.4) is based on the continuity of the map. Therefore, it also works for cubic maps. An asymmetric superstable orbit $\Sigma\overline{C}$ can be extended into a "window" by changing the last $\overline{C}$ into M or L and ordering the three words in connection with the parity of Σ:

$$\begin{aligned} &\text{If } \Sigma \text{ even}: \quad (\Sigma L, \Sigma\overline{C}, \Sigma M), \\ &\text{If } \Sigma \text{ odd}: \quad (\Sigma M, \Sigma\overline{C}, \Sigma L). \end{aligned} \tag{3.102}$$

Note that the parity of these three words attributes a signature $(+1, 0, -1)$ to the window, which coincides with the signs of the first derivatives and signals a period-doubling bifurcation at the right end of the window. The actual width of a window, i.e., its stability range, is determined by the mapping function. Symbolic dynamics alone cannot tell anything about it. In particular, a period-doubling sequence may shrink into a single point, as is the case with the surjective tent map.

In general, a symbolic description of a period-doubling sequence goes just like the unimodal maps. Write the preceding window as

$$(\lambda, \mu|_{\overline{C}}, \mu),$$

where λ and μ are symbolic sequences and $\mu|_{\overline{C}}$ denotes the sequence μ with its last letter replaced by $\overline{C}$. λ has even parity, whereas μ is odd. In addition, they differ only in the last letter, i.e., $\mu|_{\overline{C}} = \lambda|_{\overline{C}}$. The right (or upper) sequence μ goes into the next, period-doubled, lower sequence smoothly as $\mu\mu$ to ensure the correct even parity. The whole window then looks like

$$(\mu\mu, \mu\mu|_{\overline{C}}, \mu\lambda).$$

This process repeats *ad infinitum* to yield the entire period-doubling sequence. Beyond the accumulation point of the sequence, there is a period-halving sequence of chaotic bands. The point where a period 2^n band merges into a period 2^{n-1} band is described by an infinite symbolic sequence formed from the upper and lower sequences of the corresponding period 2^n window as $\mu(\lambda)^{\infty}$.

The presence of symmetric orbits, however, somewhat complicates the procedure for finding the shortest admissible sequence in between two given sequences. Postponing the complete discussion of symmetry breaking and symmetry restoration to Section 3.8, we give here the rules (Zheng, 1988d).

Suppose we are given two admissible words $\Gamma\overline{C} < \Delta\overline{C}$. We first extend both words to windows, replacing $\overline{C}$ by L and M according to the periodic window theorem, and then proceed as follows:

1. Denote the upper sequence of $\Gamma\overline{C}$ by $\Sigma^{(1)}$, i.e.,

$$\Sigma^{(1)} = (\Gamma\overline{C})_1^+ = \max(\Gamma L, \Gamma M).$$

   Similarly, let the lower sequence of $\Delta\overline{C}$ be $\Sigma^{(2)}$, i.e.,

$$\Sigma^{(2)} = (\Delta\overline{C})_2^- = \min(\Delta L, \Delta M).$$

2. If $\Sigma^{(1)} = \Sigma^{(2)}$, then $\Gamma\overline{C}$ and $\Delta\overline{C}$ are adjacent words, and there are no median sequences in between them.

3. If $\Sigma^{(1)} \neq \Sigma^{(2)}$, then denote their common leading string by Σ

$$\Sigma^{(1)} = \Sigma\mu\ldots,$$
$$\Sigma^{(2)} = \Sigma\nu\ldots,$$

   where $\mu \neq \nu$.

4. If either μ or ν is the letter L, then $\Sigma\overline{C}$ is the median sequence that we are looking for.

5. If neither μ nor ν is L, then there is a symmetric orbit $\Sigma C\overline{\Sigma C}$.

6. Keep the previous $\Sigma^{(1)}$ and let $\Sigma^{(2)} = \Sigma\mu\overline{\Sigma\mu}$, then go to 2.

7. Keep the previous $\Sigma^{(2)}$ and let $\Sigma^{(1)} = \Sigma\nu\overline{\Sigma\nu}$, then go to 2.

It is better to explain the procedure by an example.

Example. We shall construct part of the admissible sequences included in between $\Sigma_1 = R\overline{C} < \Sigma_2 = RR\overline{C}$. We shall use Σ_n to denote the nth superstable sequence that emerges from the analysis.

First of all, we will write down the upper sequence of the smaller word

$$\Sigma_1^+ = (RM)^\infty = RMRM\ldots$$

and the lower sequence of the larger word

$$\Sigma_2^- = (RRL)^\infty = RRLRRL\ldots.$$

Their common leading string is $\Sigma = R$, and there is no L following Σ. According to step 5, there is a symmetric orbit $\Sigma_3 = RCL\overline{C}$. In order to apply steps 6 and 7, we need both its lower sequence

$$\Sigma_3^- = (RMLM)^\infty = RMLMRMLM\ldots$$

and upper sequence

$$\Sigma_3^+ = (RRLL)^\infty = RRLLRRLL\ldots.$$

Comparing Σ_1^+ with Σ_3^-, we see that the common string RM is followed in one of them by a letter L, thus yielding a median orbit $\Sigma_4 = RM\overline{C}$, according to step 4. The same happens with Σ_3^+ and Σ_2^-, leading to a word $\Sigma_5 = RRL\overline{C}$.

So far, we have generated the following sequences:

n	1	4	3	5	2
Sequence	$R\overline{C}$	$RM\overline{C}$	$RCL\overline{C}$	$RRL\overline{C}$	$RR\overline{C}$

In order to find the next level of median sequences, we compare in turn

1.
$$\Sigma_1^+ = RMRM\ldots,$$
$$\Sigma_4^- = RMMRMM\ldots,$$

to get a symmetric orbit $\Sigma_6 = RMCLM\overline{C}$;

2.
$$\Sigma_4^+ = RMLRML\ldots,$$
$$\Sigma_3^- = RMLMRMLM\ldots,$$

to get another symmetric orbit $\Sigma_7 = RMLCLMR\overline{C}$;

3.
$$\Sigma_3^+ = RRLLRRLL\ldots,$$
$$\Sigma_5^- = RRLLRRLL\ldots,$$

The two sequences coincide, so Σ_3 and Σ_5 are adjacent words. In fact, they form a symmetry to symmetry-broken pair (for more details, see Section 3.8).

4.
$$\Sigma_5^+ = RRLMRRLM\ldots,$$
$$\Sigma_2^- = RRLRRL\ldots,$$

yielding again a symmetric orbit $\Sigma_8 = RRLCLLR\overline{C}$.

This process repeats itself and generates more and more sequences. For instance, in between Σ_1 and Σ_6, there is a sequence $\Sigma_9 = RMR\overline{C}$, but

$$\Sigma_1^+ = RMRM\ldots,$$
$$\Sigma_9^- = RMRMRMRM\ldots.$$

Therefore, there are no median sequences in between Σ_1 and Σ_9. In fact, the latter is the period-doubled regime of the former.

We will show part of the results thus obtained in Table 3.9, in which the parameter increases downwards.

If one cares only about windows that are capable of undergoing period-doubling, then the words could be produced by a generalization of the notion of harmonics and antiharmonics in the old-fashioned way, as we did with the unimodal maps. In Table 3.10, we list all such words and the corresponding superstable parameter values for periods $n \leq 6$, generated in this way (Zeng, 1985, 1987). The parameters are calculated, using the word-lifting method described in Section 3.2, by solving (3.91) defined with the inverse functions (3.94). In this table, an asterisk, attached to the period, indicates that the orbit is obtained after symmetry-breaking from the pattern, included

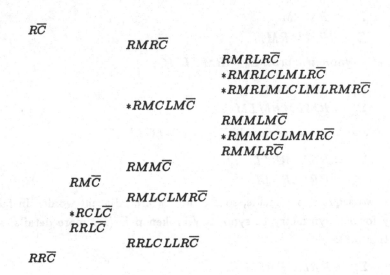

$$R\overline{C}$$

$$RMR\overline{C}$$

$$RMRLR\overline{C}$$
$$*RMRLCLMLR\overline{C}$$
$$*RMRLMLCLMLRMR\overline{C}$$

$$*RMCLM\overline{C}$$

$$RMMLM\overline{C}$$
$$*RMMLCLMMR\overline{C}$$
$$RMMLR\overline{C}$$

$$RMM\overline{C}$$

$$RM\overline{C}$$

$$RMLCLMR\overline{C}$$

$$*RCL\overline{C}$$
$$RRL\overline{C}$$

$$RRLCLLR\overline{C}$$

$$RR\overline{C}$$

Table 3.9: Part of admissible sequences included in between $R\overline{C}$ and $RR\overline{C}$. The parameter increases downwards.

in Table 3.8. The parameters for the symmetric orbits have been given separately in Table 3.8, because we do not count them, when calculating the number of different orbits for a given period (see Section 3.9.5).

In addition, a rule can be derived for the generation of all admissible words in between two given words; there exists a criterion for checking the admissibility of any word composed of L, M and R. We skip the details and only note that the periodic sequence, constructed in this way, does not depend on the particular model. It is universal for all mappings with two critical points and the antisymmetric property. Of course, the parameter values in Table 3.10 do depend on the mapping (3.81).

Table 3.10 displays clearly a regularity. In the first part of the table, every word starts with a single letter R. This can be called an R-region. Then follows the R^2-region, where every word has two consecutive R's in the leftmost part. Then comes the R^3-region, the R^4-region, etc. In fact, the words R^{n-1} play the same role as RL^{n-2} in the unimodal maps, i.e., they correspond to the last word of period n. These remarks will guide us in looking for regularities in the

i	Period	Word	Parameter	i	Period	Word	Parameter
1	2*	R	3.1213203	26	6	$RRMRR$	3.9254576
2	4	RMR	3.2628786	27	6	$RRMRM$	3.9350271
3	6	$RMRLR$	3.3340241	28	5	$RRMR$	3.9409044
4	6*	$RMMLM$	3.4632834	29	6	$RRMRL$	3.9464110
5	6	$RMMLR$	3.5282272	30	6	$RRMML$	3.9504721
6	4	RMM	3.5480858	31	5	$RRMM$	3.9553274
7	6	$RMMMR$	3.5659880	32	6	$RRMMM$	3.9597015
8	6	$RMMMM$	3.5911819	33	6	$RRMMR$	3.9637898
9	5	$RMMM$	3.6150319	34	4	RRM	3.9675403
10	5	$RMMR$	3.6662070	35	6	$RRMLR$	3.9710914
11	3	RM	3.7003155	36	6	$RRMLM$	3.9745198
12	6	$RMLRM$	3.7029894	37	5	$RRML$	3.9777816
13	5	$RMLR$	3.7339407	38	6*	$RRRLL$	3.9818990
14	5	$RMLM$	3.7753839	39	5	$RRRL$	3.9854885
15	6	$RMLMM$	3.7909088	40	6	$RRRLM$	3.9878905
16	6	$RMLMR$	3.8073689	41	6	$RRRLR$	3.9900272
17	4*	RRL	3.8398944	42	4	RRR	3.9919300
18	6	$RRLMR$	3.8610860	43	6	$RRRMR$	3.9936280
19	6	$RRLMM$	3.8734615	44	6	$RRRMM$	3.9951295
20	5	$RRLM$	3.8835860	45	5	$RRRM$	3.9964269
21	6	$RRLML$	3.8933550	46	6	$RRRML$	3.9975231
22	6	$RRLRL$	3.8982992	47	6	$RRRRL$	3.9984117
23	5	$RRLR$	3.9069063	48	5	$RRRR$	3.9991078
24	6	$RRLRM$	3.9144901	49	6	$RRRRM$	3.9996037
25	3	RR	3.9249907	50	6	$RRRRR$	3.9999009

Table 3.10: Asymmetric superstable periods in the Antisymmetric Cubic Map. An asterisk denotes a symmetry-broken orbit (see the text).

ordering of periods in the Lorenz model.

In connection with the symbolic dynamics of the cubic map, we note that Testa and Held (1983) used three letters R, M and L to name the periods seen in the cubic map (3.81) without formulating the rules to order and generate the admissible words. The number of different orbits with the same period, as given by Testa and Held (1983), unfortunately, did not agree with the actual periods listed in Table 3.10 (see Section 3.9.5 for the resolution of this puzzle).

3.7.4   Symbolic Dynamics of Four Letters

The symbolic dynamics of the sine-square map (3.11) happens to be more complicated, owing to the non-monotone dependence on a second parameter and the presence of more than two critical points. We consider instead a simpler biquadratic map (Zeng, Ding and Li, 1988)

$$x_{n+1} = -A\, x_n^4 + A\, x_n^2 - 1, \qquad x \in [-1, 1] \tag{3.103}$$

where the parameter A varies in the range $[4, 8]$. It has three critical points

$$C_1 = -1/\sqrt{2}, \quad C_0 = 0, \quad C_2 = +1/\sqrt{2},$$

and four monotone branches, denoted by the letters L, M, N and R. M and R represent monotone decreasing branches, and, consequently, their total number determines the parity of a pattern. Moreover, this is a degenerated case, similar to Fig. 3.8(c), as $I_{C_0} = L^\infty$ coincides with the smallest possible sequence L^∞ and there is no need to impose an additional admissibility condition like (3.99).

Almost all the discussions on symbolic dynamics of two and three letters may be transplanted to this case. One starts again from the natural order

$$L < C_1 < M < C_0 < N < C_2 < R \tag{3.104}$$

and distinguishes two classes of orbits. Class 1 includes orbits containing C_1, Class 2 those containing C_2. The fact that $I_{C_0} = L^\infty$ for all parameter A values further simplifies the symbolic dynamics as compared to that for the sine-square map.

The ordering of symbolic sequences is defined on the basis of the natural order (3.104) in the same manner as the unimodal or cubic maps, with the only complication that now both letters R and M have parity -1, whereas L and N parity $+1$. For example, we have

$$RC_2 < RLLC_1,$$

because the common leading string has parity -1 and $C_2 > L$.

The generation of all median words[3] is achieved by extending the periodic window theorem, which is based on continuity arguments, and by invoking further continuity considerations. Any superstable orbit may be extended into a window by replacing C_i by its two neighbouring letters in (3.104), i.e.,

$$C_1 \to M \text{ or } L,$$
$$C_2 \to R \text{ or } N.$$

The resulted words are then ordered according to their parity to match the signature $(1, 0, -1)$. We give a few examples:

$$RC_2 \quad \to \quad (RR, RC_2, RN),$$
$$RLLC_1 \quad \to \quad (RLLM, RLLC_1, RLLL).$$

Although the appearance of the letter C_0 leads immediately to the ending string $C_0 L^\infty$, in the construction of median words C_0 may also be replaced by its neighbours:

$$C_0 \to M \text{ or } N. \tag{3.105}$$

Thus, a word $\Sigma C_0 L^\infty$ generates two sequences:

$$\Sigma C_0 L^\infty \to \Sigma M L^\infty \text{ and } \Sigma N L^\infty,$$

which must be ordered by taking into account the parity of Σ.

In order to generate the shortest admissible word in between two given words, say, $P_1 = RC_2$ and $P_2 = RLLC_1$, we compare the upper sequence $RNRNRN \ldots$ of the smaller word P_1 and the lower sequence $RLLMRLLM \ldots$ of the larger word P_2. Instead of simply taking the common leading part of these two sequences, we must insert the suitable critical points according to the natural order (3.104). We present this in the form

$$
\begin{array}{lll}
R & N & RNRN\ldots \\
C_0 & L^\infty & \\
C_1 & & \\
R & L & LMRLLM\ldots.
\end{array}
$$

---

[3]The method given in the paper of Zeng, Ding and Li (1988) was based on extension of the outdated notions of harmonic and antiharmonic. The presentation here follows an unpublished note of Zheng Wei-mou.

Therefore, the shortest median word between RC_2 and $RLLC_1$ is RC_1, which extends to a window

$$(RM, RC_1, RL).$$

Now, for the generation of the median words between RC_1 and $RLLC_1$ we compare the corresponding upper and lower sequences and insert the appropriate critical points:

$$
\begin{array}{lll}
RL & R & LRL\ldots \\
 & C_2 & \\
 & C_0 & L^\infty \\
 & C_1 & \\
RL & L & MRLLM\ldots.
\end{array}
$$

Consequently, we find two median words of the same length, RLC_2 and RLC_1.

Sometimes one has to use the C_0 replacement rule (3.105) to continue the construction. Take, for example, the words $RLC_2 < RLC_1$. The only critical point, which may be inserted into the corresponding upper and lower sequences is C_0. We apply the rule (3.105), order the results, and insert critical points to get

$$
\begin{array}{lll}
RLN & R & LN\ldots \\
 & C_2 & \\
 & C_0 & L^\infty \\
 & C_1 & \\
RLN & L & LL\ldots \\
RLC_0 & L & LL\ldots \\
RLM & L & LL\ldots \\
 & C_1 & \\
 & C_0 & L^\infty \\
 & C_2 & \\
RLM & R & LM\ldots.
\end{array}
$$

Thus we have found four median words

$$RLC_2 < RLNC_2 < RLNC_1 < RLMC_1 < RLMC_2 < RLC_1.$$

The insertion rules may be written down as follow

$$(R, L) \rightarrow C_2, C_1$$
$$(R, M) \text{ and } (R, N) \rightarrow C_2$$
$$(M, L) \text{ and } (N, L) \rightarrow C_1$$
$$(M, N) \rightarrow C_0 L^\infty.$$

In fact, all the replacement and insertion rules may be read off from the natural order (3.104).

Once all the required words have been generated, their superstable parameter values may be determined by the word-lifting technique. We list in Table 3.11 the superstable words and corresponding parameter values for all periods, included in between RC_2 and $RLLC_1$ (Zeng, Ding and Li, 1988). In this table, we denote orbits of the first class by a subscript 1, e.g.,

$$RLL_1 \equiv RLLC_1,$$

and orbits of the second class, analogously, by a subscript 2.

i	k	Pattern	A	i	k	Pattern	A
1	2	R_2	7.0865566	16	4	RMN_1	7.8400428
2	4	RNR_2	7.1609685	17	4	RMN_2	7.8682263
3	3	RN_2	7.2666150	18	3	RM_2	7.8804502
4	4	RNN_2	7.3172501	19	4	RMR_2	7.8923602
5	4	RNN_1	7.4263223	20	2	R_1	7.9223487
6	4	RNM_1	7.4985938	21	4	RLR_1	7.9235784
7	4	RNM_2	7.5622701	22	4	RLR_2	7.9481516
8	3	RN_1	7.5839836	23	3	RL_2	7.9559618
9	4	RNL_2	7.6053123	24	4	RLN_2	7.9630460
10	4	RNL_1	7.6566512	25	4	RLN_1	7.9754389
11	4	RML_1	7.7009193	26	4	RLM_1	7.9851942
12	4	RML_2	7.7421327	27	4	RLM_2	7.9924663
13	3	RM_1	7.7586452	28	3	RL_1	7.9951758
14	4	RMM_2	7.7747733	29	4	RLL_2	7.9972858
15	4	RMM_1	7.8093680	30	4	RLL_1	7.9996987

Table 3.11: Superstable orbits of periods $n \leq 4$ between RC_2 and $RLLC_1$ in the biquadratic map.

In concluding this section, we make a remark concerning the admissibility condition of symbolic sequences in the case of more letters.[4]

Consider a symbolic sequence

$$\Sigma = \sigma_1 \sigma_2 \ldots \sigma_k \sigma_{k+1} \ldots .$$

We denote the string following the symbol σ_k by (σ_k), i.e.,

$$(\sigma_k) = S^k \Sigma = \sigma_{k+1} \sigma_{k+2} \ldots ,$$

where S is the shift operator, introduced in Sections 3.6.3 and 3.7.2. In our case of the biquadratic map σ may be one of the letters R, L, M, and N. In order to check the admissibility of a given sequence, we look at all possible shifts and compare them with the itineraries of the critical points I_{C_0}, I_{C_1} and I_{C_2}. The admissibility conditions may be written down by inspecting the graph of the map:

$$
\begin{aligned}
(R) &\leq I_{C_1} \\
I_{C_0} \leq (N) &\leq I_{C_1} \\
I_{C_0} \leq (M) &\leq I_{C_2} \\
(L) &\leq I_{C_2} .
\end{aligned}
$$

In general, there are as many conditions as the number of symbols. In this form the admissibility conditions may be applied to circle maps as well.

We see that the itineraries of the critical points play a special role in the admissibility conditions of the symbolic sequences. They are also called kneading sequences in the literature.

3.8 Symmetry Breaking and Restoration in Antisymmetric mappings

Symmetry breaking and symmetry restoration are common phenomena in physical systems with a certain kind of symmetry. An equation or a thermodynamical potential may possess a higher symmetry, but a particular solution or an equilibrium state may exhibit only a lower symmetry. However, all these asymmetric solutions (or states) taken together restore the original symmetry. In

---

[4]Zheng Wei-mou, unpublished.

fact, the notion of symmetry breaking has been playing an increasingly important role in the understanding of diverse problems ranging from continuous phase transitions to the origin of the Universe.

It is interesting to note that a simple form of symmetry breaking and restoration appears in the bifurcation structure of many dynamical systems. It manifests itself clearly in the bifurcation diagram of the antisymmetric cubic map (3.81) (see Fig. 1.7 in Chapter 1 for the most clearly seen symmetry breaking of the 2-cycle).

The existence of symmetric orbits, which first undergo symmetry-breaking bifurcation into asymmetric orbits, and then enjoy period-doubling, has been observed in numerical studies of many systems of ordinary and partial differential equations, as well as in laboratory experiments. It seems that for the time being the Lorenz model is in this respect the most-studied system. The systematics of periodic solutions has been associated with the symbolic dynamics of three letters, and many cases of symmetry breaking have been described explicitly in terms of symbolic sequences (Ding and Hao, 1988, and Chapter 5). In a sense, this section is a preparation for the corresponding sections in Chapter 5, devoted to the Lorenz system.

Observations have led to the conjecture that symmetry breaking must precede period-doubling, whence follows the expression "precursor" (D'Humieries *et al.*, 1982; Kumar *et al.*, 1987) to period-doubling or "suppression" (Swift and Wiesenfeld, 1984; Wiesenfeld *et al.*, 1984) of period-doubling by symmetry-breaking, etc. The phenomenon was correctly related to the symmetry of the governing equations (D'Humieries *et al.*, 1982), and a heuristic explanation based on bifurcation theory was given by Wiesenfeld *et al.* (1984). The most complete bifurcation theory analysis applied to the Poincaré mapping of flows appeared in the paper of Swift and Wiesenfeld (1984) with the conclusion that a symmetric orbit cannot undergo period-doubling directly except in extraordinary cases.

In contrast to symmetry breakings, the phenomenon of symmetry restoration has rarely been discussed in the literature. There has been, e.g., a case of mentioning "a bifurcation back to symmetry" (Knobloch and Weiss, 1981), but nothing was said about the nature of the bifurcation.

In fact, the bifurcation analysis of symmetry-breakings can be carried out simply and thoroughly for general antisymmetric mappings in much the same way as for tangent and period-doubling bifurcations, as we did in Chapter 2. Moreover, symbolic dynamics determines the precedence of symmetric orbit over the asymmetric ones, and provides a selection rule to pick up those orbits of even periods that are destined for symmetry breaking. Symbolic dynamics also furnishes a precise description for symmetry restoration. Being formulated in the language of symbolic dynamics, the essence of the entire analysis applies to differential equations as well. We shall elaborate these points in the following Sections (Zheng and Hao, 1988).

3.8.1 Symmetry Breaking Bifurcation

From now on we will consider a general map

$$x_{n+1} = f(A, x_n), \tag{3.106}$$

where the function f is antisymmetric and depends on a parameter A:

$$f(A, -x) = -f(A, x) \quad \forall \ x. \tag{3.107}$$

Notice that the k-th iterate of f is also antisymmetric:

$$f^{(k)}(A, -x) \equiv \underbrace{f \circ f \circ \cdots \circ f}_{n \ times}(A, -x) = -f^{(k)}(A, x). \tag{3.108}$$

Moreover, all odd order derivatives of f with respect to x are even functions, while even order derivatives are odd functions of x:

$$
\begin{aligned}
\left. \frac{\partial^{2k+1} f}{\partial x^{2k+1}}(A, x) \right|_{-x^*} &= \left. \frac{\partial^{2k+1} f}{\partial x^{2k+1}}(A, x) \right|_{x^*}, \\
\left. \frac{\partial^{2k} f}{\partial x^{2k}}(A, x) \right|_{-x^*} &= -\left. \frac{\partial^{2k} f}{\partial x^{2k}}(A, x) \right|_{x^*}.
\end{aligned}
\tag{3.109}
$$

Consider the iteration

$$x_{n+1} = f(A, -x_n) = -f(A, x_n). \tag{3.110}$$

Suppose that this iteration reaches an *inversion point* x^*

$$x^* = -f(A, x^*) \tag{3.111}$$

just as the iteration (3.106) may reach a *fixed point*, then x^* must be a fixed point of $f^{(2)}$, i.e., there is a symmetric 2-cycle

$$x^*, -x^*, x^*, -x^*, x^*, \ldots$$

One can check the stability of the inversion point as well as the stability of the 2-cycle. When the inversion point is no longer stable, symmetry breaking bifurcation takes place and gives birth to a pair of asymmetric 2-cycles:

$$x_1^*, -x_2^*, x_1^*, -x_2^*, \ldots,$$

and its mirror image

$$-x_1^*, x_2^*, -x_1^*, x_2^*, \ldots.$$

$(|x_1^*| \neq |x_2^*|)$. We are interested here in the formulation of the conditions when this will happen.

In general, when there is an inversion point for a certain iterate

$$x^* = -f^{(k)}(A, x^*), \tag{3.112}$$

it must be a $2k$-cycle of f, i.e., a fixed point of $f^{(2k)}$

$$\pm x^* = f^{(2k)}(A, \pm x^*). \tag{3.113}$$

Therefore, double application of the inversion relation leads to an even period. However, not all even periods may be decomposed into two consecutive inversions. The appropriate criterion will be given later, using symbolic dynamics. For the time being, we will ask the question — under what conditions can an inversion point lose stability and cause symmetry breaking bifurcation to occur.

The analysis furnishes another simple example of applying the implicit function theorem (Section 2.2.3) and proceeds much like that for period-doubling bifurcation (Section 2.6.4). The difference of these two cases consists of the fact that, while many partial derivatives encountered in the analysis of period-doubling vanish due to the condition

$$\left.\frac{\partial f}{\partial x}\right|_{x^*} = -1,$$

they now cancel each other as a result of the antisymmetry conditions (3.109). We will perform the bifurcation analysis in some detail.

We first formulate the conditions. In order to avoid clumsiness of notation, we consider only f and $f^{(2)}$; these functions may be replaced by $f^{(k)}$ and $f^{(2k)}$ everywhere below.

Proposition. Suppose $f(A, x) : \ I \ \rightarrow \ I$ is an antisymmetric map of the interval I into itself and the following conditions hold:

1. There is an inversion point x^* at the parameter value A^*

$$x^* = -f(A^*, x^*).$$

   It is also a fixed point of $f^{(2)}$.

2. The stability is marginal with the only possibility that

$$\frac{\partial f}{\partial x}\bigg|_{x^*} = \frac{\partial f}{\partial x}\bigg|_{-x^*} = 1.$$

3. The Schwarzian derivative $S(f, x)|_{x^*} < 0$. This condition holds automatically for most maps of physical interest, because they do have negative Schwarzian derivatives on the whole interval I, not only at the point x^*.

4. The mixed second derivative of the second iterate does not vanish:

$$\frac{\partial^2 f^{(2)}}{\partial x \partial \mu}\bigg|_{x^*} \neq 0.$$

Then, on one side of A^* (which side depends on the sign of the mixed derivative that figures in Condition 4), there exists a stable symmetric 2-cycle, which undergoes symmetry breaking bifurcation at $A = A^*$ into a pair of asymmetric 2-cycles, both stable but observable only in their own basins. As in Section 2.6.4, we divide the proof into several steps.

Step 1. Construct an auxiliary function

$$h(A, x) = f(A, x) + x;$$

the inversion point of f becomes a zero of h:

$$h(A^*, x^*) = 0.$$

Calculate the derivative

$$\left.\frac{\partial h}{\partial x}\right|_{x^*} = 2 \neq 0 \quad \text{(Condition 2)}.$$

Therefore, according to the implicit function theorem, there exists a function $x = x(A)$ such that

$$h(A, x(A)) = 0$$

in the vicinity of (A^*, x^*). In other words, an inversion point exists both above and below A^*, but the stability may disappear on crossing A^*. In order to check the stability, we calculate the first derivative close to (A^*, x^*):

$$
\begin{aligned}
\frac{\partial f}{\partial x}(A^* + dA, x^* + dx) &= 1 + \left(\left.\frac{\partial^2 f}{\partial x \partial A}\right|_* - \left.\frac{1}{2}\frac{\partial f}{\partial A}\right|_* \left.\frac{\partial^2 f}{\partial x^2}\right|_* \right) dA + \dots \\
&= 1 + \left.\frac{1}{2}\frac{\partial^2 f^{(2)}}{\partial x \partial A}\right|_* dA + \dots
\end{aligned}
$$

Therefore, as long as Condition 4 holds, the inversion point (and the 2-cycle formed by iterating it) will be stable on one side and unstable on the other side of A^*, depending on the sign of the mixed derivative. So far we have not studied the stability of 2-cycles by directly inspecting $f^{(2)}$. For this purpose, we turn to

Step 2. Construct another auxiliary function

$$g(A, x) = f^{(2)}(A, x) - x.$$

It is readily seen that both its first partial derivatives vanish:

$$\left.\frac{\partial g}{\partial x}\right|_* = 0, \quad \left.\frac{\partial g}{\partial A}\right|_* = 0,$$

due to the antisymmetry conditions (3.109). Therefore, $g = 0$ is not a suitable relation for the application of the implicit function theorem. Before going to the next step, we will calculate two higher derivatives for later use:

$$\left.\frac{\partial^2 g}{\partial x^2}\right|_* = 0, \quad \left.\frac{\partial^3 g}{\partial x^3}\right|_* = 2S(f, x)|_* < 0. \tag{3.114}$$

Step 3. The lack of success in choosing g was caused by the fact that $f^{(2)}$ contains the "trivial" 2-cycle, formed by the inversion point studied in Step 1. It is better to exclude it in advance. Thus, instead of g, we define

$$k(A, x) = \frac{g(A, x)}{h(A, x)}.$$ (3.115)

In order to calculate $k(A^*, x^*)$ and its derivatives, one encounters many uncertainties of the 0/0 type, which can be resolved by repeated use of L'Hospital rule, taking into account the derivatives of h and g, calculated before. The final results are:

$$k(A^*, \pm x^*) = 0,$$
$$\left.\frac{\partial k}{\partial x}\right|_{\pm *} = 0,$$
$$\left.\frac{\partial k}{\partial A}\right|_{\pm *} = \pm\frac{1}{2}\left.\frac{\partial^2 f^{(2)}}{\partial x \partial A}\right|_{\pm *} \neq 0,$$ (3.116)
$$\left.\frac{\partial^2 k}{\partial x^2}\right|_{\pm *} = \pm\frac{1}{3}S(f, x)|_{\pm *} < 0.$$

Therefore, according to the implicit function theorem, there exist functions $A = A_+(x)$ and $A = A_-(x)$ such that

$$k(A_\pm(x), x) = 0$$ (3.117)

in the vicinities of $(A^*, \pm x^*)$, respectively. One can differentiate (3.117) twice to get

$$\left.\frac{dA}{dx}\right|_{\pm *} = 0,$$
$$\left.\frac{d^2 A}{dx^2}\right|_{\pm *} = -\frac{2}{3}\frac{S(f, x)|_*}{\left.\dfrac{\partial^2 f^{(2)}}{\partial x \partial A}\right|_*} \neq 0.$$ (3.118)

Clearly, $A = A_\pm(x)$, viewed as function of x, have either a minimum or a maximum in A, depending on the sign of the mixed derivative in Condition 4. In addition, the convexity of $A = A_\pm(x)$ agrees with the sign of the dA term in the expression for the first derivative, given in Step 1, as both were determined by the same mixed derivative. The four branches of $A_+(x)$ and $A_-(x)$ form a pair of 2-cycles. There remains to check the stability of these 2-cycles by expanding the derivative $\partial f^{(2)}/\partial x$ in the neighbourhood of $(A^*, \pm x_*)$. Since

$$\frac{dA}{dx}\bigg|_{\pm*} = 0 \qquad \text{(consequence of Eq. (3.118)),}$$

$$\frac{\partial^2 f^{(2)}}{\partial x^2}\bigg|_{\pm*} = 0 \qquad \text{(consequence of Eqs. (3.109)),}$$

$$\frac{\partial^3 f^{(2)}}{\partial x^3}\bigg|_{\pm*} = 2S(f,x)|_{\pm*};$$

the terms linear in dx vanish in the expansion and we have

$$\frac{\partial f^{(2)}}{\partial x} = 1 + \frac{2}{3}S(f,x)|_{\pm*}(dx)^2 + \dots . \tag{3.119}$$

Therefore, both 2-cycles are stable. Which 2-cycle is observed for a given initial value depends on the basin; this question goes beyond the reach of the local analysis of bifurcation theory.

3.8.2 Symbolic Dynamics Analysis of Symmetry Breakings

Symbolic dynamics of the cubic map has been developed in Sections 3.7.2 and 3.7.3. Now we apply it to analyze symmetry breakings and restorations. First of all consider a *symmetric orbit* described by a sequence $\Sigma C \overline{\Sigma} \overline{C}$. Let us disturb the orbit slightly, giving up the superstability, but retaining the symmetry. The letter C may be changed to R or M by continuity, and $\overline{C}$ to L or M as the mirror image of C. The three words, thus obtained, may be ordered according to the parity of Σ:

$$\begin{aligned} \Sigma \text{ even}: &\quad (\Sigma L \overline{\Sigma} R, \Sigma C \overline{\Sigma} \overline{C}, \Sigma M \overline{\Sigma} M), \\ \Sigma \text{ odd}: &\quad (\Sigma M \overline{\Sigma} M, \Sigma C \overline{\Sigma} \overline{C}, \Sigma L \overline{\Sigma} R). \end{aligned} \tag{3.120}$$

In contrast to windows, capable of undergoing period-doubling, this window has the signature $(+1, 0, +1)$ and may be called an inversion point window, in accordance with our analysis in Section 3.8.1. It is easy to check that the inversion point window is equivalent to its mirror image (up to cyclic shifts, inessential due to periodicity). From Section 3.8.1, we know that this window undergoes a symmetry-breaking bifurcation, giving birth to an asymmetric window of the same period. The upper sequence of the window (3.120) goes smoothly into the next window, as its lower sequence. The upper sequence of the new window may be obtained by replacing the last letter to fit the -1 parity and the central word — by changing the last letter to $\overline{C}$. Now, by taking the mirror image, we get another window, not equivalent to the original one. This is nothing but the other symmetry-broken orbit. If the three words in the original window are ordered as maximal sequences, then the words in the mirror image are ordered as minimal sequences, as the orbit starts from the valley of the mapping function.

We illustrate what has been said by a few examples.

Example 1. The simplest symmetric superstable word is $C\overline{C}$. It extends to the inversion point window

$$(MM, C\overline{C}, RL). \tag{3.121}$$

This is the most clearly seen period 2 in the bifurcation diagram of Fig. 1.7. By continuing RL to the lower sequence of the next window, we construct one of the asymmetric period 2 orbits as

$$(RL, R\overline{C}, RM). \tag{3.122}$$

Its mirror image, ordered as minimal sequences,

$$(LR, LC, LM), \tag{3.123}$$

describes the other symmetry-broken orbit. By cyclic shifting, the above window may be transformed to maximal sequences

$$(RL, CL, ML). \tag{3.124}$$

Example 2. There is only one period 4 superstable orbit $RCL\overline{C}$. We write the whole bifurcation structure as:

$$(RMLM, RCL\overline{C}, RRLL) \begin{cases} (RRLL, RRL\overline{C}, RRLM), \\ (RRLL, RCLL, RMLL). \end{cases} \tag{3.125}$$

Example 3. There are two symmetric orbits among the 30 period 6 windows (Zeng 1985, 1987; Hao and Zeng, 1987) that allow symmetry-breaking bifurcation: $RMCLM\overline{C}$ and $RRCLL\overline{C}$. Among the 205 period 8 windows, only 5 can undergo symmetry-breaking, namely:

$$RMMRLM\,M\overline{C}, \quad RMLMLM\,R\overline{C}, \quad RRLRLLR\overline{C},$$
$$RRMMLLM\overline{C}, \quad RRRRLLL\overline{C}.$$

Hence we see that, although symmetry-breakings are always associated with orbits of even periods, not all even periods are capable to do so. Symbolic dynamics provides the selection rule.

3.8.3 Symbolic Dynamics and Symmetry Restoration

We have mentioned at the beginning that symmetry restoration has not been analyzed in the literature to date. In fact, symbolic dynamics provides us with a straightforward tool to accomplish this task. We recall that a periodic window

$$(\lambda, \mu|_{\overline{C}}, \mu),$$

which permits period-doubling, i.e., a window with $(+1, 0, -1)$ signature, corresponds to a band-merging point, described by the sequence $\mu(\lambda)^{\infty}$. Applied to the period 2 symmetry-broken window (3.122), we locate the ending of the symmetry-broken regime at

$$RM(RL)^{\infty}. \tag{3.126}$$

The very form of (3.126) tells the nature of this point. It corresponds to the "crisis" created by the collision of the unstable symmetric period 2 orbit, represented by $(RL)^{\infty}$, with the asymmetric chaotic attractor.

Its parameter value can be calculated from the following pair of "lifted" equations:

$$\begin{aligned} a &= R \circ M(a, z), \\ z &= R \circ L(a, z), \end{aligned} \tag{3.127}$$

where the functions R, M, and L have been defined in (3.94). The iterative solution of (3.127) yields

$$A = 3.360893769096575\ldots. \tag{3.128}$$

On the other hand, the mirror image of the window (3.122), i.e., the window (3.124), corresponds to an asymmetric chaotic band, ending at

$$ML(RL)^\infty, \qquad\qquad (3.129)$$

Referring to (3.95), it is easy to verify that the parameter value of (3.129) is the same as that given by (3.128).

The symbolic sequence, corresponding to the symmetry restoration point, should have, so to speak, a "double personality". Seen from the left, it is asymmetric; seen from the right, it ought to be symmetric. Indeed, the sequences (3.126) and (3.129) are manifestly mirror images of each other, a property of symmetry-broken pairs. On the other hand, the symmetry has actually been restored, since both sequences appear to be the $m \to \infty$ limit of the following periodic orbits, all located beyond the symmetry-restoration point (3.128):

$$RM(RL)^{m+1}ML(RL)^m = RM(RL)^m RLM(LR)^m L.$$

These orbits are nothing but the lower sequences of the inversion point window, formed by the symmetric periodic orbits

$$RM(RL)^m CLM(LR)^m \overline{C}, \qquad\qquad (3.130)$$

which are equivalent to the cyclically shifted sequences

$$LM(LR)^m \overline{C} RM(RL)^m C, \qquad\qquad (3.131)$$

as long as m remains finite. However, the $m \to \infty$ limits of (3.130) and (3.131) approach the asymmetric sequences (3.126) and (3.129), respectively.

It is interesting to investigate the rate of convergence of these periods towards the symmetry-restoration point, which may be considered as the $m = \infty$ member of the family (3.130). The corresponding superstable parameters may be determined by the word-lifting technique described in Section 3.2 and are listed in Table 3.12 (in the actual calculations, we have retained more digits).The coefficients δ_m are calculated from the formula

$$\delta_m = \frac{A_{m-2} - A_{m-1}}{A_{m-1} - A_m}.$$

These coeffcients A_m are asymptotically proportional to the length of the periods and converge much faster than the Feigenbaum period-doubling sequence, with the convergence rate $\delta = 2.96455\ldots$, which coincides with the derivative

m	Period	A_m	δ_m
0	6	3.41592 61947	
1	10	3.39439 31930	
2	14	3.37305 84369	
3	18	3.36519 01399	2.71148
4	22	3.36237 69478	2.79693
5	26	3.36139 92781	2.87745
6	30	3.36106 50252	2.92494
7	34	3.36095 16371	2.94786
8	38	3.36091 33022	2.95784
9	42	3.36090 03597	2.96193
10	46	3.36089 59924	2.96355
11	50	3.36089 45191	2.96417
12	54	3.36089 40221	2.96441
13	58	3.36089 38544	2.96450
14	62	3.36089 37978	2.96453
15	66	3.36089 37788	2.96455
16	70	3.36089 37723	2.96455
17	74	3.36089 37702	2.96455
18	78	3.36089 37695	2.96455
19	82	3.36089 37692	2.96455
20	86	3.36089 37691	2.96455
21	90	3.36089 37691	2.96455
∞	∞	3.36089 37690	

Table 3.12: Convergence towards the symmetry-restoration point.

$$u = f^{(2)'},$$

taken at the unstable 2-cycle $z^* \equiv (RL)^\infty$ of the parameter A_∞. This observation can be explained as follows[5] . By (3.130), we have

$$f^{(3)}(A_m, \overline{C}) = f^{(-2m)}(A_m, C) \equiv z_m. \qquad (3.132)$$

Denote by ϵ_m the parameter difference $A_\infty - A_m$. For m large enough,

[5]Cf., e.g., Dias de Deus and Taborda Duarte (1982).

$$|z_m - z^*| \sim \epsilon_m,$$
$$\frac{\partial}{\partial x} f^{(2)}(A_m, x)|_{z_m} \sim u + \gamma \epsilon_m, \tag{3.133}$$

where γ is a constant, independent of m. Thus, one has asymptotically

$$(u + \gamma \epsilon_m)^m \epsilon_m \sim 1,$$

which implies $\epsilon_m \sim u^{-m}$ or $\delta = u$. In contradistinction to Feigenbaum's δ, the convergence rate δ_∞ here depends on the mapping function and, therefore, cannot be universal. In fact, this kind of converging periodic sequences exist in abundance. Take, for example, the RL^m words in unimodal mappings. The convergence rate equals

$$\delta = f'(x)|_{x^* = -1} = 4,$$

the numerical value being given for the logistic map of which the unstable fixed point L^∞ at $m = \infty$ is $x^* = -1$.

Symmetry restoration for higher periods may be analyzed in a similar way. Suppose there is a symmetric orbit $\Sigma C \overline{\Sigma} C$. The inversion point window may be represented as

$$(\overline{\mu}\mu, \ \Sigma C \overline{\Sigma} C, \ \rho\lambda), \tag{3.134}$$

where ρ, λ and μ are symbolic strings made up of R, L and M, each having the parity of the corresponding letter. They are related to Σ and $\overline{\Sigma}$ in the following manner:

$$\begin{aligned} \rho &= \text{the even one of } (\Sigma R, \ \Sigma M), \\ \mu &= \text{the odd one of } (\overline{\Sigma} M, \ \overline{\Sigma} L), \\ \lambda &= \text{the even one of } (\overline{\Sigma} M, \ \overline{\Sigma} L). \end{aligned} \tag{3.135}$$

In fact, ρ and λ are mirror images, but $\overline{\mu}$ may, in general, differ from μ. Then the pair of asymmetric orbits are given by

$$\begin{aligned} (\rho\lambda, \ \rho\overline{\Sigma} C, \ \rho\overline{\mu}), \\ (\lambda\rho, \ \lambda\Sigma C, \ \lambda\mu). \end{aligned} \tag{3.136}$$

The symmetry restoration point is described by

$$\rho\overline{\mu}(\rho\lambda)^\infty \quad \text{or} \quad \lambda\mu(\lambda\rho)^\infty. \tag{3.137}$$

Period	Σ	Symmetric	Asymmetric	Restoration	δ
2		2.5	3.12132 03436	3.36089 37691	2.96455
6	RM	3.3453155584	3.4632834578	3.46851 19419	2.83903
4	R	3.8308115142	3.8398944876	3.84470 61353	2.84205
6	RR	3.9817973948	3.9818990306	3.9819541883	2.82018

Table 3.13: Symmetry breaking and restoration parameters.

They are the $m \to \infty$ limits of the sequence of symmetric orbits

$$\Sigma_m C \overline{\Sigma}_m \overline{C}, \tag{3.138}$$

where

$$\Sigma_m = \rho \overline{\mu} (\rho \lambda)^m \quad \text{and} \quad \overline{\Sigma}_m = \lambda \mu (\lambda \rho)^m. \tag{3.139}$$

It is worth mentioning that all the equations (3.134) to (3.139) may be obtained from the simplest case of symmetry breaking and restoration, namely, that associated with the sequence $C\overline{C}$ and described by (3.121) to (3.123) as well as (3.126) through (3.131), by making the following substitutions:

$$\begin{aligned}
R &\to \rho, \\
L &\to \lambda, \\
M &\to \overline{\mu}, \\
C &\to \Sigma C, \\
\overline{C} &\to \overline{\Sigma C}.
\end{aligned} \tag{3.140}$$

All the superstable and symmetry restoration parameters may be calculated by solving the "lifted" equations. We have collected a few results in Table 3.13, the first three columns of which have already appeared in Table 3.8.

3.9 The Number of Periodic Windows

We already know that each stable n-cycle lives in a finite interval on the parameter axis and gives rise to a periodic window. Somewhere in the middle of the window, there exists a superstable orbit which can be taken as a representative of the whole window. One might ask how many different superstable orbits are there for a given n and, in total, for a given type of one-dimensional mapping. This enumeration problem has many facets which can be explored more or less thoroughly for the logistic map, and can be extended to a few other mappings:

1. It is given by the number of real roots of the equation $P_n(\mu) = x_c$ within a certain interval of μ, where P_n is the composite function defined in Section 2.1.

2. It gives the number of admissible words in the symbolic dynamics of a certain number of letters.

3. As we have learnt from Section 3.5.2, it also gives the number of finite λ-auto-expansions of a given length.

4. It has some relationship to a combinatorical problem, namely, the necklace problem: how many different necklaces can one make from n pieces of stones, each of which comes in one of q colours?

5. There exists a simple recursion formula which yields these numbers.

6. They count for the number of "saddle orbits" in the process of forming a "horseshoe" (Yorke and Alligood, 1985; see Section 3.9.3 below).

We shall discuss mainly the logistic map, and only touch briefly the antisymmetric cubic map. We begin with the recursion formula.

3.9.1   A Recursion Formula for the Number of Periodic Windows

Relying upon the universal nature of the bifurcation structure of unimodal mappings, we derive a recursion formula for the number of periodic windows for a particular map, namely, the logistic map. Once obtained, it applies to all maps of the same class. Each point from the n-cycle is a fixed point of the nested function $f^{(n)}$, or, using the notation of Section 2.3.2,

$$F(n, \mu, x) = x. \tag{3.141}$$

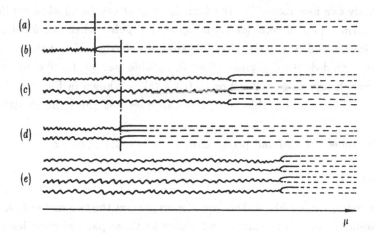

Figure 3.9: Different types of roots (schematic): wavy line — a pair of complex roots, solid line — a real stable solution, dashed line — a real unstable solution.

This is a polynomial equation of degree 2^n. Therefore, we can refer to the fundamental theorem of algebra by which "an algebraic equation of degree n has n roots (in the complex field)" as a conservation law for the number of roots. It is sufficient to sum up all possible types of roots to 2^n. This balance equation will yield the recursion formula. Let us go step by step, starting from the simplest case.

1. For $n = 1$ we have the fixed point equation

$$f(\mu, x) = x. \tag{3.142}$$

As a second degree equation, it has either a pair of complex roots or two real roots, depending on the parameter μ. For some range of μ, one or another of the real roots may become stable, and correspond to the observable fixed point. We draw this situation schematically in Fig. 3.9(a), where a dashed line represents an unstable fixed point, and a solid line a stable fixed point.

2. The $n = 2$ equation

$$F(2, \mu, x) = x \tag{3.143}$$

has $2^2 = 4$ roots among which there are always the two roots of (3.142). Hence there are only two new roots. For the logistic map, it can be checked explicitly that they form a pair of complex roots up to the point when the fixed point of (3.142) loses stability and they become simultaneously real and stable. This is just the period-doubled regime after the stable period 1. We write the balance symbolically as $2^2 = 2 + 2$ or as $(a) + (b)$, as shown in Fig. 3.9. These two roots may become unstable later, but this does not affect the counting of roots.

3. Now comes the odd and prime number $n = 3$. Its equation

$$F(3, \mu, x) = x \qquad\qquad (3.144)$$

has $2^3 = 8$ roots, among which there are always present the two roots of (3.142). The remaining 6 roots may correspond either to three pairs of complex roots or to 6 real roots. The complex roots are unobservable in numerical iterations with real numbers. The real roots are combined into three pairs, one stable and one unstable in each pair. They arise from a new phenomenon — tangent bifurcation. As we have learnt in Section 2.6, at tangent bifurcation, stable and unstable periodic points come in pairs. Denote the number of different tangent bifurcations into stable period p windows by $M(p)$; the contribution of these bifurcations to the number of roots may be expressed as $2 \times p \times M(p)$. Therefore, in our $p = 3$ case, we have the following balance $[(a) + (c)$ in Fig. 3.9]

$$2^3 = 2 + 2 \times 3 \times M(3),$$

leading to $M(3) = 1$. We might have included a $M(2)$ term in the balance equation for the roots of (3.143), but it would simply yield $M(2) = 0$.

4. Now let us consider a general prime number p. The equation

$$F(p, \mu, x) = x \qquad\qquad (3.145)$$

has 2^p roots, among which there must be the two roots of (3.142). Since p is a prime, all the other $2^p - 2$ roots must associate with tangent bifurcations into period p orbits. Suppose that there are $M(p)$ different tangent bifurcations into period p, we can then write the balance as

$$2^p = 2 + 2 \times p \times M(p),$$

which leads to the general and useful formula for the number of windows of prime period p (Metropolis, Stein and Stein, 1973):

$$M(p) = (2^{p-1} - 1)/p. \tag{3.146}$$

Thus, there are 3, 9, 93, 315, 3855, ... different windows for periods $p = 5, 7, 11, 13, 17, \ldots$, etc. We have referred to the number 93 in the beginning of Section 3.2.

5. The $n = 4$ equation

$$F(4, \mu, x) = x \tag{3.147}$$

has $2^4 = 16$, roots among which there must be that of the $n = 1$ and $n = 2$ cases [Figs. 3.9(a) and (b)]. Among the other roots, there are 4 coming from the period-doubled regime of period 2 (Fig. 3.9(d)) and those corresponding to tangent bifurcations into period 4 (Fig. 3.9(e)). Therefore, we can write the balance as

$$2^4 = 2 + (2 + 4) + 2 \times 4 \times M(4).$$

It follows from this equation that there is only the $M(4) = 1$ primitive period 4 window. We already know that it corresponds to the word RL^2.

Now the readers must have grasped the idea. For an arbitrary n, we first find all the factors d that divide n. In elementary number theory, the notation $d|n$ is used to denote this fact. We must count all the roots coming from the equations

$$F(d, \mu, x) = x,$$

with $d < n$, and those coming from possible period-doublings of shorter periods, to write down the "conservation of the number of roots":

$$2^n = 2 + \sum_{i=1}^{k} 2^i + \sum_{\{d:d|n\}}^{1<d<n} 2 \times d \times M(d) \times Q(d) + 2 \times n \times M(n).$$

Here, the first sum exists, if $n =$ some odd number $\times 2^k$ ($k \geq 1$) and the factor $Q(d)$ in the second sum is calculated by decomposing n into $d \times$ (odd number, if any) $\times 2^l$ ($l \geq 0$) and equating $Q(d) = 2^l$. $Q(d)$ takes into account the period-doubled regimes from lower periods. Therefore, we have got the formula for $M(n)$ (Zeng, 1985):

$$M(n) = \frac{1}{2n} \left[2^n - 2 - \sum_{\substack{i=1}}^{k} 2^i - 2 \sum_{\{d:d|n\}}^{1<d<n} dM(d)Q(d) \right], \qquad (3.148)$$

where the last sum runs over all divisors d of n. This is a recursion formula of variable recursion depth, which depends on the number of factors d into which n decomposes. If one prefers, the identity

$$M(2) = 0$$

may be taken as the initial condition to start the recursion.

3.9.2   The Necklace Problem — Symmetry Types of Periodic Sequences

Consider a periodic sequence

$$\ldots, a_1, a_2, \ldots, a_n, a_{n+1} = a_1, \ldots,$$

where each a_i may take one of q values. Two sequences are considered to be equivalent, if they differ only by a shift of origin. In other words, they are necklaces made of stones of q different colours. How many different necklaces of length n can be made from these stones? The reference to this necklace problem in Riordan's book[6] dates back to 1891. Clearly, the answer to this question depends on what necklaces are considered to be equivalent. In general, one tests for invariance of the sequence under a certain group G. We just have had the case $G = C_n$, C_n being the cyclic group of order n. We might allow for permutation of colours ($G = S_q$, the symmetric group of order q) or allow for both shift of origin and change of colours ($G = C_n \otimes S_q$). Fine[7], Gilbert and Riordan[8] have derived explicit formulae for the cases $G = C_n$, D_n (the dihedral group, i.e., C_n with mirror inversion added), $C_n \otimes S_q$ and $D_n \otimes S_q$.

Since the multiple kn of a period n will also be a period, it is convenient to introduce the number $F_q(n)$ of primitive period n, where the sequence does not have any period smaller than n. The total number of sequences of period n, denoted by $F_q^*(n)$, is related to $F_q(n)$ by the relationship

---

[6] J. Riordan, *An Introduction to Combinatorial Analysis*, John Wiley & Sons, New York, 1958.

[7] N. J. Fine, *Illinois J. Math.* **2**(1958),285.

[8] E. N. Gilbert, and J. Riordan, *ibid.* **5**(1961), 657.

$$F_q^*(n) = \sum_{d|n} F_q(n),$$
(3.149)

where the sum runs over different factors of n, including 1 and n. Therefore, if $n = p$ is a prime number, then

$$F_q^*(n) = F_q(n) + 1.$$
(3.150)

Formula (3.149) can be inverted to yield

$$F_q(n) = \sum_{d|n} \mu(\frac{n}{d}) F_q^*(d),$$
(3.151)

where $\mu(n)$ is the so-called Möbius function in number theory: $\mu(1) = 1$, $\mu(n) = (-1)^r$, if n is a product of r distinct primes, and otherwise $\mu(n) = 0$. Metropolis, Stein and Stein (1973) have argued that the total number $N^*(n)$ of admissible words in the symbolic dynamics of two letters corresponds to the number of symmetry types for the group $C_n \otimes S_2$:

$$N^*(n) = F_2(n) \quad \text{for} \quad G = C_n \otimes S_2.$$
(3.152)

We skip the explicit formulae derived by mathematicians for the calculation of $F_q(n)$ and $F_q^*(n)$, because it is much easier to compute them using the recursive formula (3.148) of the last section. Nevertheless, we summarize all these quantities in Table 3.14.

3.9.3 The Number of Saddle Orbits in the Horseshoe

As we have seen in Fig. 2.7, the dynamics of unimodal maps, geometrically speaking, consists of repeatedly stretching and folding back of the interval. The mathematician Smale has extracted the basics of this transformation and suggested the celebrated "horseshoe mapping", to reflect the essence of complicated dynamics in higher dimensional systems (Smale, 1967). The mapping can be loosely described as follows. Take a rectangular region in a plane and deform it into a long strip. Then fold this strip into a horseshoe-shaped figure, and then put it back into the original region. This process is repeated indefinitely. In a sense, it is an analog of the surjective logistic map.

Recently, Yorke and Alligood (1985) have introduced a parameter μ into the horseshoe transformation (see Figure 3.10). When $\mu = 0$, the deformed

n	$F_2^*(n)$	$F_2(n) = N^*(n)$	$M(n) = S(n)$	$N(n)$
1	1	1	1	1
2	2	1	0	0
3	2	1	1	1
4	4	2	1	1
5	4	3	3	3
6	8	5	4	3
7	10	9	9	9
8	20	16	14	13
9	30	28	28	27
10	56	51	48	45
11	94	93	93	93
12	180	170	165	159
13	316	315	315	315
14	596	585	576	567
15	1096	1091	1091	1085
16	2068	2048	2032	2017
17	3856	3855	3855	3855
18	7316	7280	7252	7217
19	13798	13797	13797	13797
20	26272	26214	26163	26115
21	49940	49929	49929	49911
22	95420	95325	95232	95139
23	182362	182361	182361	182361
24	349716	349520	349350	349102
25	671092	671088	671088	671079

Table 3.14: Comparison of various enumeration results for unimodal maps.

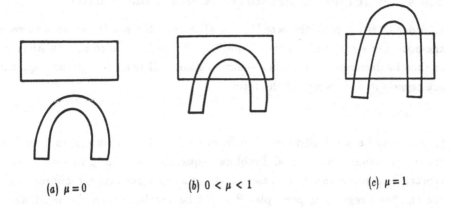

(a) $\mu = 0$ (b) $0 < \mu < 1$ (c) $\mu = 1$

Figure 3.10: Horseshoe transformation with parameter (after Yorke and Alligood).

strip is put back outside the rectangle. When $\mu = 1$, it is put entirely into the old region, as Smale did. When $0 < \mu < 1$ there is only a partial overlap between the original and deformed figure. They have shown that in the process of forming the standard horseshoe, i.e., when μ goes from 0 to 1, attracting periodic orbits of all periods must be present at a certain range of the parameter. Each of these periods then develop into a period-doubling cascade. Although their arguments are highly topological in nature, Yorke and Alligood have succeeded in tabulating the number $S(n)$ of distinct cascades for a given period n (Table 1 in the cited paper). In fact, all these numbers $S(n)$ follow from our recursion formula (3.148); thus we have

$$M(n) = S(n).$$

One sees from Table 3.14 that

$$N^*(n) = M(n) = S(n), \qquad (n \text{ odd})$$
$$N^*(n) > M(n) = S(n), \qquad (n \text{ even})$$

Futhermore, for all non-prime numbers n, we have

$$M(n) = S(n) > N(n),$$

where $N(n)$ is the number of primitive, or indecomposable, admissible words. This leads us to the new enumeration problem of the next section.

3.9.4   Number of Primitive Admissible Words

In order to select primitive words from all admissible words, we have to recall the notion of *-composition (Section 3.3.4). A word is said to be primitive, if it cannot be decomposed by using *-composition. All period-doubling sequences are represented by words of the form

$$P * R^{*k};$$

thus it may be a primitive word only when $k = 0$. If P is a blank, we have the Feigenbaum main period-doubling sequence. If P is primitive, then it represents a period-doubling cascade, originating from tangent bifurcation in the one-band region. In principle, P might be another composite word, say,

$$P = R^{*2} * RLR^2 * R * RL;$$

then $P * R^{*3}$ would describe a period 960 orbit, developed after a three-fold period-doubling (R^{*3}) from a period 3 (type RL) tangent bifurcation in the secondary 2-band (R) chaotic region, which itself is the inverse sequence coming from a period 5 (type RLR^2) tangent bifurcation in the 4-band (R^{*2}) region of the main sequence. What we have just said may be repeated recursively to produce more complicated periodic orbits.

Keeping in mind this recursive construction and what has been said before about the number of real roots at period-doublings and tangent bifurcations, it is easy to generalize the recursion formula (3.148) to that for the number $N(n)$ of primitive words. It reads (Hao and Zeng, 1987)

$$N(n) = \frac{1}{2n} \left[2^n - 2 - \sum_{i=1}^{p} 2^i - 2 \sum_{\substack{\{d:d|n\} }}^{1<d<n} Q_0 \prod_j d_j N(d_j) Q(d_j) \right],$$

where $Q_0 = 2^k$ $(k \geq 0)$ and $Q(d)$, in general, should be defined recursively in the form

$$Q(d) = 2 + \sum_{i=1}^{q} 2^i + 2 \sum_{\substack{\{c:c|d\} }}^{1<c<d} Q_0 \prod_k c_k N(c_k) Q(c_k),$$

where all that has been said with respect to n must be repeated with respect to d, and so on, and so forth. In practice, one encounters the full recursive structure only when n gets large enough.

These numbers $N(n)$ are given in the last column of Table 3.14. When $n = p$ is a prime number, it is natural to have

$$N^*(p) = M(p) = S(p) = N(p).$$

The fact that the number of primitive periodic sequences $F_2(n)$ equals $N^*(n)$, which contains non-primitive periods in the sense of *-composition, tells us that the necklace problem may be further refined to include the consideration of subpattern compositions.

3.9.5 Number of Periods in the Antisymmetric Cubic Map

Almost all the discussion in Sections 3.9.1 to 3.9.4 may be carried over to the antisymmetric cubic map, i.e., symbolic dynamics of three letters. We will first of all write down the recursive formula for the number of periodic windows (Zeng, 1985):

$$M(n) = \frac{1}{4n} \left[3^n - 3 - 2 \sum_{i=0}^{k} 2^i - 4 \sum_{\substack{\{d:d|n\} }}^{1 < d < n} dM(d)Q(d) \right], \qquad (3.153)$$

where the first sum to k exists, if $n = $ (odd number) $\times 2^k$, $(p > 1)$, and $Q(d) = 2^l$, if $n = d \times$ (odd number, if any) $\times 2^l$, $(l \geq 0)$.

The number of windows of a given period also coincides with the number of finite λ-auto-expansions, extended to the interval $(2,3)$, i.e., for $2 < \lambda < 3$ (Zeng, 1987). The generalized λ expansion provides a procedure for obtaining all admissible words of given length as well as a criterion for admissibility.

Nevertheless, as far as we are concerned with the group-theoretical approach, something subtle emerges. Metropolis, Stein and Stein (1973) have referred to the results of mathematicians for the group $C_n \otimes S_2$ and their conclusion appeared to be correct. Later on, the results for the group $C_n \otimes S_3$ have been applied to the antisymmetric cubic map (Testa and Held, 1983). However, they did not agree with the numerical findings on the number of superstable periodic windows (Zeng, 1985). What is wrong with the reasoning? Actually, the antisymmetric cubic map has a hump and a valley, and locally one must have the symbolic dynamics of two letters (R, M or M, L, each described by

Group	$C_n \otimes S_2 \otimes C_2$			$C_n \otimes S_3$	
n	$F_3^*(n)$	$F_3(n) = N^*(n)$	$M(n)$	$F_3^*(n)$	$F_3(n)$
1	1	1	1	1	1
2	2	1	0	2	1
3	3	2	2	3	2
4	7	5	4	6	4
5	13	12	12	9	8
6	34	30	28	26	22
7	79	78	78	53	52
8	212	205	200	146	140
9	549	546	546	369	366
10	1490	1476	1464	1002	992
11	4027	4026	4026	2685	2684
12	11090	11070	11040	7434	7404
13	30661	30660	30660	20441	20440
14	85490	85410	85332	57046	56992
15	239159	239144	239144	159451	159440

Table 3.15: Comparison of two groups for the cubic map.

a group S_2) in a suitable parameter range. In addition, there exists an operation of interchanging R and L, but leaving M unchanged. This corresponds to a cyclic group of order 2, and converts the two S_2 groups one into another. Therefore, instead of the group S_3, one should consider the group $S_2 \times C_2$. It is a smaller group (order 4) than S_3 (order 6), hence the larger number of periodic sequences (see Table 3.15, taken from Hao and Zeng, 1987). The working formula to calculate $F_3^*(n)$ for the group $C_n \otimes S_3$ may be modified to treat the case $C_n \otimes S_2 \otimes C_2$. It reads (Zeng, 1987):

$$F_3^*(n) = \frac{1}{4n} \sum_{(k_1, k_2)} \sum_{d|n} \phi(d) N(k_1, k_2) (R(d))^{n/d}, \tag{3.154}$$

where $\phi(d)$ is the Euler totient function (see, e.g., the book by Hua[9]),

[9]Hua Loo Keng (1982), *Introduction to Number Theory*, Springer-Verlag.

$$R(d) = \sum_{c|d} c\, k_c \leq 3,$$

and k_i satisfies $k_1 + 2k_2 = 3$, $N(3,0) = N(1,1) = 1$. As we have seen in (3.148), $c|d$ denotes an integer c that divides d. The double sum in (3.154) runs over possible combinations of (k_1, k_2), i.e., $(3,0)$ and $(1,1)$, and over distinct factors of n, including 1 and n. The results for $F_3^*(n)$ and $F(n)$ [see (3.154)] are listed in the second and third columns of Table 3.15; they agree with the results of the generalization of the λ-expansion and the recursion formula (3.153).

We see now that the cubic map has deviated from the necklace problem. Conceptually, Metropolis, Stein and Stein (1973) should have chosen the group $C_n \otimes C_2$ instead of $C_n \otimes S_2$, since all groups of order 2 are isomorphic to each other. Then the choice of $C_n \otimes C_2 \otimes C_2$ for the antisymmetric cubic map would look like a natural generalization of the unimodal case.

Chapter 4

Circle Mappings and Two-Dimensional Maps

Circle mappings and higher dimensional maps are in their own rights rich subjects; a detailed study would go beyond the scope of this book. We include this chapter in order to link one-dimensional mappings and differential equations. Therefore, instead of a comprehensive review, we shall only touch on those aspects of the problem which either are needed in the sequel or contain new contributions by the Chinese researchers.

Circle mappings, or maps of a circle into itself, provide another important paradigm in nonlinear dynamics and occupy an intermediate position between one-dimensional maps and higher dimensional systems. They give clues to understand chaotic behaviour in a much broader class of nonlinear systems, namely, those involving two or more competing frequencies. Circle mappings are the simplest models of coupled nonlinear oscillators, yet they exhibit a variety of new phenomena, e.g., quasiperiodic motion, mode-locking, and many new forms of transitions from quasiperiodic regimes to chaos. These phenomena are much more frequently encountered in higher dimensional systems and are either absent or not so transparent to be visible in one-dimensional mappings of the interval that we have studied in the preceding chapters. Nevertheless, mappings of the interval and circle mappings may be closely related through

205

the symbolic description of their orbits.

The importance of circle maps consists of the fact that it highlights the physics of chaos as one of the typical regimes of nonlinear oscillations that cannot be reduced to periodic or quasiperiodic motions. The "physics" here is centered around the phenomena of resonance and their nonlinear generalizations — mode or frequency locking and the destruction of these locked regimes. Although an elaborate description of nonlinear oscillations inevitably involves nonlinear differential equations, the main ingredients are present in circle mappings and can be studied in a fairly elementary manner. Moreover, many circle mappings are very nice mathematical creatures in their pre-chaotic regimes: they are diffeomorphisms, i.e., continuous and invertible transformations of a circle into itself; there exists a wide mathematical literature on circle mappings. However, in accordance with the spirit of this book, we shall confine ourselves mainly to symbolic description and the systematics of periods in these mappings.

4.1   The Physics of Linear and Nonlinear Oscillators

We begin the discussion of circle maps with a qualitative comparison of linear and nonlinear oscillators without the use of the language of differential equations.

A simple linear oscillation is described by $A \sin(\omega t + \phi)$. Among the three characteristics of this oscillation: the amplitude A, the phase ϕ and the frequency ω, only the last, i.e., the frequency, is an intrinsic property of the oscillator. The amplitude and phase are determined by initial conditions. There are no transients in such simple linear oscillations without friction or external forcing: a pendulum would start swinging from the point where it has been released and return to this point indefinitely many times, provided the initial conditions have put it in the linear, i.e., the small amplitude regime.

If two linear oscillators are coupled, they "interact" only under one condition, namely, that their frequencies must be the same: $\omega_1 = \omega_2$, otherwise they will just go their own way without influencing each other. Recall a middle school physics demonstration: two pendulums, hanging from a beam, will interchange energy only when they are of the same length. Even in this case, the transfer of energy from one pendulum to another and *vice versa* become possible only due to small nonlinearities, inevitably present in the system. At resonance, the amplitude of oscillations will grow without limits in the simple linear theory; in practice, it is bounded by dissipations and nonlinearities. Although dissipations can be included in oscillator models in a linear manner, e.g., by a friction term $\gamma \dot{x}$ proportional to the friction coefficient γ and velocity $\dot{x}$, they originate from interactions, i.e., nonlinearities, among a huge number of degrees of freedom. When friction is included, the resonance amplitude will be restricted to a finite peak and acquires a finite width. This width smears the sharp resonance condition and causes interactions under near-resonance conditions. Nevertheless, we repeat that friction is essentially a nonlinear and many-body effect.

There are many types of nonlinear oscillators, a frequently encountered type being a limit cycle oscillator. Contrary to a linear oscillator, the frequency, amplitude and phase of a limit cycle oscillator are closely related intrinsic properties of the system. The system will eventually reach a steady oscillatory state of motion, independently of the initial conditions. If we "couple" two nonlinear oscillators, e.g., by making the state of one oscillator dependent on the amplitude of the other, then in most cases they will try "to sing in unison": either the frequency of the stronger one takes over, or they compromise by finding a common frequency — usually at a fixed ratio of the two "free" frequencies. This is the phenomenon of *frequency-locking* or *mode-locking*. Linear oscillators simply cannot do this, because they are not capable of changing their own frequency. Even coupled, nonlinear oscillators cannot always compromise at a common periodic regime, whence quasiperiodic and chaotic motions come into play.

The nice properties of limit cycle oscillators of retaining their individuality, on the one hand, and of being entrained by external periods, on the other hand, have made them into good candidates for the modelling of biological processes. Indeed, many biological rhythms must be an intrinsic property of a species, in the first place, independently of initial conditions or small changes in the environment, and, at the same time, they should be in harmony with the rhythms of Nature (seasons, day and night, etc.) These facts might explain why researchers in the life sciences have paid so much attention and contributed so widely to the study of circle mappings (e.g., Glass and coworkers, 1982, 1983, 1984).

Nonetheless, we will begin with the simplest linear oscillator, namely, the "bare" circle map.

4.2   The Bare Circle Map

It will be quite instructive to commence our study of circle mappings with the simplest case of a rigid rotation. It is given by the linear, bare (i.e., without nonlinear interactions) map

$$\theta_{n+1} = \theta_n + A \quad (\text{mod } 1), \tag{4.1}$$

(see Fig. 4.1, where the map is represented by two segments of straight lines above and below the bisector, and a periodic orbit starting from $\theta = \theta_0$ is shown). In (4.1), (mod 1) (read "modulus one") means dropping the integer part of the numerical result of the preceding expression. In other words, we are concerned with the motion along a closed circle and count a full turn around the circle as 1 instead of calling it 2π or $360°$. The q-th iterate of this simple linear map can be written down easily:

$$\theta_q = \theta_0 + q A \quad (\text{mod } 1).$$

Therefore, if A happens to be the ratio of two integers p and q: $A = p/q$, then $q A$ drops out when taking the modulus and we have a q-cycle

$$\theta_q = \theta_0 \quad (\text{mod } 1).$$

Notice that in the figure of the bare circle map q is the total number of points, while p is the number of points located on the lower branch of the map. For instance, in Fig. 4.1, we have a 2/7 periodic orbit.

If we wish to retain the cumulative effect of the rotations, we can drop (mod 1) and write

$$\theta_q = \theta_0 + p.$$

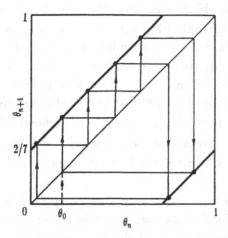

Figure 4.1: The bare circle map with an $A = 2/7$ periodic orbit shown.

In general, when a circle map is drawn without taking (mod 1), it extends to the whole $f(\theta)$ — θ plane instead of being confined to the unit square. Sometimes this is called a *lift* of the circle map. We shall define general lifts later. The lift of the bare map is an ever ascending straight line. When confined to the unit square, each of the p points on the lower branch results from a single (mod 1) operation. It will be quite useful to become accustomed to going back and forth between these two representations.

We see that when $A = p/q$, there are q points on the circle, and they are visited in p full turns. A is called the (bare) *rotation number*. In the case of a general circle map (see the next section), usually the rotation number no longer enters into the original map explicitly and has to be calculated especially. The dependence of the rotation number on the parameters is one of the central problems in the study of circle maps.

When A is an irrational number, i.e., it cannot be represented as the ratio of two integers, we will never obtain a finite periodic cycle from the bare circle map 4.1: the successive iterates will fill up the circle without ending or repeating themselves. We say that we have a *quasiperiodic motion*, characterized by the irrational rotation number A. Quasiperiodic motion is a qualitatively new phenomenon that comes into play whenever two or more basic periods are present in the system. In the case of (4.1), these periods are 1 and A.

Now we have a complete picture of the bare circle map in its parameter space $A \in (0, 1)$. When A varies from 0 to 1, the map (4.1) shows a periodic cycle at each rational A, and it exhibits quasiperiodic motion at all other values. Since rational numbers form only a countable, infinite set of measure zero in the unit interval, and irrationals fill up the line with measure 1, periodic cycles are exceptions, and quasiperiodic motion is the general rule. When one picks up an A at random, there is an overwhelming chance that a quasiperiodic orbit will be obtained. Nevertheless, the bare circle map is rather important for at least two reasons.

Firstly, inclusion of a nonlinearity in the map (4.1) tends to enhance the possibility of mode-locking, increasing thereby the chance of a periodic regime and giving it a positive measure in the parameter space. For example, if one modifies the bare map into the so-called sine circle map

$$\theta_{n+1} = f(\theta_n) = \theta_n + A + \frac{B}{2\pi} \sin(2\pi\theta_n) \quad (\text{mod } 1), \tag{4.2}$$

where the sine term is clearly a periodic and nonlinear function, the new parameter B characterizes the strength of the coupling to the nonlinearity. The previous periodic regime at one single point $A = p/q$ will be widened into a finite interval in A at nonzero B. This is a new phenomenon, called *frequency-locking* or *mode-locking*: the nonlinear oscillator, represented by $f(\theta)$, locks into the p/q regime of the bare linear map instead of exhibiting its own period 1. Figure 4.2 gives a schematic presentation of these *mode-locking tongues* or *Arnold tongues*. For different p/g, when B is small, mode-locked regions do not overlap. Therefore, at least for sufficient small B, the systematics of periodic regimes in the nonlinear map (4.2) will be given by that of the bare map.

Secondly, the bare map serves as a reference for the ordering of orbits in more general circle mappings. The point is that, whenever the map $f(\theta)$ remains monotonically increasing in θ, i.e., the slope $df/d\theta$ remains positive, the order of the visits along the unit circle in such maps is the same as that in the bare circle with the same rotation number, although the distance between points may vary. If two or more initial points are chosen, then their relative order will be preserved at subsequent iterations (this is, of course, a consequence

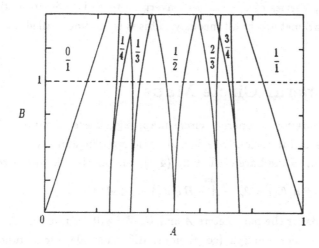

Figure 4.2: Mode-locking tongues in the parameter plane of a circle map (schematic).

of the monotonicity of the map). We shall call this kind of order a *good order*, and the corresponding map — an order-preserving map. Since the orientation of the circle may be determined by following a small "vector" from one point to its neighbouring point, order-preserving also means orientation-preserving.

Intuitively speaking, when an initial point maps successively into its images along the circle, if it always moves forward in one sense without marking time or jumping back, the motion is regular (periodic or quasiperiodic). Whenever it starts to hesitate or to step back, owing to the non-monotonicity in $f(\theta)$, the good order is no longer preserved and chaotic motion may appear. However, bad order alone does not necessarily mean chaos, since period-doubling already violates the good order. This remark may become clearer later on when we try to establish the connection between the symbolic sequences in unimodal and circle maps.

Perhaps this is the right place to note that at least two parameters [A and B in (4.2)] are needed for an exploration of the full complexity of circle mappings, because, in order to follow the transition to chaos through a specific sequence of regimes (e.g., a series of p/q values converging to a given irrational), one must adjust two parameters simultaneously. This is sometimes called a codimension two study.

4.3   General Circle Maps

Instead of studying a concrete circle map, e.g., the sine map (4.2), we will develop some useful notions relating to circle mappings in general. Coupling a nonlinear term to the bare circle map (4.1), one obtains the more general map

$$\theta_{n+1} = f(\mu, \theta_n) = \theta_n + A + B\, g(\theta_n) \quad (\text{mod } 1), \tag{4.3}$$

where μ stands for the parameters A and B. We will assume for the time being, that $g(\theta)$ is a good enough (continuous, differentiable, etc.) function. The differentiability requirement will be dropped in Section 4.5, when considering a piecewise linear circle mapping. Moreover, we will demand that $g(\theta)$ is a periodic function:

$$g(\theta + 1) = g(\theta). \tag{4.4}$$

Frequently, we will omit the parameters, when writing down the map, if no confusion can occur. Dropping (mod 1) in the definition (4.3) leads to the *lift* of $f(\theta)$: $F(\theta)$. A lift takes into account the cumulative effect of rotation. Defined by (4.3) and (4.4), the lift F has the obvious property

$$F(\theta + 1) - F(\theta) = 1.$$

This is called a *degree 1 map*. One could have changed the original definition (4.3) in order to arrive at, for example, a degree d map, i.e.,

$$F(\theta + 1) - F(\theta) = d.$$

However, degree 1 maps are the most natural and important circle mappings requiring study (see, e.g., MacKay and Tresser, 1986).

The rotation number of the circle map f is defined, with the aid of its lift F, by

$$\rho(\theta_0) = \lim_{n \to \infty} \frac{F^{(n)}(\theta_0) - \theta_0}{n}; \tag{4.5}$$

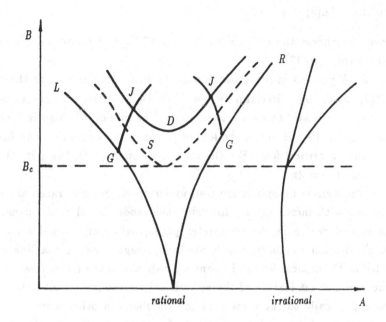

Figure 4.3: Mode-locking tongues for rational and irrational rotation numbers (schematic, after Boyland).

it measures the average rotation per iteration. In general, the above limit may not exist (corresponding to an orbit without rotation number) and ρ may depend on the initial point θ_0. Therefore, in general one should write $\rho(f, \theta)$.

For small B, the nonlinear term does not change the monotonic behaviour of the linear part. When both the function f and its inverse f^{-1} exist and are continuous, the map is, in fact, a diffeomorphism; it has been studied for a long time as a model of simple dynamical systems. In this case, ρ does not depend on the initial point θ_0 and one calls $\rho(f)$ the rotation number of the map f (or F). For an irrational rotation number $\rho = A$, a theorem due to Denjoy[1] says that the mapping f is topologically conjugate to a rigid rotation by A, i.e., there exists a function u that transforms the mapping f into a bare circle map:

[1]A. Denjoy (1932), *Comptes Rendus Sci. l'Acad.*, **195**, 478.

$$u^{-1} \circ f \circ u(\theta) = \theta + A.$$

Moreover, for almost all irrational ρ and an analytic f, the function u is analytic as well (Herman, 1977).

The lift F persists in being monotonically increasing as long as the derivative $F'(\theta) > 0$. The first time when F' is equal to zero at one or more θ values, the map is said to be a *critical map*. The monotonic map is a *subcritical map*. In the $A - B$ plane, the loci of parameters, where the map becomes critical, form a *critical line*. For the sine circle map (4.2), the critical line is the horizontal $B = 2\pi$.

Subcritical circle mappings are well understood. For any rational rotation number p/q with increasing nonlinearity, the mode-locked zone opens into a tongue-shaped region in the parameter plane, called an Arnold tongue; for irrational rotation numbers, the mode-locked region remains a line of zero width below the critical line and opens up only above the latter (see Fig. 4.3). The most important feature of the subcritical regime is that all initial values lead asymptotically to the same rotation number. In other words, all orbits are qualitatively the same for a given set of parameters. One speaks about the rotation number of the map, not of an individual orbit.

Above the critical line, the map is no longer monotonic, but monotonic maps still play an essential role in the description of the motion. This fact can be understood by inspecting Fig. 4.4, where a general non-monotonic map F is confined between two monotonic maps F^+ and F^-: F^+ is the smallest non-decreasing map that is larger than or equal to F everywhere, while F^- is the largest non-decreasing map that is less than or equal to F everywhere.[2] In other words, for all x, we have, by definition,

$$F^-(x) \leq F(x) \leq F^+(x).$$

Applying the monotonic functions F^- and F^+ to

$$F^-(x) \leq F(x) \quad \text{and} \quad F(x) \leq F^+(x),$$

respectively, and using again the above inequalities, we find

$$F^-(F^-(x)) \leq F^-(F(x)) \leq F^2(x) \leq F^+(F(x)) \leq F^+(F^+(x)).$$

---

[2]This kind of construction has been used by Kadanoff and by Hall, see Boyland (1986).

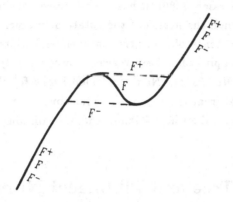

Figure 4.4: The maps F^+, F, and F^-.

This process may be repeated to yield

$$F^{-(n)}(x) \leq F^{(n)}(x) \leq F^{+(n)}(x),$$

from which it follows that the rotation number of $F(x)$ must lie between that of $F^-(x)$ and $F^+(x)$, i.e.,

$$\rho(F^-) \leq \rho(F) \leq \rho(F^+).$$

For a non-monotonic circle map the rotation number may acquire different values for a different choice of the initial points θ_0 (at one and the same parameter set), giving rise to a *rotation interval* (Newhouse, Palis and Takens, 1983). It has been proved that the rotation interval is a closed interval, i.e.,

$$\rho(F) = [\rho_1(F), \rho_2(F)]$$

(Ito, 1981). Moreover, this interval can be determined by computing the rotation numbers of the two monotonic maps F^+ and F^-:

$$\rho(F) = [\rho(F^-), \rho(F^+)]$$

(Gambaudo, Glendinning and Tresser, 1984).

The stretching of a single rotation number into an interval may be taken as a symptom of chaos in circle mappings. It happens to be a more sensible criterion than the positivity of topological entropy ("topological chaos", see, e.g., MacKay and Tresser, 1986); it has been termed *rotational chaos* by Casdegli (1988). Although the notion of the rotation interval was introduced by H. Poincaré, the existence of a rotation interval and its relevance to chaotic motion in circle mappings has been revealed only recently (Newhouse, Palis and Takens, 1983; Ito, 1981). Numerical algorithms for the computation of the rotation interval from the dynamical equations or from time series have been proposed by several authors (Gambaudo, Glendinning and Tresser, 1984; Casdegli, 1988).

4.4 Farey Tree and Fibonacci Numbers

It is quite curious that the development of nonlinear dynamics has related some well-known constructions in elementary number theory to physical reality. In order not to interrupt our discussion in the subsequent sections, we will now make a detour to recollect a few notions from the elementary theory of numbers for future reference.

4.4.1 Farey Tree: Rational Fraction Representation

There are many ways of ordering rational numbers on the unit interval. One of them, namely, the Farey construction, has attracted much attention due to a recent "experimental" observation that in between two frequency-locked regimes, described by rational fractions p/q and p'/q', respectively, the most easily observable, i.e., the widest in the parameter space, period would be given by the Farey composition

$$\frac{p}{q} \oplus \frac{p'}{q'} = \frac{p + p'}{q + q'}. \tag{4.6}$$

Applying this composition rule to the two extremes of the unit interval $0/1$ and $1/1$, one gets $1/2$. Reapplying the Farey construction to all adjacent pairs, we arrive at the *Farey tree*, shown in Fig. 4.5. We see that, forgetting the two "ancestors" $0/1$ and $1/1$ at the top (zeroth) level of the binary tree, there is only one, i.e., $2^0 = 1$ member, namely, $1/2$, while on the k-th level there are 2^k members $\{x_i = p_i/q_i\}$, the sum of which is

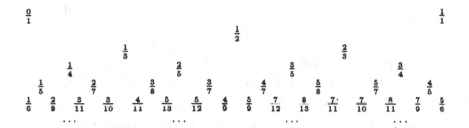

Figure 4.5: The Farey tree: rational fraction representation.

$$\sum_{i=0}^{2^k-1} x_i = 2^{k-1} \quad \forall\, k > 0.$$

Moreover, the values of two Farey members, which are located symmetrically with respect to the central vertical, yield the sum 1: they are p/q and $1-p/q = (q-p)/q$. Geometrically they differ only in the sense of rotation.

Since the Farey tree plays a key role in the understanding of the ordering of periods in circle mappings, we will present a few alternate representations, which may be more useful in one or another context.

4.4.2   Farey Tree: Continued Fraction Representation

Any rational fraction may be converted into a unique, simple *continued fraction* by applying the division algorithm, usually used to find the greatest common divisor of two integers. Take, for example, the fraction 3/11, we have P

$$\frac{3}{11} = \frac{1}{\dfrac{11}{3}} = \frac{1}{3+\dfrac{3}{2}} = \frac{1}{3+\dfrac{1}{1+\dfrac{1}{2}}} \equiv [3,1,2].$$

A *simple* continued fraction has only 1's in the successive numerators. In general, we write

$$\frac{p}{q} = \cfrac{1}{a_0 + \cfrac{1}{a_1 + \cfrac{1}{a_2 + \cfrac{1}{a_3 \cdots}}}} = [a_0, a_1, a_2, a_3, \cdots]. \tag{4.7}$$

```
  [ ]                                                                    [1]
                                        [2]
              [3]                                        [1, 2]
        [4]              [2, 2]              [1, 1, 2]              [1, 3]
    [5]    [3, 2]    [2, 1, 2]    [2, 3]    [1, 1, 3]    [1, 1, 1, 2]    [1, 2, 2]    [1, 4]
      . . .              . . .                . . .                . . .
```

Figure 4.6: The Farey tree: continued fraction representation.

The integers $a_1, a_2, \ldots$ are called the first, second, ... partial quotients of the continued fraction. A rational number is expressed by a finite (terminating) continued fraction and an irrational number by a nonterminating continued fraction. The truncations of an infinite continued fraction give a systematic approximation to an irrational number by rational fractions. In order to make the notation unique, we will never allow the last partial quotient to be 1, because

$$[a_0, a_1, \ldots, a_n] = [a_0, a_1, \ldots, a_n - 1, 1].$$

The Farey tree, shown in Fig. 4.5, can be transformed into a continued fraction representation, as is shown in Fig. 4.6. Now, at the k-th level of the tree, the Farey members are different partitions of the number $k + 2$, i.e.,

$$\sum_i a_i = k + 2.$$

One can obtain readily from the continued fraction representation of the Farey tree the rule for the generation of the next level members. Any member (mother) in the tree gives birth to two members (daughters) at the next level, according to two different rules

1. *Rule 0.*

   $$[a_0, a_1, \ldots, a_n] \rightarrow [a_0, a_1, \ldots, a_n + 1],$$

   where the length of the daughter equals that of the mother, i.e., the length increment is 0.

2. *Rule 1.*

$$[a_0, a_1, \ldots, a_n] \;\rightarrow\; [a_0, a_1, \ldots, a_n - 1, 2],$$

where the length of the daughter is longer than that of the mother by an increment 1.

Starting from $k = 1$, these rules are used in between any two levels in an alternating way, i.e.,

$$0\;1\;1\;0\;0\;1\;1\;0\;0\;1 \;\ldots\; 0\;1\;1\;0$$

with *Rule 0* being used at the two flanks of the tree. Written on the branches of the tree and read from the top $(1/2)$, these 0's and 1's assign a unique binary number to each Farey member. In this way one could have given a binary number representation of the Farey tree. However, we prefer an equivalent, but more convenient representation using so-called *Farey Addresses* (Ostlund and Kim, 1985).

In order to introduce the Farey addresses, we notice that among the two daughters one is greater and the other is smaller than the mother (in the sense of the natural order of numbers in between 0 and 1). An inspection of the continued fraction representation of the Farey tree (Fig. 4.6) shows that the smaller daughter always has an odd number of partial quotients (we call it an odd parity for short), while the elder daughter has even parity, independently of the number of partial quotients of the mother. We agree to put a -1 on the branch connecting the mother to the smaller daughter, and a $+1$ on the other branch. This is a fixed assignment with respect to a mother: -1 on the left branch and $+1$ on the right. We mention in passing that the alternating assignment of 0 or 1 to the branches is clear in terms of parity: *Rule 0* is used when the mother and daughter have the same parity, and *Rule 1* is applied otherwise.

Now let us formulate the rules for finding the parents of a Farey member from its own continued fraction representation. It is easy to see that a member $[a_0, a_1, \ldots, a_n]$ comes from the Farey composition of the following two members:

1. The mother $[a_0, a_1, \ldots, a_n - 1]$ and

2. The father $[a_0, a_1, \ldots, a_{n-1}]$.

In fact, one can forget the fathers all together and trace back only along the maternal line, remembering our convention of absorbing a 1 at the last place of a continued fraction into the next-to-the-last partial quotient. For example, we have for $3/11 = [3, 1, 2]$:

$$[3, 1, 2] \overset{-1}{\rightleftharpoons} [3, 2] \overset{+1}{\rightleftharpoons} [4] \overset{-1}{\rightleftharpoons} [3] \overset{-1}{\rightleftharpoons} [2]$$

where ∓ 1's have been stacked on the arrows, according to the relative values of daughter-mother pairs (or even simpler, according to the parity of the number of partial quotients in the daughter). Reading downwards from the top [2], we find the Farey address of the Farey member $3/11$:

$$\langle\, -1 -1 +1 -1 \,\rangle. \tag{4.8}$$

In this way we could have given an "address representation" for the Farey tree.

4.4.3   Farey Tree: Symbolic Representation

We shall try to introduce symbolic systematics for all possible orbits in the circle mapping and to establish some links to the symbolic dynamics for the mappings of the interval that we have discussed in Chapter 3. Since the rational rotation numbers represented in the Farey tree will play an essential role even for the systematics of chaotic orbits, we start with replacing each rational fraction in the tree by a symbolic sequence. In order to obtain some feeling for the connection to mappings, we draw the graph of a general circle map in Fig. 4.7 with a period 2 orbit shown explicitly. This is a periodic orbit in the $1/2$ mode-locked tongue. If we label the upper left branch by the letter L and the lower right branch — by R, this orbit may be represented by the word RL. Therefore, the rotation number $1/2$ orbit has a continued fraction representation [2] and a symbolic word RL. In contradistinction to unimodal maps, now both letters represent monotone ascending branches. This remains true as long as we deal with subcritical circle mappings (see Section 4.3). Above the critical line, there appears one or more descending branches in the map, which may require an additional letter, say, M (see Fig. 4.10 in Section 4.4.4.)

The symbolic representation of any Farey member may be obtained from its Farey address by applying the Farey transformation to the top of the tree, i.e., to the word RL (Zheng, 1988c). We will define Farey transformations as

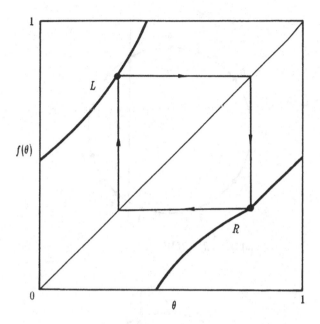

Figure 4.7: A period 2 orbit in circle mapping.

substitution rules applied to the letters R and L, analogously to the generalized composition rules discussed in Section 3.6.6.

There are two *Farey transformations*:

$$
\mathcal{T}_+ : \begin{cases} R \to R \\ L \to RL \end{cases} \quad \text{(order reversing)},
$$

$$
\mathcal{T}_- : \begin{cases} R \to RL \\ L \to L \end{cases} \quad \text{(order preserving)}.
$$

$$(4.9)$$

In general, we will write $\mathcal{T}_\epsilon$ where ϵ may be ± 1. Suppose we have a word

$$\Sigma = s_0 s_1 \ldots s_n,$$

then its Farey transformation is

$$\mathcal{T}_\epsilon(\Sigma) = \mathcal{T}_\epsilon(s_0)\mathcal{T}_\epsilon(s_1) \ldots \mathcal{T}_\epsilon(s_n).$$

Now the rule to generate the symbolic representation W of a Farey member from its Farey address $\langle b_0 b_1 \ldots b_n \rangle$ can be formulated as follows:

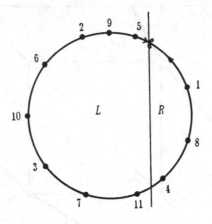

(a) 3/11 : $RL^2(RL^3)^2$

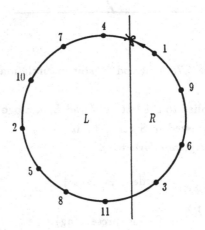

(b) 4/11 : $RL(RL^2)^3$

Figure 4.8: Assignment of symbolic sequence to rational rotation numbers.

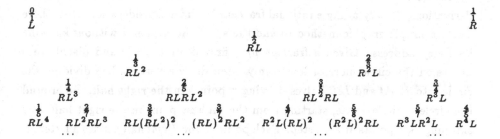

Figure 4.9: The Farey tree: symbolic representation.

$$W = T_{b_0} T_{b_1} \cdots T_{b_n}(RL). \tag{4.10}$$

The following working procedure may be adopted:

1. Transform the given rational fraction into its continued fraction representation by using the divide algorithm.

2. Trace its maternal line back to the top member [2] and write down the ± 1's to form the Farey address.

3. Apply the Farey transformation (4.10) to the Farey address to get the word W.

A few simple examples follow.

Example 1. The fraction 2/7 has a Farey address $\langle\, -1 \ -1 \ 1 \,\rangle$. Its symbolic representation is

$$\begin{aligned}
W_{\frac{2}{7}} &= T_{-1}T_{-1}T_{+1}(RL) \\
&= T_{-1}T_{-1}(RRL) \\
&= T_{-1}(RLRLL) = RLLRLLL.
\end{aligned}$$

Example 2. We have calculated the Farey address of 3/11 to be $\langle\, -1 \ - 1 \ +1 \ -1 \,\rangle$ [see Equation (4.8)]. Consequently,

$$\begin{aligned}
W_{\frac{3}{11}} &= T_{-1}T_{+1}T_{-1}T_{-1}(RL) \\
&= RLLRLLLRLLL = RL^2(RL^3)^2.
\end{aligned}$$

As we shall see later on, the Farey transformation is useful for general derivations. If only a single rational fraction is under consideration, then there exists a simple graphic method to find the symbolic sequence without knowing its Farey address. Given a fraction p/q, first draw a circle and distribute q points on the circle more or less evenly, then draw a vertical line dividing the circle into *Right* and *Left* halves, leaving p points in the right half. Go around the circle anti-clockwise, starting from the highest point in the right half, and skipping $p-1$ points at each step. Each point in the *Right* half carries a letter R, and each point in the *Left* half — a letter L. When all the points on the circle have been reached, one gets the desired symbolic sequence. Figures 4.8 show two examples of assigning symbolic sequences to rational fractions. It is readily checked that the symbolic sequences thus obtained are maximal and coincide with those built by applying the transformations (4.9) according to the Farey addresses.

Figure 4.9 is the symbolic representation of the top part of the Farey tree.

4.4.4   The W-sequence: Ordering of Rotation Numbers

Now we are in a position to introduce an ordering for all rotation numbers using the symbolic representation of the Farey tree and Farey transformations (Zheng, 1988c). This ordering is called a W-sequence, in analogy with the U-sequence of Metropolis, Stein and Stein (1973); W stands for Winding or simply for Word. We start from the case when only two letters R and L, both representing an increasing branch, are needed. It should be made clear right away that this case is not only restricted to the subcritical regime when the map is monotone and the dynamics preserves the good order (see Section 4.2). Orbits that are badly ordered, but do not require an additional letter, say, M, are also included. We shall see that the appearance of a monotone decreasing branch, requiring a third letter, does not cause any difficulty.

We have compared the ordering of all admissible words for unimodal maps to the compilation of a dictionary (see Section 3.3.2). Speaking again in terms of a dictionary, the ordering of rotation numbers leads to a dictionary of infinitely many volumes, each "entitled" by a member of the Farey tree. These volumes have a similar structure and there are similarities within each volume. All the words in a volume have one and the same winding number, namely, that of the title, but they are ordered in the sense of the natural order $L < R$ on the basis of a letter by letter comparison from left to right. To be more precise, in order to compare two patterns

$$W_1 = W^* \sigma \cdots,$$
$$W_2 = W^* \tau \cdots.$$

with a common leading part W^*, we look at the first unequal letters σ and τ. $W_1 > W_2$ if $\sigma > \tau$ in the sense of $R > L$, and *vice versa*. In contradistinction to unimodal maps, no parity is checked, since both letters represent a monotone increasing function.

Let us first describe the fundamental volume for the rotation number $1/2$ which will serve as a templet in the construction of other volumes. On the title page, the word RL is inscribed as the first word in the volume. Different words in the volume may correspond to different periods, but they all have the same rotation number. In order to keep the rotation number equal to $1/2$ there must be an equal number of R's and L's in a word. We include in the volume only the words composed from segments $(RL)^m$ or $(LR)^n$ or both. We shall represent an infinite periodic symbolic sequence Σ^∞ only by its shortest basic unit Σ, and write it in the maximal form. The last word in the volume is $R(RL)^\infty$. A general word in between must be of the form

$$R(RL)^{m_1}(LR)^{n_1} \ldots (RL)^{m_k}(LR)^{n_k} \ldots L.$$

In order to ensure maximality, we must require either $m_1 > m_k$ for all k, or $m_1 = m_k$, but $n_1 < n_k$. Therefore, we have in this volume the following sequence

$$RL, R(RL)L, R(RL)^2(LR)L, \ldots, \quad R(RL)^2 L, \ldots,$$
$$R(RL)^3, \ldots, R(RL)^\infty. \tag{4.11}$$

The sequence (4.11) can be obtained from that part of the period-doubling sequence of unimodal mappings that is included in between the start of period 2 window $[(RR)^\infty]$ and the merging point of the 2-band region $[RL(RR)^\infty]$ by "turning over" the letters according to the following rule (Procaccia, Thomae and Tresser, 1987). Suppose we have a pattern $P = P^*\sigma \ldots$ from the U-sequence of unimodal mappings and we have turned its leading part P^* into a pattern W^* in the W-sequence of circle mappings. In order to get the next letter in W we look at the R-parity of P^*. If it is even, we just copy σ into W without change; if it is odd, we append to W^* a $\overline{\sigma}$ obtained from σ by interchanging R and L. This rule is applied recursively, starting from copying the first letter of P to W, since, by definition, a blank leading part is an even pattern. The procedure can be reversed to get a word in the U-sequence from a symbolic rotation number in the W-sequence, remembering, however, only the parity of the *resulted* leading part is to be checked.

We shall demonstrate this rule by two examples. The period 2 window in the U-sequence starts from RR (see Example 1 in Section 3.6.7). Thus we have

$$U: \quad RRRRRRRRR\ldots,$$
$$W: \quad RLRLRLRLRL\ldots .$$

The result is nothing but the first word RL in the sequence (4.11). The $2 \to 1$ band-merging point in the U-sequence is $RL(RR)^\infty$ (see Example 4 in Section 3.6.7). Therefore, we get

$$U: \quad RLRRRRRRRR\ldots,$$
$$W: \quad RRLRLRLRLRL\ldots .$$

This is just the last pattern $R(RL)^\infty$ in the sequence (4.11). Any admissible word included in between these two may be turned into a pattern in the sequence (4.11). However, only the two extreme patterns correspond to good rotational order, all the other words being badly ordered. In particular, the period 4 orbit, generated by a doubling from RR, is badly ordered, as can be seen by closing the interval into a circle and tracing the locations of the periodic points.

Now let us look at the volume denoted by the rotation number 2/3. On the title, one has the word RRL. It has been obtained from the fundamental RL by using the Farey transformation $R \to R$ and $L \to RL$. Applying the same transformation to the last word $R(RL)^\infty$ in the fundamental sequence (4.11), we get the last pattern in the 2/3 volume, namely, $R(RRL)^\infty$. All the intermediate words in (4.11) may be transformed in this way. The similarity between these two volumes is obvious from the construction. This volume, however, cannot be obtained from the period 3 window in the U-sequence, because, strictly speaking, the "turning over" procedure described above applies only to the rotation number 1/2.

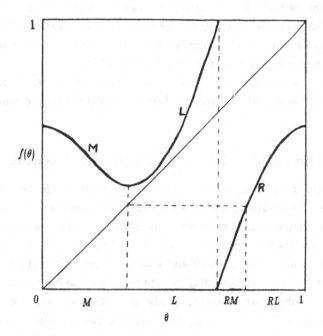

Figure 4.10: Circle mapping with a descending branch denoted by M.

All other "rational" volumes may be constructed in a similar way by first calculating its Farey address and then applying the corresponding Farey transformation to the fundamental sequence (4.11). An irrational number has an infinite Farey address and its volume contains a single infinitely long word. In order to construct this word, one first writes down the fundamental segment RL, then steps down the Farey tree towards the irrational number. One stops *en route* at each Farey member and copies the symbol of its father as a new word segment. One prefixes or suffixes the segment to the existing word according to whether the father is larger or smaller than the daughter.

Next, we will make allowance for a descending branch in the map, which requires a third letter M (see Fig. 4.10). The interval $(0, 1)$ is divided into several subintervals, denoted by the letter(s) which will lead any sequence that originates from the corresponding subinterval. We have clearly the ordering

$$M < L < RM < RL < R.$$

It can be shown that, if there exist two maximal sequences $\Sigma RL\Delta$ and $\Sigma L\Delta$, then $\Sigma RM\Delta$ must be maximal and be ordered in between the first two. Generally speaking, the rotation number of the latter orbit, if it exists, is equal to the rotation number of one of the two former orbits.

So far we have mentioned only the "well-behaved" words. A certain kind of "misbehaved" words which imply a rotation interval has been discussed (Zheng, 1988c).

4.4.5   The Golden Mean and Fibonacci Numbers

Quasiperiodic motion with two or more incommensurable frequency components cannot be identified with the help of a finite number of digits, shown by a measuring instrument or represented by words in computer memory, because irrationals are infinitely long real numbers. Any truncation of an irrational number yields a rational number and leads to a periodic regime. Fortunately, however, irrational numbers may be approximated systematically by means of rational numbers, and a quasiperiodic orbit may be well approached by a sequence of periodic orbits. If properly chosen, these periodic orbits may exhibit a certain scaling property that permits renormalization group treatment. There have been some achievements along this direction to arrive at an understanding of the transition from a quasiperiodic regime to chaos (Shenker, 1982; Feigenbaum, Kadanoff and Shenker, 1982; Rand, Ostlund, Sethna and Siggia, 1982; Ostlund, Rand, Sethna and Siggia, 1983; Zheng, 1986 a and b, 1987b). Nevertheless, we shall confine ourselves only to the approximation of irrational numbers, which is required in the sequel.

Truncation of the infinite continued fraction representation of an irrational number leads naturally to a systematic approximation of the number. This is realized by taking a certain a_i in the simple continued fraction. It is clear that the larger the a_i's in the partial quotient, the smaller the errors caused by discarding the higher quotients. The least favourable case is encountered, when all the a_i's equal 1. This corresponds to a particular irrational number, namely, the reciprocal of the *golden mean*:

$$w \equiv [1, 1, 1, \ldots] = (\sqrt{5} - 1)/2 = 0.618\ldots \ .$$

The truncation of this continued fraction leads to the following sequence of rational fractions

$$w_0 = [1] = 1/1,$$
$$w_1 = [1, 1] = 1/2,$$
$$w_2 = [1, 1, 1] = 2/3,$$
$$w_3 = [1, 1, 1, 1] = 3/5,$$
$$\cdots$$

In general, the n-th member of this sequence is given by the ratio of two *Fibonacci numbers*:

$$w_n = F_n/F_{n+1},$$

which in turn are defined by the recursion relation:

$$F_0 = F_1 = 1,$$
$$F_n = F_{n-1} + F_{n-2}, \quad n = 2, 3, \ldots .$$

The golden mean and the Fibonacci numbers occur in many contexts in nonlinear dynamics. In particular, the renormalization group analysis for the transition from quasiperiodicity to chaos in circle mappings has been based on approaching the golden mean rotation number by ratios of Fibonacci numbers (see references, cited above). However, we will put aside this interesting development and turn to a special circle map, which may be entirely treated by analytical means.

4.5 How the Arnold Tongues Become Sausages: a Piecewise Linear Map

From the study of the tent map (Derrida, Gervois and Pomeau 1978; see Section 3.5) and a similar construction for the antisymmetric cubic map (Zeng, 1987), we know that a lot of instructive results may be obtained analytically by means of piecewise linear maps. In this section, we discuss a piecewise linear circle map, replacing the $\sin(2\pi\theta)$ term by a properly normalized sawtooth function (Yang and Hao, 1987). We write

$$\theta_{n+1} = f(\theta_n) \equiv \theta_n + A - B\, g(\theta_n) \quad (\text{mod } 1), \tag{4.12}$$

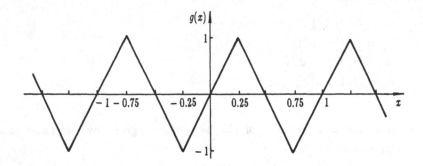

Figure 4.11: The piecewise linear function $g(\theta)$.

where

$$g(\theta) = g_k(\theta) \equiv (-1)^k(4\theta - 2k) \quad \text{for} \quad \theta \in [(2k-1)/4, (2k+1)/4].$$

The function $g(\theta)$ is given in Fig.4.11. Although it is only a C^0 function, there are just two undifferentiable points in $[0,1)$. If necessary, we can smoothen it carefully to make it closer to the sine map without drastically changing its shape. The purely piecewise linear case (4.12) can be studied analytically and the mode-locking structure in the parameter space is illustrated in Fig. 4.12. The most striking feature of this figure is that all the Arnold tongues of rotation numbers p/q with $q > 3$ have become "sausages": their widths have shrunk to zero at certain values of B. We shall consider the critical line and the subcritical behaviour separately.

4.5.1   The Critical Line

We first explore the mode-locking structure along the critical line. It is clear that the map remains monotonic as long as $0 < B < 1/4$, and $B = 1/4$ determines the critical line. Figure 4.13 shows the map (modulus 1) at $B = 1/4$. There are two regions where the map has different slopes. Written down explicitly, we have

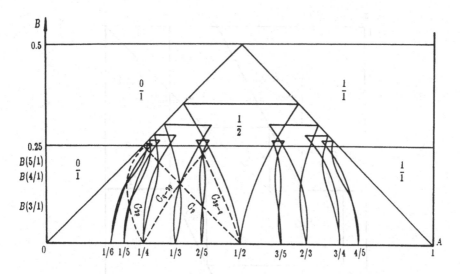

Figure 4.12: The mode-locking structure in the $A - B$ plane. For C_{ap+bq} and $B(q/m)$, see text.

in Region 1 : $\quad f_0(\theta) = f_2(\theta) = A,$

in Region 2 : $\quad f_1(\theta) = 2\theta + A - 1/2, \text{ if } 1/4 \leq \theta < (3 - 2A)/4, \quad (4.13)$

$\qquad\qquad f_3(\theta) = 2\theta + A - 3/2, \text{ if } (3 - 2A)/4 \leq \theta < 3/4.$

Since the slope in Region 1 is 0, any initial point θ_0 chosen in this region would lead to one and the same point A in Region 2 where $f'(\theta) = 2$. Therefore, in order to make a q-cycle stable, there must be just one point in Region 1 and $q - 1$ points in Region 2. In other words, mode-locking can be realized by keeping one point of a cycle in Region 1. Using $f_1(\theta)$ from (4.13) for the map in Region 2 and taking the (mod 1) operation at the end of the calculation, we get for θ_i, $0 < i < q$, all being in Region 2, the expression

$$\theta_q = 2^{q-1}A + (2^{q-1} - 1)(A - 0.5) = (2^q - 1)A + 2^{q-2} - 0.5 \quad (\text{mod } 1).$$

In order to find the width $\triangle A_{p/q}$ of the p/q (p and q being coprimes) mode-locking interval, we must keep θ_q in Region 1, i.e., take either $0 \leq \theta < 1/4$ or $3/4 < \theta < 1$, i.e., $\triangle\theta_q = 1/2$. For this reason the width happens to be independent of p:

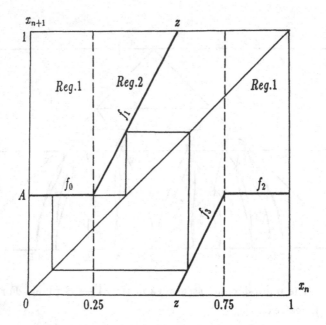

Figure 4.13: The piecewise linear map drawn in the unit square at $A = 0.375$, $B = 0.25$.

$$\triangle A_{p/q} = \triangle A_q = \frac{\triangle \theta_q}{2^q - 1} = \frac{1}{2^{q+1} - 2}.$$

The mode-locking intervals for larger q turn out to be exponentially small. However, this does not mean that the mode-locked regimes will occupy very small parts of the critical line. On the contrary, the dimension of the quasiperiodic set along the critical line happens to be exactly zero instead of the "universal" value $D_0 = 0.87 \ldots$. A heuristic proof of D_0 being zero has been given in Yang and Hao (1987); a rigorous proof for piecewise linear circle mappings may be found in Alstrøm (1986).

4.5.2   Structure of the Mode-Locking Tongues

Below the critical line $B = 1/4$, the map behaves monotonically and the periodic orbits appear in "good order", i.e., in the same order as they appear in the bare circle map (4.12). In order to determine the boundaries of a mode-locking interval, we introduce the notions of *forward map* and *backward map* (Arnold, 1965). Since $g(0) = 0$ in (4.2) or (4.12), at fixed B, the map as a whole shifts upwards when A increases, and downwards when A decreases. The same remark applies to $f^{(n)}(\theta)$, the difference being that the form of the map changes with A slightly when B is small enough. In Fig.4.14, we draw the situation of $f^{(2)}(\theta)$ for a general circle map $\theta_{n+1} = f(\theta)$ in and at the boundaries of the 1/2 tongue. The maps are drawn in the unit square, but the discussion proceeds more easily with the lifted map, i.e., without applying (mod 1). At the tongue boundaries, the p/q-periodic orbit loses stability. When the parameter A crosses the right boundary of the p/q tongue, the map ensures $f^q(\theta) - p - \theta \geq 0$ for all θ [see Fig. 4.14(b)]. This is called the forward map. In the opposite case, when A reaches the left boundary of the tongue, we have the backward map: $f^q(\theta) - p - \theta \leq 0$ for all θ [see Fig. 4.14(c)]. These situations of marginal stability resemble those of tangent bifurcations (cf. Section 2.6.2 and Fig. 2.16).

In order to apply these definitions to the piecewise linear map (4.12), it is more convenient to draw it in the unit square, taking (mod 1) in both directions. Then the map is given by four linear segments (cf. Fig. 4.13):

$$
\begin{array}{lll}
f_0(\theta) = \theta + A - 4B\theta, & (k = 0) & 0 \leq \theta < 1/4, \\
f_1(\theta) = \theta + A + 4B(\theta - 1/2), & (k = 1) & 1/4 \leq \theta < z, \\
f_3(\theta) = f_1(\theta) - 1, & & z \leq \theta < 3/4, \\
f_2(\theta) = \theta + A - 4B(\theta - 1) - 1, & (k = 2) & 3/4 \leq \theta < 1,
\end{array} \tag{4.14}
$$

where

$$
z = (2B - A + 1)/(4B + 1).
$$

A stable periodic orbit, corresponding to the rotation number p/q, must have p points on the lower branch f_2 or f_3, while the total number of points is q. It follows from a comparison of the local behaviour near the periodic points that the forward map must contain $\theta = 1/4$ as its periodic point, whereas the backward map goes through $\theta = 3/4$. Carefully keeping the "good order" at given p/q and checking the stability of the orbit, we arrive at the implicit equations for the tongue boundaries. Take, for example, the $p/q = 1/3$ tongue.

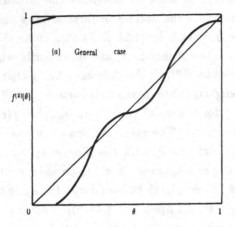

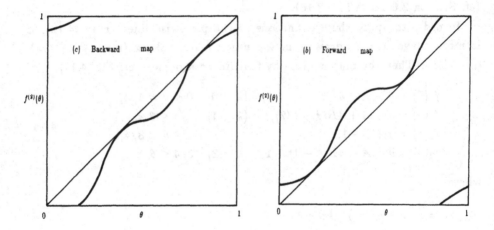

Figure 4.14: The circle map at fixed B.

The right-hand boundary is given by (for the shrinking point $B = 0.1545\ldots$ see below)

$$f_2 \circ f_1 \circ f_0(1/4) = 1/4 \quad (B \leq 0.1545\ldots),$$
$$f_3 \circ f_1 \circ f_0(1/4) = 1/4 \quad (B \geq 0.1545\ldots), \tag{4.15}$$

and the left-hand boundary by

$$f_1 \circ f_0 \circ f_2(3/4) = 3/4 \quad (B \leq 0.1545\ldots),$$
$$f_1 \circ f_1 \circ f_2(3/4) = 3/4 \quad (B \geq 0.1545\ldots). \tag{4.16}$$

Note that, by definition, $f_0(1/4) = f_1(1/4)$ and $f_2(3/4) = f_3(3/4)$. Since A enters (4.14) linearly and nonlinearity associates with B only, Equations (4.15) and (4.16) can be solved explicitly, to yield the tongue boundaries in the $A - B$ plane. All the answers will be rational fractions in B. As one goes to rotation numbers with larger and larger p and q, the calculation, although elementary, becomes more and more tedious. However, this task may be accomplished easily by using any algebraic manipulation language such as REDUCE or MACSYMA on a computer. We collect the results in Table 4.1.

Note that the boundaries of tongues with $q > 2$ comprise more than one section, divided by shrinking points, where the width of the Arnold tongue contracts to zero and only a single value of A allows the p/q mode-locking to occur. We calculate these shrinking points in the next section.

4.5.3 Shrinking Points of the Arnold Tongues

At a shrinking point the right and left boundaries of an Arnold tongue intersect. Therefore, the forward condition $f^q(\theta) - p - \theta \geq 0$ and the backward condition $f^q(\theta) - p - \theta \leq 0$ must hold simultaneously. Thus, the $f^q(\theta) - \theta$ map must be a straight line from $(0,0)$ to $(1,1)$ in the unit square, and any θ_0 would lead to a q-cycle with rotation number p/q. In other words, the $f^q(\theta) - \theta$ map coincides with the bisector. Consequently, we have the following condition for the width of the Arnold tongue to shrink to zero:

$$\frac{d}{d\theta} f^q(\theta) = 1. \tag{4.17}$$

Since the slopes of the piecewise linear map (4.12) in Regions 1 and 2 are

$$(1 - 4B) < 1 \quad \text{and} \quad (1 + 4B) > 1,$$

W	R/L	A Boundaries	Range of B
$\frac{0}{1}$	R	B	
$\frac{1}{1}$	L	$1-B$	
$\frac{1}{2}$	R	$\dfrac{4B^2+2B+1}{2(2B+1)}$	
	L	$\dfrac{-4B^2+2B+1}{2(2B+1)}$	
$\frac{1}{3}$	R	$\dfrac{16B^3+4B^2+B-1}{16B^2+4B-3}$	≤ 0.15451
	L	$\dfrac{-16B^3+4B^2-B-1}{16B^2-B-3}$	≤ 0.15451
	R	$\dfrac{16B^3+12B^2+3B+1}{16B^2+12B+3}$	≥ 0.15451
	L	$\dfrac{-16B^3+4B^2+5B+1}{16B^2+12B+3}$	≥ 0.15451
$\frac{1}{4}$	R	$\dfrac{64B^4+32B^3-1}{4(16B^3+8B^2-1)}$	≤ 0.20982
	L	$\dfrac{64B^4+4B+1}{4(16B^3+4B+1)}$	≤ 0.20982
	R	$\dfrac{64B^4+64B^3+24B^2+4B+1}{4(16B^3++16B^2+6B+1)}$	≥ 0.20982
	L	$\dfrac{-64B^4+24B^2+8B+1}{4(16B^3+16B^2+6B+1)}$	≥ 0.20982
$\frac{1}{5}$	R	$\dfrac{256B^5+64B^4-32B^3-3B+1}{256B^4+64B^3-32B^2-16B+5}$	≤ 0.09734
	L	$\dfrac{-256B^5+64B^4+32B^3+3B+1}{256B^4-64B^3-32B^2+16B+5}$	≤ 0.09734
	R	$\dfrac{256B^5+192B^4+32B^3-8B^2-B-1}{256B^4+192B^3+32B^2-8B-5}$	$0.09734 \sim 0.23189$
	L	$\dfrac{-256B^5-64B^4-16B^3-7B-1}{256B^4+64B^3-64B^2-32B-5}$	$0.09734 \sim 0.23189$
	R	$\dfrac{256B^5+320B^4+160B^3+40B^2+5B+1}{256B^4+320B^3+160B^2+40B+5}$	≥ 0.23189
	L	$\dfrac{-256B^5-64B^4+96B^3+56B^2+11B+1}{256B^4+320B^3+160B^2+40B+5}$	≥ 0.23189
$\frac{2}{5}$	R	$\dfrac{256B^5+64B^4-24B^2-3B+2}{256B^4+64B^3-64B^2-8B+5}$	≤ 0.09734
	L	$\dfrac{-256B^5+64B^4-24B^2+3B+2}{256B^4-64B^3-64B^2+8B+5}$	≤ 0.09734
	R	$\dfrac{256B^5+192B^4+64B^3-8B^2-9B-2}{256B^4+192B^3-24B-5}$	$0.09734 \sim 0.23189$
	L	$\dfrac{-256B^5+64B^4+96B^3+16B^2-7B-2}{256B^4+192B^3+32B^2-16B-5}$	$0.09734 \sim 0.23189$
	R	$\dfrac{256B^5+320B^4+160B^3+56B^2+13B+2}{256B^4+320B^3+160B^2+40B+5}$	≥ 0.23189
	L	$\dfrac{-256B^5-64B^4+96B^3+72B^2+19B+2}{256B^4+320B^3+160B^2+40B+5}$	≥ 0.23189
$\frac{1}{6}$	R	$\dfrac{1024B^6+512B^5-64B^4-64B^3+4B^2-2B+1}{2(512B^5+256B^4-32B^3-32B^2-6B+3)}$	≤ 0.15451
	L	$\dfrac{-1024B^6+192B^4+32B^3+12B^2+6B+1}{2(512B^5-96B^3+16B^2+18B+3)}$	≤ 0.15451
	R	$\dfrac{1024B^6+1024B^5+320B^4-20B^2-2B-1}{2(512B^5+512B^4+160B^2-10B-3)}$	$0.15451 \sim 0.24149$
	L	$\dfrac{-1024B^F6-512B^5-64B^4-64B^3-44B^2-10B-1}{2(512B^5+256B^4-96B^3-96B^2-26B-3)}$	$0.15451 \sim 0.24149$
	R	$\dfrac{1024B^6+1536B^5+960B^4+320B^3+60B^2+6B+1}{2(512B^5+768B^4+480B^3+160B^2+30B+3)}$	≥ 0.24149
	L	$\dfrac{-1024B^6-512B^5+320B^4+320B^3+100B^2+14B+1}{2(512B^5+768B^4+480B^3+160B^2+30B+3)}$	≥ 0.24149

Table 4.1: Boundaries of mode-locking tongues.

respectively, Condition (4.17) implies

$$(1 - 4B)^m (1 + 4B)^{q-m} = 1, \quad m = 1, 2, \ldots, (q + 1)/2. \tag{4.18}$$

This is an equation, which does not depend on A and p. It should be solved for

$$0 < B < 1/4.$$

It has a solution for each $m = 1, 2, \ldots, (q + 1)/2$, the last one being $B = 0$, corresponding to the mode-locking point at rational p/q for the bare circle map $\theta_{n+1} = \theta_n + p/q$. Rewriting Equation (4.18) in the form

$$B = \frac{1}{4(1 - \dfrac{1}{(1 + 4B)^{q/m-1}})}, \tag{4.19}$$

it becomes clear that B depends on q/m only. Equation (4.19) can be solved easily by iteration

$$B_{n+1} = \frac{1}{4(1 - \dfrac{1}{(1 + 4B_n)^{q/m-1}})}.$$

We shall call $B(q/m)$ the m-th shrinking point of the p/q tongue. All period q tongues will shrink at the same $B(q/m)$ values. All $B(q/m)$ values for $q \leq 9$ are listed in Table 4.2.

There is an alternative point of view for the study of the shrinking of mode-locking tongues. We know that the boundaries contain either $\theta = 0.25$ (right-hand boundaries) or $\theta = 0.75$ (left-hand boundaries) as a periodic point in the cycle. Therefore, if both $\theta = 0.25$ and $\theta = 0.75$ are periodic points of a cycle, the two boundaries must cross each other, giving birth to a shrinking point. Elaborating this point of view, we can define infinitely many curves in the A — B plane, joining various shrinking points belonging to different tongues.

Consider, for example, any cycle with $\theta_0 = 3/4$ and $\theta_1 = 1/4$ as the first two points. This leads immediately to the condition $1/4 = f_3(3/4)$, i.e., $A = 1/2 - B$. This straight line is labelled C_p in Fig. 4.12 and Table 4.3. Suppose that the cycle under consideration belongs to a p/q tongue, then, according to the "good order" in the subcritical regime, there must be $(p + 1)$

$q \backslash m$	1	2	3	4	5
1	0				
2	0				
3	0.154508	0			
4	0.209822	0			
5	0.231890	0.097348	0		
6	0.241487	0.154508	0		
7	0.245896	0.188719	0.070459	0	
8	0.247991	0.209822	0.119836	0	
9	0.249008	0.223205	0.154508	0.055099	0

Table 4.2: Shrinking points of the Arnold tongues.

points in Region 1, including the two end points θ_0 and θ_1. As we have shown before, at the shrinking point, the $f^q(\theta)$ map degenerates into a straight line and an arbitrary θ_0 gives rise to a q-cycle, we can make θ_0 slightly larger than 3/4 and thus keep p points in Region 1. Now, each point in the cycle has a well-defined slope, and we have for this q-cycle the stability condition (4.17):

$$(1 - 4B)^p (1 + 4B)^{q-p} = \frac{d}{dx} f^q(\theta) = 1. \qquad (4.20)$$

Comparing (4.20) with (4.18), we see that the curve C_p joins the p-th shrinking point in the p/q tongue.

Similar analysis, again using an algebraic manipulation language, can be carried out for other cycles containing 3/4 and 1/4 as two periodic points. The results for the simplest cases are summarized in Table 4.3. Moreover, these curves may be combined into families. For instance, the curves C_p, C_{2p}, C_{3p} in Table 4.3 belong to the C_{kp} family, joining the kp-th shrinking point in different p/q tongues. Equations for these C_{kp} curves can be written in the form

$$C_{kp}: \quad A = \frac{1}{2Q_k(B)} - B,$$

where the $Q_k(B)$ are defined recursively by

$$Q_1(B) = 1,$$
$$Q_k(B) = (1 - 4B)Q_{k-1}(B) + 1.$$

Curve	Cycle		Equation	Which points it joins
C_p	075 025		$A = 1 = 2\,B$	p-th (q 2p)
C_{q-p}	025 075		$A = 1 = 2 + B$	(q p)-th (q 2p)
C_{2p}	075 f$_0$ 025		$A = \dfrac{1}{4-8B}\,B$	2p-th (q 4p)
C_{q-2p}	025 f$_1$ 075		$A = \dfrac{1}{4+8B} + B$	(q 2p)-th (2p q 4p)
C_{3p}	075 f$_{0\,0}$ 025	$A =$	$\dfrac{1}{\dfrac{02B^3-24B+6}{8B+3}}$	3p-th (q 6p)
C_{3p-q}	075 f$_{1\,3}$ 025	$A = 1$	$\dfrac{1}{\dfrac{32B^2+24B+6}{}}$	(3p q)-th (2p q 3p)

Table 4.3: Curves joining the shrinking points.

In summary, we have shown that the shrinking of Arnold tongues occurs at the intersections of horizontal lines $B = B(q/m)$ and the C_{ap+bq} curves, where a and b are integers. Therefore one can fix the subcritical mode-locking structure by drawing these curves without calculating the tongue boundaries explicitly.

The shrinking of mode-locking tongues is related to a problem in surface physics. A monoatomic layer adsorbed on a crystal surface may be described by the so-called Frenkel-Kontorova model (see the beginning of Section 4.7). Griffiths and Chou[3] studied a modified Frenkel-Kontorova model and discovered shrinking of tongues in the corresponding phase diagram by means of a variational numerical approach.

4.6 The Hénon Map

We devote the rest of this chapter to a brief review of two-dimensional maps, in order to introduce a new, geometrical point of view into the study of chaotic behaviour. So far, periodic and chaotic orbits in one-dimensional mappings have been studied with the help of symbolic dynamics. Although it is topological in nature, this might be called an algebraic approach, since we have been dealing with symbols and combinatorics. The geometric picture of chaotic motion, suppressed to a large extent in one dimension, may be better exposed in a two-dimensional phase plane. We shall see that the transverse intersections of stable and unstable invariant manifolds, so-called homoclinic and heteroclinic

[3]R. B. Griffiths, and Weiren Chou, *Phys. Rev. Lett.*, **56**(1986), 1929. We thank Griffiths for a discussion on this point.

points, play the role of organizing centers of chaotic motion. In fact, these two approaches are closely related and their combination furnishes the most rigorous, but often not practical, definition for chaos. Anyway, researchers in the physical sciences now feel the necessity to become acquainted with a number of notions that two decades ago belonged to the realm of mathematics. The whole geometric approach is now backed by the well-shaped mathematical theory of dynamical systems. However, we shall not tend to mathematical rigour and shall continue to insist on a practical way of presentation.

One-dimensional mappings often appear as projections of dynamical processes evolving in higher dimensional spaces. It is a general rule that unfolding of low-dimensional processes in higher dimensions, either by adding new parameters or by introducing a larger phase space, may bring about a more complete understanding. On the other hand, some qualitatively new features, e.g., the crossover between conservative and dissipative processes, or more complicated basin dependence of the dynamics, may appear only in higher dimensional phase space. Two-dimensional mappings arise as models in many physical problems. Higher dimensional flows may be visualized as two dimensional mappings by taking Poincaré sections (see Chapter 6). Therefore, these mappings deserve detailed study on their own. However, two-dimensional mappings are not our main concern in this book. We shall use them to introduce some important notions that will be needed when working on differential equations. Accordingly, instead of attempting an overall review, we shall mention only a few selected topics.

Hénon (1976) first studied the two-dimensional map of the plane

$$
\begin{aligned}
x_{n+1} &= 1 - \mu\, x_n^2 + y_i, \\
y_{n+1} &= b x_i.
\end{aligned}
\qquad (4.21)
$$

The original motivation for the introduction of this map was "to find a model problem which is as simple as possible, yet exhibits the same essential properties as the Lorenz system". However, it does not enjoy the invariant property when changing the signs of the variables which the Lorenz system possesses. Nevertheless, the Hénon map has become a subject of research on its own. Being a representative of a large class of two-dimensional mappings and simple enough in structure, it has served as a touchstone for many new ideas and numerical procedures. There exists a wide literature on this map (Feit, 1978;

Curry, 1979; Simó, 1979; Marotto, 1979; Franceschini and Russo, 1981; just to name a few early ones).

The map (4.21) transforms the (x, y) plane into itself. The old and new area elements are related by the Jacobian

$$J = \frac{\partial(x_{n+1}, y_{n+1})}{\partial(x_n, y_n)} - \begin{bmatrix} -2\mu x_n & 1 \\ b & 0 \end{bmatrix}.$$

When the determinant $|J| = b = 1$, it preserves the area and thus imitates a conservative dynamical process. If $|b| < 1$, the contraction of the area may be thought to be due to the presence of dissipation. In the extremely dissipative limit $b = 0$, we recover the logistic map. This is why one-dimensional mappings usually represent simple dissipative dynamics. Nonlinear one-dimensional mappings are often said to be non-invertible, although, in fact, they have multivalued inverse branches which may be precisely described by means of symbolic dynamics, as we have been doing in the two preceding chapters. Non-invertibility represents certainly a remarkable distinction of the logistic map from the Hénon map which can be inverted as long as $b \neq 0$, but one should not confuse it with dissipation which is an essentially physical phenomenon.

4.6.1 Fixed Points and Their Stability

In order to look for the fixed points of the Hénon map (4.21), we solve

$$x^* = 1 - \mu x^{*2} + y^*,$$
$$y^* = b x^*$$

and find

$$x^* = (-(1 - b) \pm \sqrt{(1 - b)^2 + 4\mu})/2\mu,$$
$$y^* = bx^*. \tag{4.22}$$

We shall follow the notation of Simó (1979) and denote these two fixed points by H_+ and H_-, according to the choice of sign in (4.22). In order to remain in the field of real numbers, we must demand that

$$\mu > \mu_0 = -(1 - b)^2/4. \tag{4.23}$$

As usual, the next question to ask relates to the stability of these fixed points. In the vicinity of the fixed point (x^*, y^*), one writes

$$x_n = x^* + \xi_n,$$
$$y_n = y^* + \eta_n,$$
(4.24)

where ξ and η are small deviations from the fixed points. Inserting (4.24) into the map (4.21) and taking into account the fixed point condition, we get in linear approximation

$$\begin{bmatrix} \xi_{n+1} \\ \eta_{n+1} \end{bmatrix} = \begin{bmatrix} -2\mu x^* & 1 \\ b & 0 \end{bmatrix} \begin{bmatrix} \xi_n \\ \eta_n \end{bmatrix}.$$
(4.25)

The 2×2 matrix in (4.25) is the *tangent map* or the linearized map at (x^*, y^*). Equations (4.25) realize a linear transformation in the tangent plane. By solving the characteristic equation

$$\begin{bmatrix} -2\mu x^* - \lambda & 1 \\ b & -\lambda \end{bmatrix} = 0,$$

we get two eigenvalues

$$\lambda_\pm = -\mu x^* \pm \sqrt{\mu x^{*2} + b}$$
(4.26)

at each fixed point. In the two eigen-directions, small deviations change independently according to

$$\xi_{n+1}^+ = \lambda_+ \xi_n^+,$$
$$\eta_{n+1}^- = \lambda_- \eta_n^-.$$
(4.27)

(The quantities ξ_n^+ and η_n^- are linear combinations of the above ξ_n and η_n, obtainable by applying the similarity transformation that diagonalizes the tangent map, but we skip the simple arithmetic.) Several possibilities arise, according to the absolute values of the eigenvalues:

1. If both $|\lambda_\pm| < 1$, we have a stable fixed point which is also called a stable node or a sink in the plane.

2. If both $|\lambda_\pm| > 1$, we have an unstable fixed point, also called an unstable node or a source in the plane.

3. If one eigenvalue $|\lambda_>| > 1$, the other $|\lambda_<| < 1$, a small vector in one eigen-direction will be stretched, while in the other direction it will be contracted. This is the interesting case of a saddle point that we shall study in more detail in Section 4.6.3.

4. Either or both eigenvalues have an absolute value 1. This happens only as an exceptional case, because it requires that additional conditions are satisfied exactly.

In the first three cases, the system is said to be *hyperbolic*. The last case violates hyperbolicity and occurs when the system undergoes a transition between two different hyperbolic regimes: a *bifurcation* takes place. In general, the eigenvalues may be complex numbers as well, leading to new types of fixed points (stable and unstable foci). Having imposed the condition for x and y to be real numbers, we exclude complex eigenvalues for the time being. We shall return to this topic when dealing with differential equations in Chapter 6.

The marginal condition $\lambda_\pm = +1$ coincides with $\mu = \mu_0$, μ_0 being defined in (4.23). The other border $\lambda_\pm > -1$ leads to the new condition

$$\mu < \mu_1 = 3(1-b)^2/4. \tag{4.28}$$

The fixed point H^- is a saddle for all values of μ, whereas H^+ remains a sink as long as $\mu < \mu_1$. At $\mu = \mu_1$, the eigenvalue λ reaches -1 and period-doubling bifurcation takes place upon μ crossing μ_1, just as it happens in the case of a unimodal map. Beyond μ_1, a stable 2-cycle comes into existence, but H^+ itself becomes a saddle. We shall look at the dynamics close to these saddle points later on.

4.6.2 Stable Periodic Orbits for the Hénon Map

Since the ordering of periodic windows along the $b = 0$ line in the $b \sim \mu$ parameter plane is well-understood, it is natural to look for their extensions into the whole parameter plane. This problem has been studied by El Hamouly and Mira (1982a and b) numerically. In fact, there exists an algebraic method for the determination of the location of these stable periods, and analytical results can be obtained for orbits up to period 6 (Huang, 1985, 1986). We sketch this method in what follows.

For the sake of convenience, we rescale the map (4.21) by substituting

$$X_n = \mu x_n,$$
$$Y_n = \mu y_n,$$

and write the two-dimensional map as a second order difference equation in one variable. If the points $\{X_i, i = 1, 2, \ldots, N\}$ form an N-cycle for the map, they must satisfy the closed chain of equations

$$
\begin{aligned}
X_2 &= \mu - X_1^2 + bX_N, \\
X_3 &= \mu - X_2^2 + bX_1, \\
X_4 &= \mu - X_3^2 + bX_2, \\
&\ldots \qquad \ldots \\
X_1 &= \mu - X_N^2 + bX_{N-1}.
\end{aligned}
\tag{4.29}
$$

On the other hand, one may imagine these N real numbers to be the N roots of an N-th order algebraic equation

$$
X^N + a_1 X^{N-1} + a_2 X^{N-2} + \cdots + a_{N-1}X + a_N = 0,
$$

the coefficients of which are related to the $\{X_i\}$ by the Viéte formulae

$$
\begin{aligned}
-a_1 &= X_1 + X_2 + \ldots + X_N, \\
a_2 &= X_1 X_2 + X_1 X_3 + \ldots + X_{N-1} X_N, \\
&\ldots \qquad \ldots \\
(-1)^N a_N &= X_1 X_2 \ldots X_N.
\end{aligned}
$$

These formulae lead to the equivalent relations

$$
\begin{aligned}
2a_2 + \sum_{i=1}^{N} X_i^2 &= a_1^2, \\
3a_3 + a_1 \sum_{i=1}^{N} X_i^2 + \sum_{i=1}^{N} X_i^3 &= a_1 a_2, \\
&\ldots \qquad \ldots \\
Na_N + a_{N-2} \sum_{i=1}^{N} X_i^2 + \ldots + \sum_{i=1}^{N} X_i^N &= a_1 a_{N-1}.
\end{aligned}
$$

The fact that the N roots also satisfy Equations (4.29) imposes some constraints on the a_i. For example, subtracting the second equation in (4.29) from the first, then the third from the second, etc., and ordering the results, we arrive at a constraint

$$
\begin{bmatrix}
X_1 + X_2 + b & 1 + b & b & \cdots & b \\
-b & X_2 + X_3 & 1 & \cdots & 0 \\
\cdots & & \cdots & & \cdots \\
0 & 0 & 0 & \cdots & 1 \\
-1 & -1 & -1 & \cdots & X_{N-1} + X_N - 1
\end{bmatrix} = 0.
$$

Using similar conditions, one can eliminate all a_i for $i \leq 2$, leaving only one equation for a_1. Since at the beginning of each window a stable and an unstable orbits emerge simultaneously, the double root condition for the equation of a_1 must be fulfilled.

In order to check the stability of the N-cycle, we study the Jacobian

$$J_N \equiv J(X_N)J(X_{N-1})\dots J(X_1) \equiv \begin{bmatrix} W_{N,1} & W_{N,2} \\ W_{N,3} & W_{N,4} \end{bmatrix}.$$

The matrix elements $W_{N,i}$, $i = 1$ to 4, may be calculated from a set of recurrence relations. The characteristic equation of J_N simplifies due to the relation

$$W_{N,1}W_{N,4} - W_{N,2}W_{N,3} = (-b)^N,$$

which may be verified easily. When a stable period N orbit is created at $\lambda = 1$, the characteristic equation yields the window edge condition

$$W_{N,1} + W_{N,4} + \lambda - (-b)^N = 0.$$

This relation coincides with the double root condition. The window loses stability at $\lambda = -1$, where the characteristic equation provides an additional condition. Skipping the detailed derivation (Huang, 1985, 1986), we list the results for a few short periods (see Figure 4.15). The period 1 orbit appears at the curve (4.23):

$$\Lambda_{(1)_0} : \quad \mu = -(1-b)^2/4.$$

It destabilizes along the line (4.28), where the stable period 2 appears:

$$\Lambda_{(2)_0} : \quad \mu = 3(1-b)^2/4.$$

Then the period 4 orbit replaces period 2 at (Simó, 1979):

$$\Lambda_{(2^2)_0} : \quad \mu = (1-b)^2 + (1+b)^2/4.$$

This period 4 orbit remains stable until μ reaches the curve

$$\Lambda_{(2^3)_0} : \quad \mu = \frac{3}{4}\{(1+b)^2 + [(1-b)(1+b)^2]^{2/3}\}.$$

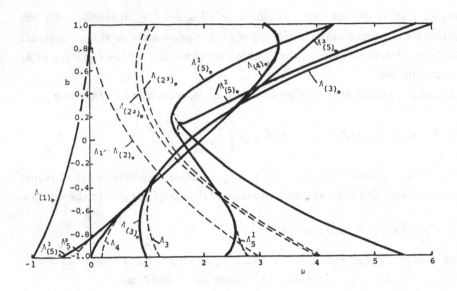

Figure 4.15: The location of stable periodic orbits in the parameter plane of the Hénon map (after Huang, 1985).

The method gives the location of another period 4 orbit (the RL^2 orbit at $b = 0$) in the parameter plane. It is remarkable that the boundary of its existence zone (the curve $\Lambda_{(4)_0}$ in Fig. 4.15 and its destabilizing curve, not shown in the figure) reveals an inflection point when approaching the $b = 1$ straight line. Discovered numerically by Hamouly and Mira (1982 a and b), this phenomenon has been confirmed by analytical means. Similar but more complicated results were obtained for all orbits of periods 3 through 6 (Huang, 1986). Curves for periods less than 5 are shown in Fig. 4.15.

Note that the logistic map corresponds to the central horizontal line $b = 0$ in Fig. 4.15. We see that the ordering of periods, as given by symbolic dynamics of two letters, holds only in a narrow strip along this horizontal line.

4.6.3 Stable and Unstable Invariant Manifolds

A two-dimensional mapping realizes a transformation of the plane into itself. Are there any objects in the plane that remain unchanged under the transformation? Obviously, fixed points are invariant under the mapping. Sinks and sources are isolated invariant points, as any neighbouring point will get closer to or farther away from the fixed point when the transformation is applied repeatedly. Saddle points distinguish themselves significantly in this respect: they sit at the intersection of two invariant curves.

Indeed, the eigenvector

$$\begin{bmatrix} \xi \\ 0 \end{bmatrix},$$

that corresponds to the eigenvalue that is larger than 1 (denoted by λ_u hereafter) only stretches by a factor λ_u, but it remains in the same direction, positive or negative, depending on the sign of λ_u [see Equations (4.27)]. Similarly, the other eigenvector

$$\begin{bmatrix} 0 \\ \eta \end{bmatrix},$$

corresponding to $|\lambda_s| < 1$, remains in the contracting direction, possibly with a sign change. The stretching direction is called the *unstable direction* (or the *outset*) of the saddle, while the contracting direction — the *stable direction* (the *inset*) of the saddle. What has been said holds in the linearized regime, as long as the map may be replaced by the tangent map acting on small deviations from the saddle.

What happens with these small invariant straight lines when they get further away from the saddle? In principle, they can develop into invariant curves, called in genaral *invariant manifolds*; these curves may either terminate at other fixed points or continue to wander in the plane.

There are at least two methods for the visualization of these invariant manifolds: a numerical and an analytical one. For the sake of clarity, we proceed in a slightly formal manner. Suppose that we have a two-dimensional map M:

$$\begin{aligned} x_{n+1} &= f(x_n, y_n), \\ y_{n+1} &= g(x_n, y_n). \end{aligned} \qquad (4.30)$$

It has a nonsingular Jacobian, so that also the inverse map M^{-1} exists. Suppose further that we have found a saddle type fixed point (x^*, y^*), the unstable direction ξ^+ corresponding to $\lambda_>$, and the stable direction η^- corresponding to $\lambda_<$.

Numerically, one can take a tiny segment $(0, \epsilon)$, corresponding to the saddle, along the unstable direction, and divide it into, say, $n = 1000$ subsegments: $\xi_i = i \times \epsilon/n, i = 1, \ldots, n$. Then the map M is applied to each of the end points of these subsegments (transformed back to x and y coordinates) in order to get n images. If the map reverses the direction of vectors, as the Hénon map does, one should apply M^2 instead of M to remain on the continuation of one of the two opposite unstable directions. The n images will outline a part of the unstable invariant manifold $\mathcal{M}^u$. In order to get more points on $\mathcal{M}^u$ one takes the first subsegment ξ_1, divides it again into n points and then applies to them the map M (or M^2, if the map reverses direction). This process may be repeated with more and more points. By taking initial points closer and closer to the saddle, we ensure the validity of the linear approximation, and hence the precision of the reconstructed invariant manifold. If the points on $\mathcal{M}^u$ are distributed unevenly, then one has to compensate for the exponential separation of the points under the mapping. It is sufficient to divide the initial segment also in an exponential way, adjusting the numerical factor by trial and error. In order to construct the stable invariant manifold $\mathcal{M}^s$, one starts with the stable direction of the saddle and applies the inverse map M^{-1} (or M^{-2}) to the initial points. It is understandable that one can get a general (usually fairly good) idea regarding the shape of the invariant manifolds, but should not overdo it, due to an unavoidable accumulation of numerical errors.

Analytically, a part of the invariant manifolds can be obtained as a series expansion in terms of small deviations from the saddle (Simó, 1979; Franceschini and Russo, 1981). Suppose the invariant curve is given by $y = \phi(x)$. The saddle point (x^*, y^*), any initial point (x_n, y_n) and its image (x_{n+1}, y_{n+1}) should all lie on this curve, i.e.,

$$y^* = \phi(x^*), \qquad y_n = \phi(x_n), \qquad y_{n+1} = \phi(x_{n+1}). \tag{4.31}$$

Substituting the map (4.30) into both sides of the last equation of (4.31) and using the second one, we find

$$g(x_n, \phi(x_n)) = \phi(f(x_n, \phi(x_n))). \tag{4.32}$$

Since x_n appears everywhere in (4.32), we can drop the subscript. In the case of the Hénon map (4.21), Equation (4.32) reads

$$b\,x = \phi\big(1 - \mu\,x^2 + \phi(x)\big). \qquad (4.33)$$

This is a functional equation for $\phi(x)$. In the vicinity of the saddle one can write

$$x = x^* + \xi,$$
$$y = y^* + \eta,$$

and present ϕ as a series in ξ

$$y = y^* + \eta = \phi(x^* + \xi) = \phi(x^*) + \sum_{k \leq 1} \beta_k \xi^k. \qquad (4.34)$$

Substituting (4.34) back into (4.33) and eliminating x^* and y^*, in accordance with the fixed point condition, one gets a set of recursive relations for determining the coefficients β_k (Simó, 1979). Later on, Franceschini and Russo (1981) used 100 terms in a similar expansion. Equipped with any algebraic manipulation language, it would not be difficult to get hundreds of terms. Referring the reader to the original papers for pictures of the stable and unstable manifolds of the Hénon map, we shall show some pictures for differential equations in Chapter 5.

We have discussed the invariant curves emerging from saddle type fixed points. Similar situations exist for unstable periodic orbits. It is sufficient to consider the fixed points of the nth iterate of the map, i.e., the map M^n, when the stable and unstable invariant manifolds can be reconstructed. We shall denote these manifolds as $\mathcal{M}_{nP}^s$ and $\mathcal{M}_{nP}^u$.

The interrelations between stable and unstable manifolds play a central role in the development of chaotic motion. The unstable manifold of a saddle point, in getting away from the latter, may intersect the stable manifold of the same saddle point. This is called a *homoclinic intersection*. If one homoclinic intersection takes place, then there must be infinitely many such intersections, because the homoclinic point belongs to both invariant manifolds and all its images and pre-images under the mapping necessarily stay in the manifolds. As a rule, a homoclinic intersection first appears as a *homoclinic tangency* when one invariant manifold touches the other tangentially. Then, at a further change of the parameter, the tangent point develops into a *transversal* intersection. The M^s and M^u from different unstable fixed points or unstable periodic points may intersect as well, leading to *heteroclinic intersections*.

We cannot do better than cite the first two paragraphs of §397 from the third volume of H. Poincaré's *"Les Méthodes Nouvelles de la Mécanique Céleste"*, which appeared in 1899:[4]

"Let us attempt to imagine the figure formed by these two curves and their infinite number of intersections, each of which corresponds to a double asymptotic solution. These intersections form a kind of trellis, a fabric, a lattice with an infinitely dense mesh; each of the two curves must never cross itself, but must fold back upon itself in a very complex fashion, in order to cross through the whole infinite number of lattice sites.

The complexity of this figure, which I shall not even attempt to draw, is truly striking. Nothing else would be more appropriate in giving us an idea of the complications of the three-body problem and, in general, of all the problems in dynamics where there is no uniform integral ... ".

When chaotic motion appears as the result of a finite or infinite number of bifurcations, where some periodic regimes loss their stability, the final "chaotic attractor" has a close relation to M^s_{nP} and M^u_{nP} of the unstable periods. Geometrically, it is easy to imagine that at least one branch of M^u_{nP} will approach the attractor, while M^s_{nP}, traced backwards, will outline the boundary of the basin of attraction.

4.7   The Dissipative Standard Map

Let us consider an old problem of surface physics. Take an infinite chain of point "atoms", connected by springs, the equilibrium spacing between the atoms being a. Now put this chain in a periodical potential of period p. What is the new equilibrium spatial distribution of the atoms? The system may be described by the following phenomenological "potential energy":

$$F = \frac{1}{2} \sum_n (x_n - x_{n-1} - a)^2 - \frac{B}{(2\pi)^2} \sum_n \cos\left(\frac{2\pi x_n}{p}\right).$$

[4]We thank Dr. Wu Ling-An for the English translation from the French original.

Without any loss of generality we shall take $p = 1$. The strength of the potential K has been written with some numerical factors, in order to bring later results into a more convenient form. The stationary positions of the x_n are determined from

$$\frac{\partial F}{\partial x_n} = 0,$$

which yields an infinite system of difference equations

$$x_n - x_{n-1} - (x_{n+1} - x_n) + \frac{B}{2\pi} \sin(2\pi x_n) = 0.$$

This is the well-known Frenkel-Kontorova model[5], frequently used to describe an adsorbed atom layer on a crystal surface. The two extreme solutions of this system may be guessed from the physics. When $B \ll 1$, the atom chain ignores the potential and keeps its original equilibrium spacing a. When B is strong enough, the atoms will fall into the valleys of the potential, in order to gain potential energy at the cost of higher elastic energy. For intermediate B, various commensurable and incommensurable phases, including chaotic ordering, may occur (see, e.g., Bak, 1982; Aubry, 1983). Although the problem has been formulated in terms of spatial arrangement, the situation is identical to temporal evolution with two competing frequencies.

In order to see this, let us denote $x_n - x_{n-1}$ by ρ_n and change the remaining x_n to θ_n. We then end up with a two-dimensional mapping:

$$\begin{aligned}
\theta_{n+1} &= \theta_n + \rho_{n+1}, \\
\rho_{n+1} &= \rho_n + \frac{B}{2\pi} \sin(2\pi\theta_n).
\end{aligned} \tag{4.35}$$

This is nothing but the *standard map* (Chirikov, 1979) which has been widely used to model conservative Hamiltonian systems. Since in this book we are concerned with dissipative systems, we modify the map (4.35) by introducing a coefficient $0 \le b \le 1$ and adding a constant A. Then we are led to the *dissipative standard map* (Feigenbaum, Kadanoff and Shenker, 1982):

$$\begin{aligned}
\theta_{n+1} &= \theta_n + \rho_{n+1}, \\
\rho_{n+1} &= b\rho_n + \frac{B}{2\pi} \sin(2\pi\theta_n).
\end{aligned} \tag{4.36}$$

The Jacobian of this map is

[5]Y. I. Frenkel, and T. Kontorova, *Zh. Eksp. Teor. Fiz.* 8(1938), 1340.

$$J = \frac{\partial(\theta_{n+1}, \rho_{n+1})}{\partial(\theta_n, \rho_n)} = b.$$

In the extremely dissipative limit $b = 0$, Equation (4.36) reduces to the circle map (4.2) of Section 4.2. In a more rigorous way, the circle map may be obtained at the extremely dissipative limit of a general annular map.

We close with a remark relating to a statement that one encounters ocassionally in the literature. Both the standard map and its dissipative counterpart are invertible (provided $b \neq 0$). On the contrary, one-dimensional maps such as the logistic map or the circle map (in the supercritical regime) are said to be non-invertible. This is not very adequate, as the inverse functions are merely multivalued, and symbolic dynamics helps to resolve the multivaluedness. Futhermore, it might even be misleading, if non-invertibility is thought to be the reason for one-dimensional mappings to be dissipative systems. One-dimensional mappings do appear as extremely dissipative limits of two-dimensional mappings, but this fact has nothing to do with non-invertibility.

Chapter 5

Chaos in Ordinary
Differential Equations

Many interesting models in the physical sciences are described by systems of ordinary differential equations. While bifurcation and chaos in discrete mappings can be studied more or less thoroughly by both analytical and numerical means, similar tasks for differential equations may encounter great difficulty. Analytical tools can be of some help only in very limited cases, e.g., when the criterion of Melnikov (see, e.g., Chapter 4 of Guckenheimer and Holmes, B1983) or Silnikov (see, e.g., Gaspard and Nicolis, 1983) is applicable. These methods usually only provide a criterion for the existence of chaotic motion and give an estimate for the threshold of transition to chaos, but cannot help with the exploration of the global structure of the parameter space. Although numerical studies are very time-consuming in most cases, nevertheless, our present understanding of chaotic behaviour in differential equations relies heavily on numerical work. Therefore, we shall basically devote this chapter to numerical studies of ordinary differential equations.

We shall first give an overview of the most frequently used numerical methods without going into programming details, and then summarize the results of two case studies in Sections 5.7 and 5.8. In particular, we shall demonstrate the method of elementary symbolic dynamics in an empirical way, to achieve a better understanding of the bifurcation and chaos "spectra" of ordinary differential equations.

The strategy of a study of chaos in ordinary differential equations involves initially an identification of periodic solutions and their systematics, and is then followed by a characterization of the attractors. We already know that symbolic dynamics may play an essential role in the first stage. This explains why we shall pay much attention to the numerical determination of periodic orbits. Regarding the second stage, the computation of various dimensions and entropies as well as of Lyapunov exponents has become the center of concern, although power spectrum analysis has been the main tool in the earlier days of numerical study. In anticipation, we point out that the success in the characterization of chaotic attractors in computer studies nowadays relies significantly on an extension of the relevant space by inclusion of the tangent space (see Section 6.4.1).

5.1 Three Kinds of Ordinary Differential Equations

In this chapter, we shall confine ourselves to the following types of ordinary differential equations.

1. Autonomous differential equations, i.e., equations without explicit time dependence on their right-hand sides, when one requires at least three independent variables, in order to observe period-doubling or chaotic transitions. A classical and much-studied example of autonomous equations is the Lorenz model (Lorenz, 1963) obtained from a three-mode truncation of the thermal convection problem:

$$\dot{x} = \sigma(y - x),$$
$$\dot{y} = rx - y - xz, \tag{5.1}$$
$$\dot{z} = xy - bz,$$

where σ, b and r are control parameters.

2. Non-autonomous systems with at least two independent variables. It is well known that by addition of one or more independent variables a non-autonomous system can be transformed into an autonomous one. The most

important class of non-autonomous systems are those driven by periodic external force. In numerical studies, an availability of control frequency creates the possibility of reaching very high frequency resolution. In this chapter, we shall frequently refer to results for the periodically forced Brusselator which has been introduced in Section 1.5 (for a recent review on this model, see Hao, 1987):

$$\dot{x} = A - (B+1)x + x^2 y + \alpha \cos(\omega t),$$
$$\dot{y} = Bx - x^2 y. \tag{5.2}$$

The "free" part of (5.2) describes the kinetics of a three-molecular chemical reaction; it is capable of developing a limit cycle type oscillation when $B > A^2 + 1$.

3. Time-delayed differential equations. Formally, it is enough to have a single independent variable, in order to display complicated bifurcating and chaotic behaviour. However, time-delayed equations are, in essence, functional equations with an infinite number of degrees of freedom. This can be seen easily by rewriting the time delay as a sum of high order derivatives:

$$f(t-T) = \exp\left(-T\frac{d}{dt}\right) f(t) = \sum_{i=0}^{\infty} (-T)^i \frac{d^i}{dt^i} f(t) \tag{5.3}$$

or by considering the dependence of the solution on an initial function instead of on an initial point. Time-delayed equations have been used, for example, in ecological models and in the description of optical bistable devices, one of the simplest cases being (Li and Hao, 1985)

$$\tau \dot{x}(t) + x(t) = 1 - \mu[x(t-T)]^2. \tag{5.4}$$

Since time-delayed differential equations will not be studied further in this book, we make a few remarks on the case (5.4).

Firstly, in the long delay limit $T \gg \tau$, one can neglect the time derivative and, measuring time in units of T, transform (5.4) into a mapping. However, no matter how detailed is the knowledge which one possesses for this mapping, it cannot be extrapolated into the $\tau - \mu$ parameter plane for small τ, because there are an infinite number of "linear" modes which are excitable near $\tau = 0$. In other words, we have a singular perturbation case, since the vanishing small parameter removes the only derivative and changes the nature of the equation drastically.

Secondly, in the opposite small delay limit, although one may use a truncated expansion (5.3) to transform (5.4) into a system of finite order ordinary differential equations, the resulting system may resemble the original equation only for small enough t. Therefore it is not of much help in the study of the asymptotic $t \to \infty$ behaviour. In fact, this is another extreme of singular perturbation, since the truncation always means neglect of higher order derivatives.

Thirdly, the solution of (5.4) depends on the choice of the initial function. One might try to use a one-parameter family of functions and look for the dependence on that parameter. Our experience has shown that sensitive dependence on the parameter appears when the asymptotic regime is chaotic. For example, in some region of the parameter space there are two coexisting chaotic attractors, distinguishable by different characteristic frequencies in the power spectra. The system jumps at random between these two attractors while the parameter in the initial function varies continuously.

5.2   On Numerical Integration of Differential Equations

A prerequisite for an application of any method, that we are going to discuss, is that a good algorithm is available for the integration of the equations. Gone is the time when a scientist must write his own mathematical routines, since so many tested programs have accumulated that surpass any amateur's handiwork in stability, precision and efficiency. All one has to do is to choose an appropriate subroutine from one of the existing libraries (e.g., IMSL, NAG, CACM, etc.) which should be available on-line at most computing centers. Therefore, we shall confine ourselves to a few comments.

In order to fix the notation, we write a general autonomous system of nonlinear ordinary differential equations in the standard vector form

$$\frac{d\mathbf{x}}{dt} = \mathbf{F}(\mathbf{x}). \tag{5.5}$$

It should be supplemented with the initial condition

$$\mathbf{x}(t=0) = \mathbf{x}_0.$$

The solution of this initial value problem is an integral curve passing through $\mathbf{x}_0$. All possible integral curves taken together constitute a *flow* Φ in the phase space and the solution of the above problem picks up a particular curve

$$\mathbf{x}(t) = \Phi_t(\mathbf{x}_0), \tag{5.6}$$

which satisfies, of course, the condition

$$\mathbf{x}_0 = \Phi_0(\mathbf{x}_0).$$

The system (5.5) can be linearized at a point, say $\mathbf{x}_1$, by letting

$$\mathbf{x} = \mathbf{x}_1 + \mathbf{W},$$

where $\mathbf{W}$ is a small vector, tangential to the integral curve at $\mathbf{x}_1$, and satisfies the linearized equation

$$\frac{d\mathbf{W}(t)}{dt} = \mathbf{J}(\mathbf{x}_1)\mathbf{W}(t), \tag{5.7}$$

with

$$\mathbf{J}(\mathbf{x}_1) = \frac{\partial \mathbf{F}(\mathbf{x})}{\partial \mathbf{x}}\Big|_{\mathbf{x}=\mathbf{x}_1},$$

an $n \times n$ matrix. We sometimes will denote its elements by a_{ij}.

Since it is a linear system, the solutions of (5.7) may be expressed by means of a linear evolution operator $\mathbf{U}(t) \equiv \mathbf{U}(t,0)$:

$$\mathbf{W}(t) = \mathbf{U}(t)\mathbf{W}(0).$$

It is readily seen that $\mathbf{U}(t)$ satisfies the same equation (5.7)

$$\frac{\mathbf{U}(t)}{dt} = \mathbf{J}(\mathbf{x}_1)\mathbf{U}(t)$$

with the initial conditions

$$\mathbf{U}(t)\,|_{t=0} = \mathbf{I},$$

where $\mathbf{I}$ is the unit matrix.

Numerical algorithms for the integration of ordinary differential equations can be subdivided into two classes. The first class is based on Taylor expansion and step by step progress from the initial point. The second class employs some numerical quadrature method over a small interval. As a rule, algorithms in the first class only require the result of the previous step and are called one-step or single-step methods. The second class leads to multi-step algorithms; in particular, it is quite useful for the treatment of time-delayed problems, when the initial function is known over an interval, for example, we have used the fourth order Adam's method to study (5.4) (Li and Hao, 1985). As a rule, multistep methods require less arithmetic operations per step than single-step algorithms, but at the start they demand initial values at several points. On the other hand, multi-step schemes are usually derived for equidistant partitions of the integration interval, whereas single-step methods are flexible enough to allow for varying step lengths.

A very frequently used single-step method is the Runge-Kutta algorithm. A fourth order Runge-Kutta scheme has a local truncation error of the order h^5, h being the integration step. It measures the difference between the discretized iteration equations and the original differential equations at one single step, hence the adjective "local". However, a small truncation error at each step does not reveal very much about the degree of global approximation to the system of differential equations. In addition, our concern with a chaotic regime implies a sensitive dependence on initial values as well as on local truncation errors. Nevertheless, as we shall see later, a number of global characteristics of the motion such as the Lyapunov exponents, dimensions and entropies, remain quite insensitive to the algorithm or local truncation, provided small enough integration steps are used.

In connection with the Runge-Kutta method we would like to point out that when the differential equations represent a conservative system, the resulting difference scheme may become dissipative. This can be checked on the example of a simple linear oscillator

$$\dot{x} = \omega y,$$
$$\dot{y} = -\omega x.$$

In this case, the corresponding Runge-Kutta difference scheme may be written down explicitly and there appears a dumping factor as well as a shift of frequency. It is much better to use the so-called symplectic difference scheme[1] in

---

[1]See, e.g., Feng Kang, in *Proceedings of the Beijing Symposium on Differential Geometry and*

dealing with numerical integration of Hamiltonian systems.

A last, but not least important point to be mentioned concerns so-called stiff equations, when the eigenvalues of the matrix $\mathbf{J}$ in (5.7) differ in orders of magnitude. One should consult the literature, e.g., the book of Lambert[2], to avoid numerical quirks.

5.3   Numerical Calculation of the Poincaré Maps

In a study of bifurcation and chaos in ordinary differential equations, one does not deal with a single system, but has to treat a family of equations, varying the control parameters. Furthermore, on a computer with finite word length and, of course, finite run time, it is impossible to distinguish, say, a very long periodic orbit from a quasiperiodic or chaotic trajectory, if one observes the trajectory in the phase (configuration) space $\mathcal{R}^n$ only. However, by invocation of the tangent space, side by side with the phase space, one can calculate with confidence such quantitative characteristics of the motion as the Lyapunov exponents which, in turn, enable us to distinguish between chaotic and non-chaotic behaviour. In addition, many systems display multistable solutions for one and the same set of parameter values, i.e., different attractors, trivial and strange ones, may coexist. There appears a basin dependence: the destiny of the motion depends on the initial values chosen.

In summary, we see that a full-scale numerical study of a system of ordinary differential equations would require a scanning of the product space $\mathcal{R}^n \otimes \mathcal{R}^n \otimes \mathcal{R}^n \otimes \mathcal{R}^m$, where n is the dimension of the phase (hence the initial value and tangent) space, and m is the dimension of the control parameter space. For the simplest autonomous system, one should take at least $n = 3$, $m = 2$ (a codimension 2 study). This would be a job which exceeds the capability of many present-day supercomputers. One has to be satisfied with the knowledge of a few sections of this huge product space. Fortunately, this happens to be

Differential Equations, ed. by Feng Kang, Science Press, Beijing, 1985; M.-Z. Qin, D.-L. Wang, and M.-Q. Zhang, "Explicit symplectic difference schemes for separable Hamiltonian systems", Preprint 1988, *J. Computational Mathematics*, to appear.

[2]J. D. Lambert, *Computational Methods in Ordinary Differential Equations*, Wiley, 1973.

sufficient in many cases, and the Poincaré sections are the most important ones to study.

The Poincaré section is a low-dimensional (usually two-dimensional, but not necessarily so) intersection of the phase space, chosen in such a way that the trajectories intersect it transversally, i.e., do not touch it tangentially. The choice of a Poincaré section must be preceded at least by a linear stability analysis of the system (5.5), in order to ensure that all qualitatively interesting trajectories actually intersect it. Once the section has been chosen, we focus on consecutive intersecting points in the Poincaré section and view the motion as a point-to-point mapping in the section itself. Only in a local sense can this map sometimes be written down approximately. In general, one has to resort to numerical calculations. Anyway, adoption of the Poincaré sections reduces the description of the dynamics significantly and it still reflects the essential features of the motion. For example, a simple periodic orbit would become a single fixed point in the map; a periodic orbit with two commensurable frequency components would give rise to a finite number of points; a quasiperiodic trajectory would draw a closed curve in the Poincaré section, and chaotic motion would show off as erratically scattered points, etc.

5.3.1   Autonomous Systems: Hénon's Method

A simple-minded approach to the calculation of a Poincaré map would involve step by step integration of the equations and a test of the sign changes of a certain component, say z, when the $z = 0$ plane is used as the Poincaré section. However, in order to locate the intersection point with a sufficient precision to match that of the integration algorithm, one must use a high order interpolation scheme, which requires saving and updating of several consecutive points at each integration step. M. Hénon (Henon, 1982) has described a clever method for the determination of the intersection point at one shot. We explain the idea by means of the following example:

$$\dot{x} = f_1(x, y, z),$$
$$\dot{y} = f_2(x, y, z), \qquad\qquad (5.8)$$
$$\dot{z} = f_3(x, y, z),$$

with $z = 0$ being the Poincaré section. Suppose that we have found at the n-th step

$$t_n, \quad x_n, \quad y_n, \quad z_n < 0$$

and find at the next step

$$t_n + \Delta t, \quad x_{n+1}, \quad y_{n+1}, \quad z_{n+1} > 0.$$

The intersection must occur between these two steps. Now, interchange the role of z and t, divide the first two equations by the third and invert the third equation:

$$
\begin{aligned}
\frac{dx}{dz} &= \frac{f_1(x, y, z)}{f_3(x, y, z)}, \\
\frac{dy}{dz} &= \frac{f_2(x, y, z)}{f_3(x, y, z)}, \\
\frac{dt}{dz} &= \frac{1}{f_3(x, y, z)}.
\end{aligned}
\tag{5.9}
$$

Integrating these equations in the new independent variable z backward by one step $\Delta z = -z_{n+1}$, using the initial values

$$x_{n+1}, \quad y_{n+1}, \quad t_n + \Delta t,$$

as initial values, or, integrating forward by one step $\Delta z = -z_n$ from the $z_n < 0$ point (in practice, the second way works a bit better), one reaches exactly the $z = 0$ plane.

Hénon's method can be used to locate the intersecting point of an orbit with a general surface

$$S(x, y, z) = \text{const.} \tag{5.10}$$

It is sufficient to introduce an additional function

$$u = S(x, y, z) - \text{const},$$

and to derive the differential equation for u using the original system (5.8):

$$\frac{du}{dt} = f_1 \frac{\partial S}{\partial x} + f_2 \frac{\partial S}{\partial y} + f_3 \frac{\partial S}{\partial z}.$$

Thus, we return to the old problem of looking for the intersection with the $u = 0$ plane for a system with one more equation.

In practice, due to the presence of transients, which may be very long when one approaches a bifurcation point ("critical slowing down", see Section 7.1), it is a time-consuming task to calculate a Poincaré map with high precision, even with the help of Hénon's procedure. However, in the case of periodic solutions, there exists a very efficient method for the location of the exact periodic orbit, starting from an approximate one. We shall devote Section 5.4 to the technique of locating periodic orbits.

5.3.2   Non-autonomous Systems: Subharmonic Stroboscopic Sampling

We shall only consider the case of periodically driven systems, where the external frequency ω provides us with a reference period $T = 2\pi/\omega$. The extended phase space, i.e., the $x - y$ plane plus time t in the case of Equations (5.2), can be considered as closed in the t -direction and thus to form a torus or a truncated cylinder of height T. The calculation of Poincaré maps now reduces to a sampling of x, y values at multiples of T (up to an unessential shift of the starting time). This procedure corresponds exactly to the stroboscopic sampling techniques used by experimentalists. We should emphasize that the widespread opinion on the uselessness of stroboscopic methods in studies of ordinary differential equations applies only to autonomous systems when there is no constant characteristic period at hand. For periodically driven systems, however, the stroboscopic technique becomes a most powerful method for the exploration of subtle dynamic details.

The subharmonic stroboscopic sampling method is a very simple, yet quite an effective extension (Hao and Zhang, 1982a, 1983) of the stroboscopic sampling idea mentioned above: besides sampling at the driving period T, one also samples at multiples of the fundamental period, i.e., at pT, where p is a correctly chosen integer. When properly used, this method may provide very high frequency resolution at the cost of longer computing time. A period-doubling cascade up to the 8192th subharmonic has been resolved in this way, and the hierarchy of chaotic bands as well as the systematics of the periodic windows,

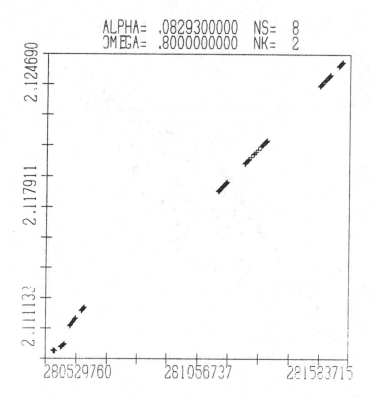

Figure 5.1: Subharmonic stroboscopic portraits for a 2^6 chaotic band, sampled at $8T$.

according to the two-letter symbolic dynamics, studied in Chapter 3, have been observed (Hao and Zhang, 1982a; Hao, Wang and Zhang, 1983; see also Section 5.7.3 in this chapter). Similar resolution can usually be attained only for one-dimensional mappings, not for ordinary differential equations.

When sampling at pT, p being an integer, one has the freedom of shifting the starting time by qT, $q = 0, 1, \ldots p - 1$, thereby picking up one of the p components of the attractor. However, as a method of discrete sampling, the subharmonic stroboscopic sampling technique has the same demerit as the discrete Fourier transform (see Section 5.6 below), namely, non-uniqueness in its interpretation and the impossibility of a resolution of frequency components

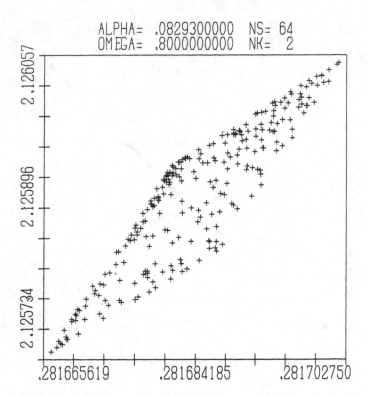

Figure 5.2: Subharmonic stroboscopic portraits for a 2^6 chaotic band, sampled at $64T$.

which are higher than the sampling frequency. Suppose the actual period T^* is related to the sampling period T by

$$T^* = \frac{n}{m}T,$$

where n and m are coprime integers, then sampling at multiples of T will always yield n points (or clusters) in the map, for all $m \geq 1$. If n is the product of two integers k and l, then one can change the sampling interval to lT or kT, increasing the frequency resolution correspondingly. However, if one has misused a k which was not a factor of n, then there would appear a spurious factor k in the measured period. For the sake of safety, one must go gradually from the fundamental frequency to subharmonics, compare the results with power spectrum analysis whenever available.

In addition, there are a number of subtle points in the execution of the subharmonic stroboscopic sampling analysis. We discuss these points in the next section by showing a few actual Poincaré maps.

5.3.3 An Atlas of Maps

All the maps which we are going to use as examples in this subsection were obtained for the periodically forced Brusselator, studied in detail in Section 5.7. In order to integrate them into this book, we have indicated the parameter values.[3] NS in the figures denotes the multiplier to the fundamental period yielding the sampling interval, while the integer NK denotes the shift of the starting point in units of T. We have denoted them by p and q in the last Section 5.3.2. In practice, in order to obtain higher resolution, it is better to let the scales of the x and y coordinates vary independently and change from map to map.

An orbit of period nT will appear as n isolated points in the map. A quasiperiodic orbit will eventually fill up a closed curve, but, in practice, there may be gaps left, if not enough points are drawn. A chaotic orbit may show off as a number of disconnected islands, when points are scattered erratically in each island. We skip the maps of these simple cases.

By means of a sufficiently long sampling period, one can distinguish a periodic orbit from chaotic islands. However, due to the inhomogeneous structure of the attractors, there exist always sparse regions in a map of chaotic bands which aggravate the determination of the true "period". A few more trials with a longer period will resolve the problem. For instance, Fig. 5.1 is a stroboscopic portrait, sampled at $8T$. There are 8 clusters of points, but it is still hard to tell whether they are periodic points or chaotic islands. Since we have got now a factor 64, we can sample at $64T$ to get the map shown in Fig. 5.2. It bears the characteristic signature of a chaotic orbit. If one suspects that there might be one or two gaps in the islands, one might try sampling at $128T$ and $192T$. Since both maps just repeat the same picture, we obtain a good evidence for

---

[3]The parameters A, B, α and ω will appear in (5.26) below. Usually, we fix $A = 0.4$ and $B = 1.2$, α and ω are written on top of each figure. The abscissae are x and ordinates — y, if not indicated otherwise.

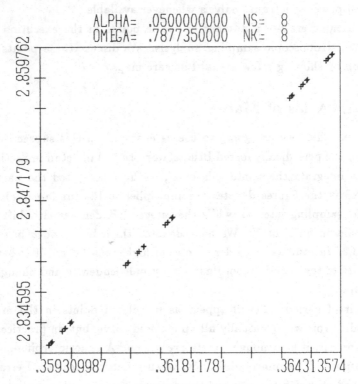

ALPHA= .0500000000   NS= 8
OMEGA= .7877350000   NK= 8

Figure 5.3: The manifestation of round-off errors. Sampling at 8T.

a 64-band chaotic regime. For the time being, subharmonic stroboscopic sampling provides the most effective means for the recognition of chaotic bands of very long "period" [or "semi-period" as it was called by Lorenz (1980)].

When the period under study gets longer and longer, the difference among various points in the subharmonic stroboscopic portrait will show up only in less and less significant digits, and round-off errors may stretch the image of a point into a strip which may be mistaken as a chaotic island. For example, sampling at 8T gave a map like that shown in Fig. 5.3. Apparently there are 8 clusters of points. However, sampling at 64T led to Fig. 5.4. The 8 strips seen in this figure are typical for round-off errors. One could even estimate the numerical difference between two neighbouring points; it agreed well with what

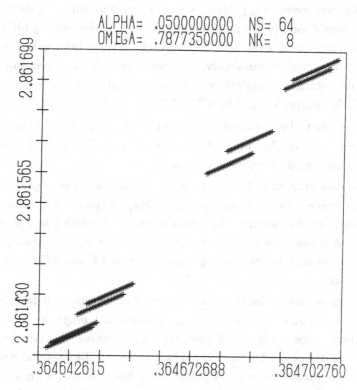

ALPHA= .0500000000 NS= 64
OMEGA= .7877350000 NK= 8

Figure 5.4: The manifestation of round-off errors. Sampling at 64T.

was expected from the computer word length. By going to double (or quadruple
if necessary and available) precision these strips shrink to points, so we have
in fact a 512T periodic orbit. With some experience, one can distinguish this
kind of systematic errors without resorting to time-consuming higher precision
arithmetic.

Another phenomenon, which is distinct from round-off errors, is a converging transient. Especially, when the parameter is close to a bifurcation point,
the transient may become very long due to critical slowing down (Hao, 1981;
and Section 10.1), and one will hardly be able to see the transient dying away
within a finite computing time. However, the round-off errors usually lead to
more or less evenly spaced points, as we have just seen in Fig. 5.4, whereas tran-

sients show a clear converging pattern. One can even determine the distribution function of these points in the neighbourhood of a period-doubling bifurcation point (Wang and Chen, 1984; and Section 6.7). With a little practice, these cases can be identified unmistakably. Transients close to the borderline separating different stationary regimes may display a rich variety of patterns. We shall show a few examples in Figs. 5.7 to 5.9.

The last point to be discussed in relation to an application of the subharmonic stroboscopic sampling method is the necessity to distinguish intermittent transitions from round-off errors or transients.

The Poincaré map close to an intermittent transition has a quite characteristic appearance (Wang, Chen and Hao, 1983). Figure 5.5 is such a map near the border of the tangent bifurcation to period 3 orbits (see Fig. 5.20 in Section 5.7). Although there are three clusters in this map, there is no period 3 component in the motion. Stroboscopic samplings at $1T$ and $3T$ give one and the same picture.

An effective procedure for the identification of an intermittent transition to period k is to plot x_{n+k} versus x_n (or y_{n+k} versus y_n) maps, known also as the k-th return maps.. Figure 5.6 presents a y_{n+3} versus y_n map, based on the same set of data as in Fig. 5.5. It outlines some one-dimensional structure near three accumulation regions of points along the bisector, with a number of points scattered in between. These are the corridors that give rise to the intermittent behaviour (cf. Figs. 2.15 and 2.16).

Transients may bring about additional structures in the Poincaré map. Figures 5.7 and 5.8 give two different types of transients to period 2 close to a quasiperiodic regime (Hao, 1982). Both upper graphs display the beginning stage of the transients, which look just like quasiperiodic loops, with possible condensations of points near the future periodic regime. The central graphs demonstrate the final stage of transients, when there has appeared a period 2 component. Now, stroboscopic sampling at $2T$ leads to quite different patterns of the manner in which the transients approach stationary states.

These two figures differ from the case of transients near a fixed point (period 1), shown in Fig. 5.9. Now the quasiperiodic loop may grow gradually from a loop of zero radius, an analog of second order phase transitions (Machlup and Sluckin, 1980). In this sense, the two previous figures may be compared with

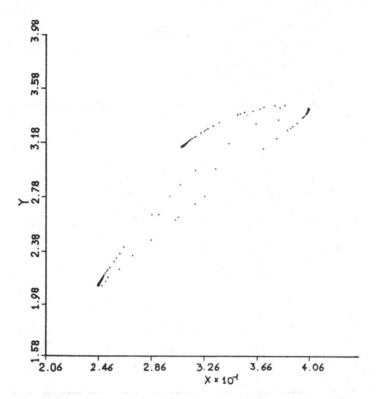

Figure 5.5: A Poincaré map near the intermittent transition to period 3.

first order phase transitions, since the new period 2 component comes into being abruptly at a certain stage of the transient process.

We shall return to the problem of transient behaviour in Chapter 7.

5.3.4   Rotation Numbers and Symbolic Dynamics

If one glance at a final Poincaré map of a periodic orbit, the immediate information gained is the period, i.e., the number of points in the plot. However, only a little more effort is required for the extraction of such useful information as the rotation number or the word describing the period in certain symbolic dynamics, provided, of course, one stays in the right part of the parameter space.

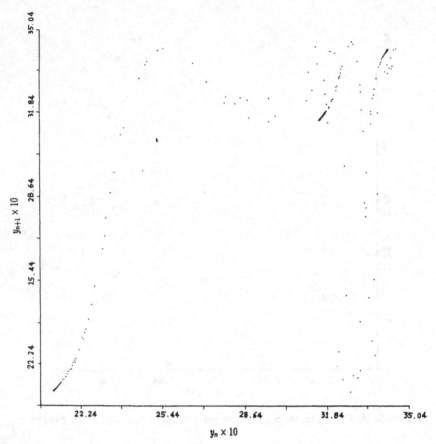

Figure 5.6: The y_{n+3} versus y_n map near the intermittent transition to period 3 (same parameters as Fig. 5.5).

We have become acquainted with the notion of the rotation number when studying the example of the bare circle map in Section 4.2:

$$\theta_{n+1} = \theta_n + \Omega \pmod 1.$$

If $\Omega = p/q$, where p and q are integers, then the q points seen in the map must be made in p turns. Therefore, one can measure p in the process of plotting the Poincaré map and define the rotation number as the ratio p/q. As we have learnt in Chapter 4, the rotation number is a very useful quantity, when dealing with periods related to frequency-locking regimes. For example,

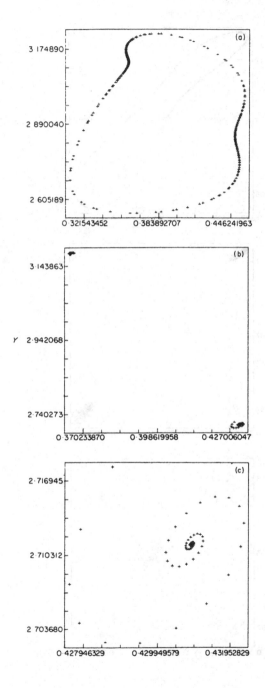

Figure 5.7: Transients to period 2 close to a quasiperiodic regime.

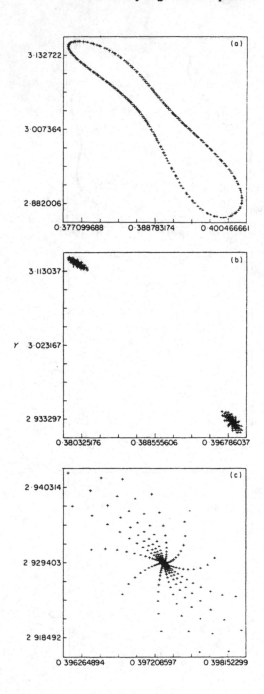

Figure 5.8: Another type of transients to period 2 close to a quasiperiodic regime.

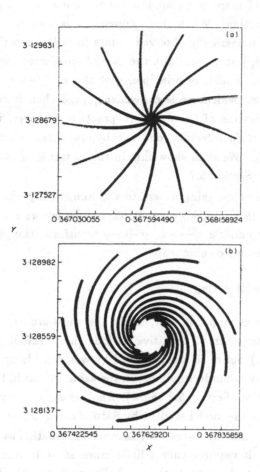

Figure 5.9: Transients near a fixed point.

if one has followed a period-doubling sequence from period 2 to 4 to 8, with the rotation numbers changing from 1/2 to 2/4 to 4/8, we are quite confident about our being in the supercritical part of the corresponding frequency-locked tongue.

Next, we will make some comments on symbolic dynamics. In dissipative systems, the Poincaré maps often approach one-dimensional objects, due to phase volume contraction. When this happens, it is sometimes possible to assign letters to the numerically observed points in order to arrive at a word describing the period, just as we did in the case of one-dimensional mappings. For this purpose, one should, in principle, plot the $x_{n+1} - x_n$ map (or the $y_{n+1} - y_n$ map, etc., whichever is convenient), and then introduce a one-parameter parametrization of the map. In practice, however, if the map is stretched along a certain direction, one can bypass these constructions and assign letters directly. We shall show this in the example of the periodically forced Brusselator in Section 5.7.

Another example of assigning letters to the numerically observed periods is the Lorenz model (5.1). In this case, at least three letters are required due to the antisymmetry under a $x \to -x$, $y \to -y$ transformation, that makes it resemble the antisymmetric cubic map

$$x_{n+1} = Ax_n{}^3 + (A - 1)x_n,$$

rather than the Hénon map. In fact, along the most-studied b line in the parameter space, altogether 47 primitive periods (i.e., excluding the period-doubled regimes, etc.) out of 53 are described by words made up of the letters R, M and L, and they are ordered almost in the same way as in the cubic map (Ding and Hao, 1988; see Section 5.8 below). Another system that may require three letters might be the model studied by Sato, Sano and Sawada (1983).

In other words, we must try to extract as much information as possible from each computer run. It requires only a little more effort to find the rotation number and, if one is lucky enough, the symbolic word from the Poincaré map of a periodic orbit.

5.4   Technique for Location of Periodic Orbits

We have emphasized repeatedly that the systematics of periodic solutions brings about essential knowledge of the global structure of the parameter space, including the location and nature of possible chaotic and quasiperiodic regimes. An effective numerical technique for the location of periodic orbits saves much computer time and guarantees high precision. We consider autonomous and non-autonomous systems separately.

5.4.1 Non-autonomous Systems

Despite its great efficiency, the subharmonic stroboscopic sampling method is still rather time-consuming, when it comes to the determination of a periodic orbit from scratch, because one has to throw away at least a few hundreds of transient periods to reach the stationary solution. However, the technique of locating periodic orbit simplifies drastically for periodically driven systems, since there is no need to calculate the period. In fact, one has to solve only 6 equations (2 nonlinear plus 4 linear) in order to locate a periodic orbit for (5.2).

Let us describe the method in more general terms. Suppose we have got a point $\mathbf{P}_n$ [$\mathbf{P}_n = (x_n, y_n)$ for (5.2)] in the stroboscopic portrait. Integrating the nonlinear system for a period T (it must be done in, say, $L = 1024$ steps), we find another point $\mathbf{P}_{n+1}$. The points $\mathbf{P}_n$ and $\mathbf{P}_{n+1}$ are quite close, whence we suspect that they are not far from an exact period $\mathbf{P}$. We may write

$$\begin{aligned}\mathbf{P}_n &= \mathbf{P} + \mathbf{W}_n, \\ \mathbf{P}_{n+1} &= \mathbf{P} + \mathbf{W}_{n+1}.\end{aligned} \tag{5.11}$$

Although $\mathbf{P}$, $\mathbf{W}_n$ and $\mathbf{W}_{n+1}$ are unknown, $\mathbf{W}_{n+1}$ must be evolved from $\mathbf{W}_n$, according to the linearized equation, i.e.,

$$\mathbf{W}_{n+1} = \mathbf{U}(T)\mathbf{W}_n. \tag{5.12}$$

Subtracting the two equations (5.11), we find

$$\mathbf{W}_n = \left(\mathbf{U}(T) - \mathbf{I}\right)^{-1} \left(\mathbf{P}_{n+1} - \mathbf{P}_n\right). \tag{5.13}$$

Every term on the right-hand side of (5.13) can be calculated: $\mathbf{P}_{n+1}$ from $\mathbf{P}_n$, according to the original nonlinear equations, $\mathbf{U}(T)$ from the linearized system. Note that $\mathbf{U}(T) \equiv \mathbf{U}(T, 0)$ is the product of L factors

$$\mathbf{U}(T) = \mathbf{U}(T, (L-1)h)\mathbf{U}((L-1)h, (L-2)h) \ldots \mathbf{U}(2h, h)\mathbf{U}(h, 0) \tag{5.14}$$

($h = T/L$ is the time step of integration), each factor being calculated from the system, linearized at the end point of the previous step. Being the solution of a linear system, there are two ways to compute $\mathbf{U}(kh, (k-1)h)$: either numerically by adding the linear equations for $\mathbf{U}$ to the linear ones with the initial values

$$\mathbf{U}(t,t) = \mathbf{I}$$

at each step, or analytically by using the closed form (when $n = 2$)

$$\mathbf{U}(t,0) = e^{\lambda t} \begin{pmatrix} C(t) + (a_{11} - a_{22})S(t)/2 & a_{12}S(t) \\ a_{21}S(t) & C(t) - (a_{11} - a_{22})S(t)/2 \end{pmatrix},$$

where

$$C(t) = \begin{cases} \cosh(st), & D > 0, \\ 1, & D = 0, \\ \cos(st), & D < 0, \end{cases} \qquad S(t) = \begin{cases} \sinh(st)/s, & D > 0, \\ t, & D = 0, \\ \sin(st)/s, & D < 0, \end{cases}$$

and

$$\lambda = (a_{11} + a_{22})/2, \quad D = \lambda^2 - \det(J), \quad s = \sqrt{|D|}.$$

Note that the a_{ij} are elements of the matrix $\mathbf{J}$ introduced in (5.7).

The decision regarding the approach to be adopted depends on the complexity of the equations involved; it may be determined by trial. The periodic orbit-following technique works equally well for both stable and unstable solutions, provided the initial data are good enough. Therefore, when using the method, it is desirable to monitor the Lyapunov exponents (see Section 6.4), in order to become aware of the stability of the orbit under study.

5.4.2   Autonomous Systems

Suppose that in the calculation of a Poincaré map by Hénon's method we have come across a suspected periodic orbit, i.e., starting from a point $\mathbf{x}_0$ in the Poincaré section, we return to its neighbourhood $\mathbf{x}_1$ after integrating the system for some time T. We suspect that there might be an exact periodic point $\mathbf{x}^*$ with period T^*. How do we find the unknown $\mathbf{x}^*$ and T^*, using our knowledge of $\mathbf{x}_0$, $\mathbf{x}_1$ and T?

The answer is based on an extension of Newton's method (method of tangents) to ordinary differential equations. This idea was first applied by O. Lanford to the Lorenz Equations (5.1), and then discussed by J. H. Curry

(1980). We follow closely the description in Appendix E in the book of Sparrow (B1982).

We shall find it quite convenient to use the concise vector notations of Section 5.2. Thus, we have, in terms of the flow $\Phi_\tau(\mathbf{x})$, an exact periodic solution

$$\mathbf{x}^* = \Phi_T(\mathbf{x}^*) \tag{5.15}$$

with both $\mathbf{x}^*$ and T^* unknown, and a nearby trajectory

$$\mathbf{x}_1 = \Phi_T(\mathbf{x}_0) \tag{5.16}$$

with $\mathbf{x}_0$ known and $\mathbf{x}_1$, T calculable by integration of the system. As we are close to the true periodic solution, we can write

$$\begin{aligned} \mathbf{x}^* &= \mathbf{x}_0 + d\mathbf{x}, \\ T^* &= T + dT, \end{aligned} \tag{5.17}$$

and insert these values into the periodic flow (5.15):

$$\mathbf{x}_0 + d\mathbf{x} = \Phi_{T+dT}(\mathbf{x}_0 + d\mathbf{x}). \tag{5.18}$$

Now, expanding the right-hand side of the above equation and using Relation (5.16) as well as the original system (5.5), we find

$$\begin{aligned} \mathbf{x}_0 + d\mathbf{x} &= \Phi_T(\mathbf{x}_0) + \frac{d}{d\tau}\Phi_T(\mathbf{x}_0)\,dT + D_{\mathbf{x}}\Phi_T(\mathbf{x}_0)\,d\mathbf{x} \\ &= \mathbf{x}_1 + \frac{d\mathbf{x}_1}{dt}\,dT + D_{\mathbf{x}}\Phi_T(\mathbf{x}_0)\,d\mathbf{x}, \end{aligned}$$

that is

$$\mathbf{x}_0 - \mathbf{x}_1 = \mathbf{F}(\mathbf{x}_1)\,dT + \left(D_{\mathbf{x}}\Phi_T(\mathbf{x}_0) - \mathbf{I}\right)d\mathbf{x}, \tag{5.19}$$

where $\mathbf{F}(\mathbf{x}_1)$ is the right-hand side of the original system (5.5) calculated at $\mathbf{x}_1$.

Equation (5.19) represents actually a system of equations which we will write down explicitly for the case $n = 3$, i.e.,

$$\mathbf{x} = (x, y, z), \quad \mathbf{F} = (f_1, f_2, f_3),$$

when it assumes the form

$$\begin{pmatrix} x_0 - x_1 \\ y_0 - y_1 \\ z_0 - z_1 \end{pmatrix} = \begin{pmatrix} \dfrac{\partial x}{\partial x_0} - 1 & \dfrac{\partial x}{\partial y_0} & \dfrac{\partial x}{\partial z_0} & f_1(\mathbf{x}_1) \\ \dfrac{\partial y}{\partial x_0} & \dfrac{\partial y}{\partial y_0} - 1 & \dfrac{\partial y}{\partial z_0} & f_2(\mathbf{x}_1) \\ \dfrac{\partial z}{\partial x_0} & \dfrac{\partial z}{\partial y_0} & \dfrac{\partial z}{\partial z_0} - 1 & f_3(\mathbf{x}_1) \end{pmatrix} \begin{pmatrix} dx \\ dy \\ dz \\ dT \end{pmatrix}.$$

The three unknown displacements dx, dy and dz must be confined to the surface of the section, i.e., subjected to the condition

$$dS = \frac{\partial S}{\partial x} dx + \frac{\partial S}{\partial y} dy + \frac{\partial S}{\partial z} dz = 0,$$

where S is the left-hand side of the intersecting surface (5.10). In the simplest case $S = z$, we have $dz = 0$ and

$$\begin{pmatrix} dx \\ dy \\ dT \end{pmatrix} = \begin{pmatrix} \dfrac{\partial x}{\partial x_0} - 1 & \dfrac{\partial x}{\partial y_0} & f_1(\mathbf{x}_1) \\ \dfrac{\partial y}{\partial x_0} & \dfrac{\partial y}{\partial y_0} - 1 & f_2(\mathbf{x}_1) \\ \dfrac{\partial z}{\partial x_0} & \dfrac{\partial z}{\partial y_0} & f_3(\mathbf{x}_1) \end{pmatrix}^{-1} \begin{pmatrix} x_0 - x_1 \\ y_0 - y_1 \\ z_0 - z_1 \end{pmatrix}. \quad (5.20)$$

Hence, the exact periodic orbit $\mathbf{x}^*$ and the period T^* can be determined from (5.20) and (5.17). In practice, one has to iterate a few times to reach sufficient precision. The convergence of this method is quadratic, just as Newton's method usually is, provided one has good enough initial estimates for $\mathbf{x}_0$ and T.

Now we can summarize the orbit locating procedure:

1. Get an initial estimate of $\mathbf{x}_0$ and T by using, e.g., Hénon's method.

2. Integrate the original system to get the next return point $\mathbf{x}_1 = \Phi_T(\mathbf{x}_0)$, and compute T and $\mathbf{F}(\mathbf{x}_1)$.

3. Integrate the linearized system to calculate the $(n - 1) \times n$ (in the case of $z = $ const) or $n \times n$ (in the case of a general surface) components of $\partial x_i/\partial x_{0j}$:

4. Use these values of $\mathbf{x}_0$, $\mathbf{x}_1$, $\mathbf{F}(\mathbf{x}_1)$ and $\partial x_i/\partial x_{0j}$ to calculate dx and dT and, consequently, $\mathbf{x}^*$ and T^*.

5. If necessary, use these as new estimates for $\mathbf{x}_0$ and T and start anew.

This method is equally well applicable to the location of stable and unstable orbits. If one is interested in stable orbits only, then the stability must be checked by, e.g., calculating the largest Lyapunov exponent (see Section 6.4).

5.4.3   Summary of Periodic Orbit Locating Technique

We summarize the periodic orbit locating technique by a comparison of the numerical work involved in terms of the number of differential equations to be solved at each step (Table 5.1).

	Nonlinear Eqs.	Linear Eqs.	Total
Autonomous:			
$z = a$ section	3	6	9
$S = \text{const surface}$	3	9	12
General case	n	n	$n(n+1)$
Driven:			
Planar	2	4	6
General	$n-1$	$(n-1)^2$	$n(n-1)$

Table 5.1: Comparison of periodic orbit locating technique.

5.5   Visualization of the Dynamics

The construction of the Poincaré maps is undoubtedly important for the visualization of the complicated dynamics of a nonlinear system. Nevertheless, in this section, we intend to discuss a few other means of visualization, including the simplest one of just looking at the trajectory.

We have shown the Poincaré map and the x_{n+k} versus x_n map near an intermittent transition in Section 5.3.3. In fact, the transition manifests itself in the case of differential equations most clearly as irregular bursts interrupting the more or less regular time variation of the dynamical variables. If one varies the parameter from a chaotic region towards the tangent bifurcation, then longer and longer regular intervals will be observed. Figure 5.10 shows a $y(t)$ versus t curve for the forced Brusselator, close enough to the transition points.

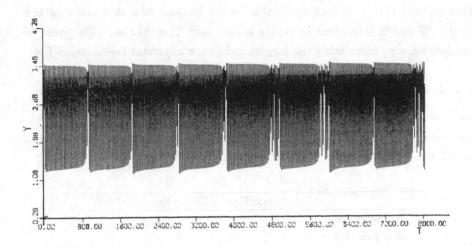

Figure 5.10: The $y(t)$ versus t curve near an intermittent transition in the forced Brusselator.

We refer the reader for more details on intermittent transitions in the forced Brusselator to Wang, Chen and Hao (1983).

There are a few cases when the difference in dynamical regimes cannot be recognized by merely looking at Poincaré maps. For instance, the period-doubled regime in the 1/1 mode-locked tongue, i.e., the 2/2 locking, may overlap with the 1/2 tongue and there may exist a borderline in the parameter plane. Nonetheless, these two different kinds of period 2 orbits can hardly be distinguished in the Poincaré sections. Their time variations, however, do differ clearly (Wang, 1983, gave an example for the forced Brusselator).

The trajectories are usually observed in projections in various planar sections of the phase space as *phase portraits*. A better view may be reached by drawing three-dimensional stereoscopic projections of a curve on a two-dimensional screen. By adjusting the projection angles, one chooses the best view of the trajectory. Continuously changing the projection angles would produce a much better effect, but it requires more computing power and sophis-

THET= 0.0 PHI= 0.0

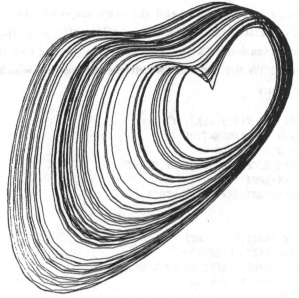

Figure 5.11: Stereoprojection of an attractor $(\theta = 0, \phi = 0)$.

THET=-30.0 PHI= 0.0

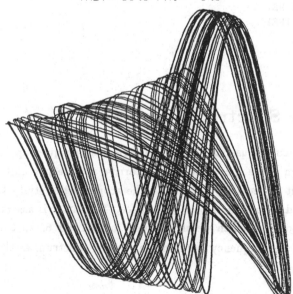

Figure 5.12: Another projection of the same attractor $(\theta = -30, \phi = 0)$.

ticated equipment. Nevertheless, one could do a good job even on personal computers taking "snapshots", or, even better, generating animated pictures.

Figures 5.11 and 5.12 show one and the same trajectory in two different projections. The calculated three-dimensional orbits $x(t)$, $y(t)$, $z(t)$ were projected onto a two-dimensional XX and YY plane by adjusting two angles θ and ϕ (given in degrees). We list here the two simple subroutines which were used to produce these figures:

```
      SUBROUTINE PROJT(X,Y,Z,XX,YY)
C STEREOPROJECTION TRANSFORMATION
C BY COURTESY OF DR. S. OHNO
      COMMON /ANGLE/CPHI,SPHI,CTHET,STHET
      XX=Y*CPHI-X*SPHI
      YY=Z*STHET-CTHET*(X*CPHI+Y*SPHI)
      RETURN
      END
      SUBROUTINE PROJI(PHI,THET)
C STEREOPROJECTION INITIALIZATION
      COMMON /ANGLE/CPHI,SPHI,CTHET,STHET
      RAD(X)=3.14159265*X/180.0
      PHI=RAD(PHI)
      THET=RAD(THET)
      CPHI=COS(PHI)
      SPHI=SIN(PHI)
      CTHET=COS(THET)
      STHET=SIN(THET)
      RETURN
      END
```

5.6   Power Spectrum Analysis

So far we have been discussing various numerical techniques applied in the time domain and in phase space. In order to arrive at a deeper understanding of the dynamics, it is necessary to extend the scope of the study to the frequency domain and to include the tangent space. The latter is associated with the Lyapunov exponents which we shall study in Chapter 6. In this section, we concentrate on the frequency domain, i.e., power spectrum analysis of the dynamics.

In a sense, power spectrum analysis is the immediate task one has to carry out after drawing the trajectories and the Poincaré maps, because it provides us with the simplest method for distinguishing between chaotic and quasiperiodic motion, the spectra of the former being noisy broad bands and the latter discrete lines without simple frequency interrelationship. Even for periodic regimes, the fine structure of power spectra can furnish useful information on our whereabouts in the parameter space. However, there are some subtleties in the performance of power spectrum analysis. We will discuss a few of these.

5.6.1 Preliminaries

Suppose we are given a time series

$$x_1, x_2, \ldots x_n, \ldots,$$

sampled at equal time intervals τ, which arise in some computer or laboratory experiment. Taking a finite subsequence

$$x_1, x_2, \ldots, x_N$$

from the series and imposing periodic boundary conditions

$$x_{N+j} = x_j \quad \forall j,$$

one calculates the correlation function

$$c_j = \frac{1}{N} \sum_{i=1}^{N} x_i x_{i+j} \tag{5.21}$$

and then performs a discrete Fourier transform to obtain the Fourier coefficients

$$p_k = \sum_{j=1}^{N} c_j \exp\left(\frac{2\pi k j \sqrt{-1}}{N}\right). \tag{5.22}$$

Roughly speaking, p_k represents the contribution of the kth frequency component to $\{x_i\}$. This was the original definition of a "power" spectrum.

Since the rediscovery of the fast Fourier transform (FFT) algorithm in 1965, power spectra have been calculated more efficiently by directly transforming $\{x_i\}$ into Fourier coefficients

$$a_k = \frac{1}{N} \sum_i x_i \cos(\pi ik/N),$$

$$b_k = \frac{1}{N} \sum_i x_i \sin(\pi ik/N), \qquad (5.23)$$

and then computing

$$\bar{p}_k = a_k^2 + b_k^2. \qquad (5.24)$$

A number of subsequences of $\{x_i\}$ will yield several sets of $\{\bar{p}_k\}$. By forming their average, one obtains the power spectrum $\{p_k\}$. In general, this is necessary whenever one deals with genuine aperiodic processes. In the case of ordinary differential equations, when there is a nice periodic substructure, a single $\{\bar{p}_k\}$ may yield quite a good spectrum and one does not need to average all the way.

We mention again that nowadays it is not necessary to write the FFT program by oneself. There is a large choice of subroutines in existing libraries. The most important step, however, in doing power spectrum analysis is to "design" the spectra before applying the FFT.

5.6.2 "Design" of the Spectrum

There are two aspects in "designing" a good spectrum: the choice of correct sampling parameters and filtering of the data (or smoothing the results). We start with the first problem. Let the sampling interval be τ and the total sampling time for a single spectrum be $L = N\tau$, where N is the number of sampled points. The quantities τ and N determine two frequencies: $f_{max} = 0.5/\tau$ is the maximal frequency one can measure, using the given sampling interval, and $\triangle f = 1/L$ is the frequency difference between two adjacent Fourier coefficients. In order to eliminate effectively the aliasing phenomenon, i.e., the appearance of spurious high frequency components in the spectrum (see, e.g., the book by Rayner[4]), one has to take $f_{max} = kf_0$, where f_0 is the fundamental frequency of the physical system and k is a multiplier of the order 4 to 8, and to retain only the lower part of the spectrum. Furthermore, we aim at resolving the pth subharmonic of f_0 and wish the subharmonic peak to be formed by s points in the spectrum, i.e., $f_0/p = s\triangle f$. Combining all these relations, we find

[4]J. H. Rayner, *An Introduction to Spectral Analysis*, Pion Limited, London, 1971, p.74.

$$p = \frac{N}{2ks}.$$

Note that this relation is independent of τ and f_0. Take, for instance, $N = 8192$, $k = 4$, and $s = 8$, when $p = 128$. This gives the resolution limit in the power spectrum analysis on a medium-size computer. Nevertheless, a broadband noisy spectrum is still the most practical and convenient criterion for the identification of chaotic motion in laboratory and computer experiments.

As regards the filtering of data, it may turn out to be more important when one deals with really aperiodic processes with noisy background. In a numerical study of ordinary differential equations, however, it is not so essential to have the data filtered before aplying a FFT. In most cases, simple smoothing of the spectrum will be sufficient. Anyway, the filtering of data and the smoothing of spectra are equivalent operations, which are interrelated by a convolution transformation. There arise a few subtleties in the performance of filtering or smoothing. Whenever the necessity should arise, the literature on time series analysis, which is numerous, should be consulted.

5.6.3 Symbolic Dynamics and Fine Structure of Spectra

When one happens to be in a chaotic regime of the parameter space, where the periodic windows are ordered according to the symbolic dynamics of a certain number of letters, the fine structure of power spectra provides useful information about one's exact location in the parameter space. We have discussed this problem in connection with the physical meaning of the *-composition in Section 3.3.4 and have shown two schematic spectra in Fig. 3.5. Now we give two real spectra taken from the periodically forced Brusselator.

We already know that the fine structure is essentially a question of factorizing non-prime numbers. Take, for example, period 24. The possible decompositions include the following two:

(1). $24 = 2^3 \times 3$, corresponding to the word

$$R * R * R * RL \rightarrow RLRRRLRLRLRRRLRRRLRRRLR\sigma,$$

i.e., the period 3 window embedded in the 8-band chaotic zone. Actually, there is no reason to present it as a superstable sequence, so the last letter

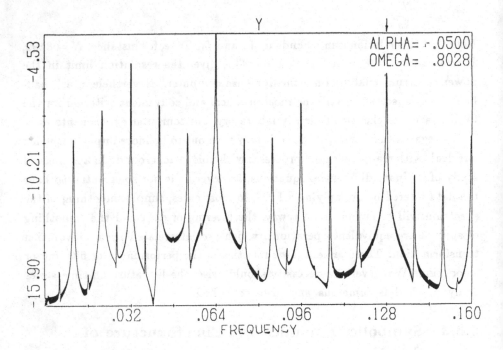

Figure 5.13: Fine structure of power spectrum: type $2^3 \times 3$, taken from the forced Brusselator at $A = 0.4$, $B = 1.2$, $\alpha = 0.05$, and $\omega = 0.83116$.

σ may be either of L, C and R according to the periodic window theorem (Section 3.6.4). The power spectrum is shown in Fig. 5.13, where the location of the fundamental frequency has been indicated by an arrow. The parameters of this regime are $A = 0.4$, $B = 1.2$, $\alpha = 0.05$, and $\omega = 0.83116$.

   (2). $24 = 2^3 \times 3 \times 2$, corresponding to the word

$$R * R * RL * R = RLRRRLRLRLRLRLRRRLRLRLR\sigma,$$

i.e., the period-doubled regime of the period 3 window, embedded in the 4-band chaotic zone. The spectrum is shown in Fig. 5.14 for the parameters $A = 0.4$, $B = 1.2$, $\alpha = 0.05$, and $\omega = 0.8028$.

   In order to complete our discussion of power spectra, we compare a typical period-doubling spectrum with its broad-band counterpart in the chaotic zone. Again Figs. 5.15 originate from the numerical study of the forced Brusselator

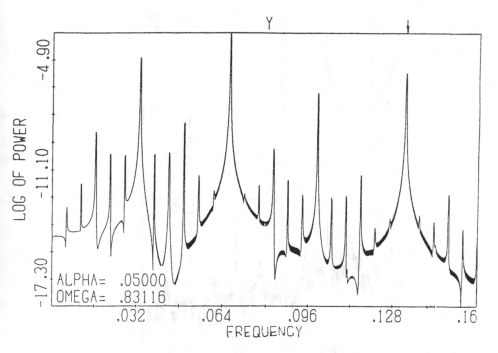

Figure 5.14: Fine structure of power spectrum: type $2^2 \times 3 \times 2$, taken from the forced Brusselator at $A = 0.4$, $B = 1.2$, $\alpha = 0.05$, and $\omega = 0.8028$.

(Hao and Zhang, 1982b). The upper figure is a period 2 orbit, while the lower figure is the 2-band chaotic regime. The arrow indicates the fundamental frequency. Note the sharp peaks in the broad band. This is characteristic for a chaotic regime in the inverse period-halving sequence of chaotic bands, and it is in contrast to the spectra corresponding to a transition to chaos from a quasiperiodic regime (e.g., Fig. 5.24).

5.7 Case Study 1: the Forced Brusselator

In the last two sections of this chapter we collect some detailed results for ordinary differential equations, in order to show how, in practice, the numerical methods work. In a sense, the choice of models does not mean very much, as

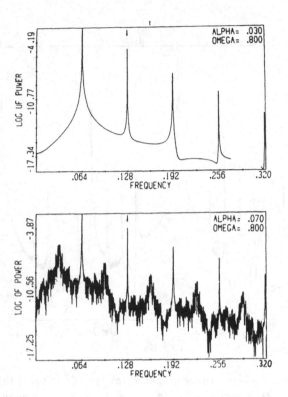

Figure 5.15: Power spectra of a period 2 orbit (upper) and a 2-band chaotic orbit (lower).

the chaotic behaviour in physically reasonable models is expected to be, to a certain extent, universal. We start with the simplest case of a periodically forced limit-cycle oscillator, where high frequency resolution can be reached by applying the subharmonic stroboscopic sampling technique described in Section 5.3.2.

The model has been described in Section 1.5. For the sake of convenience, we rewrite the equations. The free Brusselator

$$\dot{x} = A - (B + 1)x + x^2 y,$$
$$\dot{y} = Bx - x^2 y \tag{5.25}$$

describes the kinetics of a tri-molecular reaction. The control parameters A

and B are concentrations of certain chemicals, and x, y are the concentrations of the intermediate products, all being dimensionless.

With addition of a periodic driving force added, the system

$$\dot{x} = A - (B + 1)x + x^2 y + \alpha \cos(\omega t),$$
$$\dot{y} = Bx - x^2 y \tag{5.26}$$

does not have at all fixed points.

We prefer to write Equations (5.26) in the apparently autonomous form

$$\dot{x} = A - (B + 1)\, x + x^2 y + \alpha u,$$
$$\dot{y} = B\, x - x^2 y,$$
$$\dot{u} = -\omega z,$$
$$\dot{z} = \omega u,$$

with fixed initial conditions $u(0) = 1$, $z(0) = 0$. This version of the system has several merits. Firstly, since it is autonomous, there is no essential difference from other autonomous ordinary differential equations, except for the presence of a cubic nonlinearity which, as we shall see in what follows, does not cause any special features. Secondly, it represents a coupled system of one linear and one nonlinear oscillator, allowing for a more intuitive interpretation of the various regimes of motion. Thirdly, the four control parameters A, B, α and ω enter the system on an equal footing, whereas α and ω may seem to perform different roles in (5.27). In addition, one may save a few percent of computer time, owing to the absence of the cosine function on the right-hand side of the equations.

Complicated bifurcation and chaotic behaviour has been observed in Equations (5.26) by a numerical study for various parameter ranges. At this time, we have a fairly good knowledge about the structure of the parameter space. Before presenting the numerical results, however, let us see what can be learnt from a linear stability analysis of the system.

5.7.1 Linear Stability Analysis

First of all, we note that no analytical solution is available for the free oscillator (5.25). Even when more parameters are introduced into the system, no integrable point in the parameter space could be found. If one looks at the symmetry group of Equations (5.25), the only generator is the time transla-tion $\partial/\partial t$, a trivial result for any autonomous system of ordinary differential equations. In other words, since there is little hope for the development of a systematic approximate analytical approach to the free and forced Brusselator, one has to resort to numerical methods.

In a study of any nonlinear evolution equation, the first step is a search for possible steady states: simple fixed points and periodic solutions. We find fixed points by setting the derivatives $\dot{x}$ and $\dot{y}$ on the left-hand side of (5.25) to zero, when the remaining algebraic equations easily yield a unique solution $x = A$, $y = B/A$ under the reasonable assumption that $A \neq 0$. Thus, the system (5.25) has only one fixed point $P(A, B/A)$, at which the eigenvalues of the linearized system are

$$\lambda_{\pm} = \frac{1}{2}(k \pm \sqrt{k^2 - 4A^2}), \tag{5.27}$$

where $k = B - 1 - A^2$. Sometimes, for a better presentation of the results, we measure the real and imaginary parts of $\lambda_{\pm}$ in units of ω, the external control frequency, i.e., we write (5.27) in the form

$$\lambda_{\pm} = \gamma\omega \pm \beta\omega i \ . \tag{5.28}$$

In Equation (5.28), the quantity β is the ratio of the limit cycle frequency in the linearized regime to the driving frequency, an analog of the parameter A in the circle mapping (4.21), whereas the sign of γ determines the linear stability of the free oscillator. If

$$\gamma < 0 \quad \text{or} \quad B < A^2 + 1, \tag{5.29}$$

the fixed point $P(A, B/A)$ is stable. At $\gamma = 0$, there occurs Hopf bifurcation, a limit cycle develops for $\gamma > 0$. No other fixed points exist for the free Brusselator. When the coupling constant α *in (5.26) acquires a non-zero value, the fixed point P disappears. One has to examine the existence and stability of periodic solutions. To this end Fan Yin-shui (1987) succeeded in proving some global results. Since the original paper appeared only in Chinese, we reproduce here part of Fan's derivation in the next section.

5.7.2 The Existence of Periodic Solutions

As regards the global stability of Equations (5.26), Fan Yin-shui (1987) has proved that, when $|\alpha| < A$, there exists in the $x - y$ plane a closed curve Γ, which contains the point P and which has no tangential contact with the flows of (5.26), such that when $t > 0$ all integral curves starting from $S = \Gamma \otimes I(t \geq 0)$ must eventually enter S.

We give an outline of some of Fan's construction. Shifting the origin of coordinates in (5.26) to the fixed point P, i.e., letting $x' = x - A$, $y' = y - B/A$ and dropping the primes hereafter, we find

$$
\begin{aligned}
\dot{x} &= -x - G(x,y) + \alpha \cos(\omega t), \\
\dot{y} &= G(x,y),
\end{aligned}
\tag{5.30}
$$

where $G(x,y) = -(x+A)[xB/A + (x+A)y]$. First of all, we see that the curve

$$
C: \quad xB/A + (x+A)y = 0
$$

divides the region $x \geq -A$, $y \geq -B/A$ into two regions I and II (see Fig. 5.16). In region I, $G(x,y) \leq 0$, while in region II $G(x,y) \geq 0$. Next, we construct a closed curve Γ, which consists of several pieces, and show that the flow of (5.26) has no tangent contact with Γ under the condition that $|\alpha| < A$ (which holds in all numerical studies carried out so far):

1. The y-axis $x = -A$, on which

$$
\frac{dx}{dt}\Big|_{x=-A} = A + \alpha \cos(\omega t) > 0.
$$

2. Take

$$
0 < \eta \leq \frac{A - |\alpha|}{2B+1}, \quad y_0 = \frac{B(A-\eta)}{A\eta},
\tag{5.31}
$$

and draw a segment $\overline{B'C'}$ of the straight line $y = y_0 + x + A$. It intersects the y-axis at $B'(-A, y_0)$ and the curve C at

$$
C'\Big(-A - \frac{B}{2\eta} + \frac{\sqrt{B^2/\eta^2 + 4B}}{2}, \ \frac{B}{2\eta} - \frac{B}{A} + \frac{\sqrt{B^2/\eta^2 + 4B}}{2}\Big).
$$

Letting dv_2/dt be the inner product of the vector field (5.30) and the outer normal vector of $\overline{B'C'}$, one has

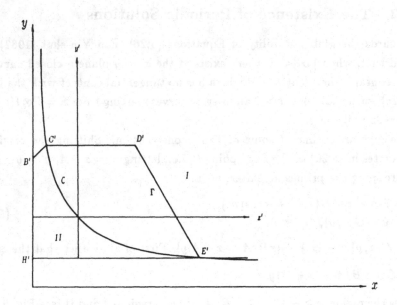

Figure 5.16: The contour Γ for the attracting region (after Fan Yin-shui).

$$
\begin{aligned}
x + 2G(x,y) &\leq x + 2B(x + A) \\
&\leq -A + (2B + 1)\left(-\frac{B}{2\eta} + \frac{\sqrt{B/\eta + 4B}}{2}\right) \\
&< -A + (2B + 1)\eta.
\end{aligned}
$$

Thus

$$
\left|\frac{dv_2}{dt}\right| = x + 2G(x,y) - \alpha\cos(\omega t) < -A + (2B + 1)\eta + |\alpha| \leq 0.
$$

3. Starting from the point C', draw a line $\overline{C'D'}$ parallel to the abscissa. It intersects the bisector $y = x$ at $D'(\xi, \xi)$, where

$$
\xi = \frac{B}{2\eta} - \frac{B}{A} + \frac{\sqrt{B^2/\eta^2 + 4B}}{2}.
$$

On $\overline{C'D'}$, one has

$$\frac{dy}{dt} = G(x, y) \le 0.$$

The equality sign holds only at the point C'.

4. Starting from D', draw a segment of the straight line $x + y = 2\xi$. It intersects the curve C at the point $E'(\mu, \nu)$, where

$$\mu = \xi + \tfrac{1}{2}\left[B/A - A + \sqrt{(A - 2\xi - B/A)^2 + 8A\xi}\right],$$

$$\nu = \xi - \tfrac{1}{2}\left[B/A - A + \sqrt{(A - 2\xi - B/A)^2 + 8A\xi}\right].$$

If η is sufficiently small such that the relation $x \ge A > |\alpha|$ holds on $\overline{D'E'}$, then

$$\left|\frac{dv_4}{dt}\right| = -x + \alpha\cos(\omega t) \le -x + |\alpha| < 0,$$

where dv_4/dt is the inner product between the vector field (5.30) and the outer normal vector of $\overline{D'E'}$.

5. Now close the contour Γ by drawing from the point E' a parallel to the abscissa, intersecting the y-axis at the point H'. On the segment $\overline{E'H'}$, we have

$$\frac{dy}{dt} = G(x, y) \ge 0,$$

where dy/dt equals 0 only at the two end points E' and H'.

Now we have constructed a closed curve $\Gamma(B'C'D'E'H'B')$ which does not have tangent contact with the flows of (5.30). Therefore, the cylinder

$$S = \Gamma \otimes I(t \ge 0)$$

is a surface without tangent contact with the flows of the equations: all integral curves, departing outside S, must enter S when t increases. It takes a little more effort to prove that S contains periodic solutions of period $2\pi/\omega$ of (5.30).

5.7.3　Hierarchy of Chaotic Bands and the U-sequence

It is instructive to proceed in historical order, in order to demonstrate the knowledge relating to the forced Brusselator which has accumulated to date. In their numerical study of the $\alpha - \omega$ plane, Tomita and Kai (1978, 1979 a and b) discovered a small chaotic region, surrounded by an apparently period-doubling cascade, with a number of isolated frequency-locked "bubbles" immersed in quasiperiodic regimes in the lower part of the $\alpha - \omega$ diagram. They noticed the nested structure (we have dropped their term "window structure" in order to avoid a possible confusion with periodic windows) in the appearance of some periodic regimes, e.g., the sequence of periods 128, 24, 48, 96, ..., 44, ..., 12, 24, 48, 192, ..., 96, 48, 24, 12, ..., 40, ..., 128. With the help of the subharmonic stroboscopic sampling technique, combined with extensive power spectrum analysis, the search was carried out with a frequency resolution as high as one-thousandth of the fundamental frequency (Hao and Zhang, 1982b) and extended to other sections of the parameter space, spanned by the four parameters A, B, α and ω.

Figure 5.17 is an enlargement of the surroundings of the chaotic region in the $\alpha - \omega$ plane (at fixed $A = 0.4$, $B = 1.2$). Surrounded by period-doubling cascades in all directions, the chaotic region itself is divided into nested bands of "period" 2^n in reverse order. In this and subsequent figures, the abbreviation nP stands for Periodic motion with period n, and nI for Inverse bands (or Islands in the Poincaré sections) of "period" n.

Two lines crossing the chaotic region were examined in detail, i.e., the lines $\alpha = 0.05$ and $\omega = 0.80$. Both lines missed the central $1I$ region, because they were chosen before the global picture of Fig. 5.17 was clarified. Along both lines one sees a direct period-doubling sequence, going into an inverse sequence of the chaotic bands. The analogy with the bifurcation diagram of one-dimensional mappings is apparent. Only further study of the homoclinic and heteroclinic crossings of the invariant manifolds will reveal a few differences, say, in the nature of some crises, unknown to the logistic map (see Section 5.7.5). In every chaotic band of the primary sequence, there are secondary bifurcation sequences with their own direct period-doubling and inverse band-merging parts. Within chaotic bands of the secondary sequences there are embedded tertiary bifurcation sequences, again with their direct and inverse

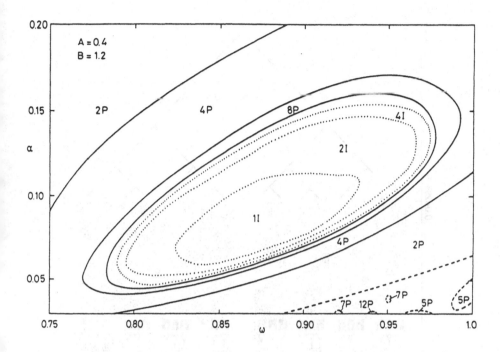

Figure 5.17: The $\alpha - \omega$ phase diagram $(A = 0.4, \quad B = 1.2)$.

parts.

This hierarchical structure is shown in Fig. 5.18, where the bifurcations at fixed frequency $\omega = 0.80$ are indicated schematically. Since the only parameter which varies now is the coupling constant α between the linear and nonlinear oscillators in (5.26), we have an intuitive interpretation of Fig. 5.18 in terms of changing regimes of nonlinear oscillations. When α is very small, the periodic driving force cannot entrain the system and the limit cycle retains its own identity. There exist two independent frequencies and one observes only a beating regime (not shown in Fig. 5.18). With α increasing, the system passes through a sharp beating-entrainment transition (horizontal dashed line in Fig. 5.18) and a period-doubling sequence comes into play, starting with period 2. Its inverse sequence merges with that of another period-doubling sequence, which is

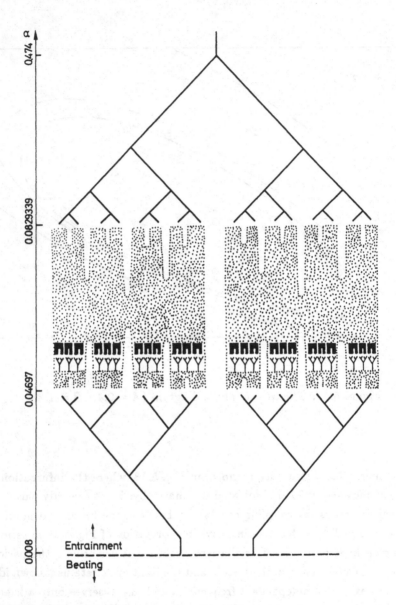

Figure 5.18: Hierarchy of period-doubling bifurcation sequences in the forced Brusselator (schematic, not to scale). Only one secondary sequence embedded in the 8I band is shown, cf. Fig. 5.19.

the manifestation of the gradual taking over of the external driving frequency. Finally, for big enough α, the latter takes over and the sequence ends with only one period of the external force. Therefore the chaotic behaviour appears as a compromise between two trends: at small α, the limit cycle tries to show itself off, at large α, the external force dominates, and chaos exists in between, as a new regime of nonlinear oscillations. This is a physical understanding of period-doubling bifurcation and the associated chaotic behaviour.

Figure 5.18 shows only one of the many secondary bifurcation sequences in the $8I$ band. A magnification of this part is given in Fig. 5.19, again schematically [the precise parameter values have been given in Table III of Hao and Zhang (1982b)]. In this figure, we see a secondary bifurcation sequence of period $8 \times 3 \times 2^n$, for $n = 0$ to 5, and its inverse sequence of chaotic bands of period $8 \times 3 \times 2^n$ for $n = 3$ to 0. At least four segments of tertiary bifurcation sequences are found embedded in the secondary $24I$, $48I$ and $96I$ chaotic bands. It is only a question of computer time and resolution power of the numerical algorithm to discover more details of this hierarchical structure. This nested bifurcation structure bears a strong resemblance to that in one-dimensional unimodal mappings. Although expected on a general theoretical basis, it is believed that Equations (5.26) represent the first case, where the existence of the hierarchy of chaotic bands in ordinary differential equations has been demonstrated by direct numerical integration.

In order to understand the global nature of the bifurcation and chaos "spectrum", we must extend the study to larger regions of the parameter space. Figure 5.20 shows the $B = 1.2$, $\alpha = 0.05$ section of the (A, B, α, ω) space, of which only a small part, namely, the upper left corner, has been shown before (Hao and Zhang, 1982b). In Fig. 5.20, the solid lines denote boundaries between periodic regions, dotted solid lines — boundaries that show intermittent transitions, dashed lines — boundaries between periodic and quasiperiodic regions, ticked solid lines — period-doubling sequences, dash-dotted lines — boundaries of unclear nature. The numbers in the figure indicate the periods. Dotted regions are chaotic with embedded periodic windows, mostly not shown explicitly. The $\alpha - \omega$ phase diagram, shown in Fig. 5.17, happens to be a perpendicular cut of the parameter space through the $A = 0.4$ line in Fig. 5.20.

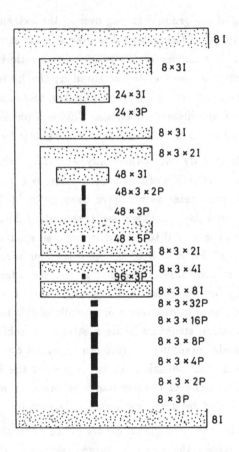

Figure 5.19: Secondary and tertiary bifurcation sequences embedded in the 8I chaotic band.

Going from the upper left corner to the lower right corner of Fig. 5.20, one encounters period-doubling sequences, beginning with periods 2, 3, 4, ... up to 8. They can be compared to the R, RL, RL^2, ... up to RL^6 words of the universal sequence of unimodal maps shown in Table 3.1 (page 118). A detailed search along the slanting line $A = 0.46 - 0.2\omega$ has exposed all but one window with periods less than or equal to 6, and they are ordered in the same way as in the logistic map. We collect in Table 5.2 the locations and widths of all these periodic windows.

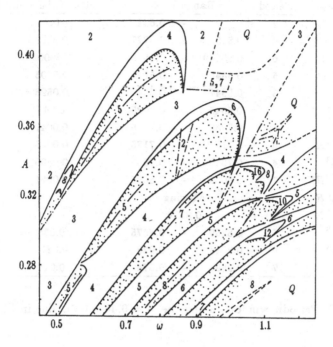

Figure 5.20: The $A — \omega$ phase diagram ($B = 1.2$, $\alpha = 0.05$). For an explanation of the symbols and lines, cf. the text.

In order to convince oneself that these periods indeed conform to the corresponding words in the U-sequence, one may proceed in the following manner. Take, for example, period 5. A detailed search along the slanting line $A = 0.46 - 0.2\omega$ in the $A — \omega$ plane has located three different period 5 orbits; we want to find the symbolic description for each of them. Since all points in the subharmonic stroboscopic sampling portrait spread along the y direction, the stretch along the x direction being much smaller, we can consider only the y_i values as the output from some unknown one-dimensional iterations. Identifying the largest y_i as the rightmost R, and attributing y_{i-1} to the central point C, because all admissible words actually begin with the letters RL, where R corresponds to the rightmost point in the interval. The central point

Word	Period	Range in ω			Width of the window
R	2		-	0.544	0.2
RLR	4	0.555	-	0.5777	0.0227
RLR^3	6	0.58249	-	0.58251	0.00002
RLR^2	5	0.5845	-	0.5848	0.0003
RL	3	0.5947	-	0.654	0.0593
RL^2RL	6	0.6545	-	0.7025	0.048
RL^2R	5	0.7068	-	0.7115	0.0047
RL^2R^2	6	0.718	-	0.7185	0.0005
RL^2	4	0.7325	-	0.792	0.0595
RL^3RL	7	0.8035	-	0.8056	0.0021
RL^3R	6	Missing			
RL^3R^2	7	0.8194			
RL^3	5	0.8259	-	0.8675	0.038
RL^4	6	0.9015	-	0.923	0.0215
RL^5	7	0.959	-	0.974	0.015

Table 5.2: Periodic windows along the $A = 0.46 - 0.2\,\omega$ line in Fig. 5.20.

may equally well be called R or L owing to the periodic window theorem (Section 3.6.4). Then all the other y_i acquire a unique assignment of letters R or L. In this way we have assigned a word to each of the observed periods. Table 5.3 shows this procedure for the period 5 orbits. This is, of course, a lucky case. In general, one has to introduce a one-parameter parametrization along the attractor to achieve a correspondence with the unimodal mappings. In order to get an idea with regard to the underlying mapping, one can plot the (y_i, y_{i+1}) or (x_i, x_{i+1}) map at a nearby parameter value, where chaotic behaviour has been known to exist. We shall see a similar construction in the next Section 5.8, when studying the Lorenz equations.

The type RLR^3, RL^2RL, RL^2R^2 and RL^4 orbits of period 6 are determined in the same way. We will now comment on the missing period 6 window, which must be of the type RL^3R. Since the widths of the three period 5 windows differ in orders of magnitude (0.038, 0.0047 and 0.0003 from Table 5.2), the width of the missing RL^3R might be much narrower than that of RLR^3, i.e.,

ω	Adjacent periodic points (x_i, y_i)					Word
0.5847	0.433,3.749	0.225,2.284	0.317,3.521	0.336, 3.548	0.285,3.248	
	R	L	R	R	C	RLRR
0.7095	0.294,3.537	0.323,3.982	0.194,1.843	0.230,3.034	0.294,3.858	
	C	R	L	L	R	RLLR
0.8480	0.211,3.364	0.245,4.036	0.285,4.333	0.190,1.478	0.187,2.492	
	L	C	R	L	L	RLLL

Table 5.3: Assignment of letters to the periodic points (the case of $5P$).

0.00002. In order to look for the missing period 6, we have located the two adjacent periods 7, but still could not find the RL^3R, using a $\triangle\omega = 0.00001$ mesh.

There is no simple relationship between a window and its width. For instance, in searching for the missing period 6, we have encountered a period 17 window ranging from $\omega = 0.8147$ to 0.81495, i.e., it was much wider than the narrowest period 6 window. In the process, we have not seen any periods between 13 and 16, which must be present in abundance nearby, due to their even narrower widths.

A striking feature of Fig. 5.20 is the bending and penetration of one period-doubling cascade into the next chaotic region. At least 5 cases of bending and three cases of penetration are resolved in Fig. 5.20. This complicated bifurcation scheme demonstrates the necessity of enlarging the parameter space while carrying out similar research. As most of the earlier studies of (5.26) were restricted to the $A = 0.4$ horizontal line segment in Fig. 5.20, which passes near the bending top of the main 2×2^n bifurcation sequence, the results could not be understood fully in the light of a general classification scheme. Even when taking a lower horizontal, say, the $A = 0.35$ line, one comes across a larger part of the primary sequences, but still cannot recognize the periods as members of the U-sequence. Now, we see clearly that the nested appearance of periods, first noticed by Tomita and Kai (1979b), is caused by cutting the bending top of the first period-doubling sequence of type R^{*n}. The bending and intersection of different period-doubling sequences will be better understood by a comparison with the mode-locking structure of circle mappings in the following Section 5.7.4, where we express the "phase diagram" in terms of a new combination of parameters (see Fig. 5.25).

The paper by Hao and Zhang (1982b) left open the question of the role and origin of the large period 3 region observed in the $A - \omega$ phase diagram. It was suspected that it might be linked to the cubic nonlinearity in the system (5.26), i.e., it might be a specific feature of the particular model. Now, it appears to be the only admissible $3P$, namely, the RL orbit, in the general scheme of the U-sequence. Looking at the isolated "bubble" labelled 3 in the original $\alpha - \omega$ phase diagram of Tomita and Kai (1979a), how should one have realized that it is connected to the unique period 3 in the framework of symbolic dynamics of two letters?

From the study of one-dimensional mappings, we know that period-doubling and intermittency are twin phenomena. In Fig. 5.20, regions in between any pair of type RL^{n-1} and RL^n periods correspond to a chaotic regime with many high order periodicities embedded in them. If one enters a given chaotic region from the RL^{n-1} side, then the transition takes place via an accumulation of period-doubling bifurcations. If one approaches the same chaotic region from the RL^n boundary, then intermittent transitions are observed (Wang, Chen and Hao, 1983). Intermittency associated with tangent bifurcations into not very long period k can easily be distinguished from ordinary chaotic or transient behaviour by their specific Poincaré maps or by plotting the (P_i, P_{i+k}) return maps. We have shown some sample maps in Section 5.3.3. The measured maximal "laminar" time and the ratio of "turbulent" time to the total time fit well into the general theory of intermittency (see Section 2.6.3) with the "mean field" value 1/2 for the corresponding exponents. Since both period-doubling and intermittency are well understood by now, we shall not go into details here.

The last question we would like to touch on briefly in this section is the coexistence of different attractors at one and the same parameter value, or, in other words, the dependence of the asymptotic behaviour of (5.26) on the initial values. Most of the results shown in Figs. 5.17 through 5.20 were obtained by integrating the differential equations from an initial point at or close to the unstable fixed point $(A, B/A)$, because we aimed to stay in one basin as long as we could. A scan of the initial value dependence of the asymptotic regimes has revealed interwoven periodic and quasiperiodic regimes. Take, for example,

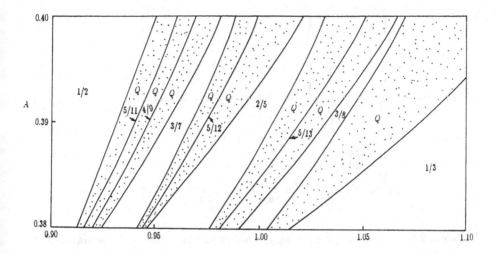

Figure 5.21: A blow-up of a small region in Fig. 5.20 calculated by using different initial values (after Yu *et al.*, 1987).

the penetrating tail of the R^{*n} cascade into the $(RL)^{*n}$ zone in Fig. 5.20. A suitable choice of initial values led to Fig. 5.21, which clearly shows a mode-locking structure, according to the Farey composition rule (Yu *et al.*, 1987). This naturally calls for a study of the transition from quasiperiodicity to chaos in the forced Brusselator.

5.7.4   Transitions from Quasiperiodicity to Chaos

Transitions from quasiperiodic regimes to chaos have been studied mainly on various circle mappings (see Chapter 4). With a detailed knowledge of the bifurcation scheme in several sections of the parameter space, the forced Brusselator (5.26) provides a suitable candidate for the study of such transitions in differential equations. We have seen in Fig. 5.20 that the RL^n bands do not extend to very large n: they get immersed in the sea of the quasiperiodic regime. Intuitively speaking, this is just the location to look for transitions to chaos. A comparison of the forced Brusselator (5.26) with the circle map [see,

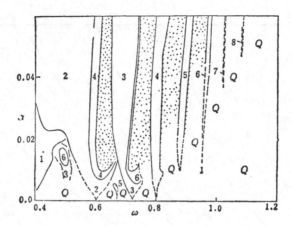

Figure 5.22: An $\alpha - \omega$ phase diagram along the $A = 0.48 - 0.2\omega$ slanting line in Fig. 5.20.

e.g., Equation (4.2)] implies that it is more appropriate to study the $\alpha - \omega$ phase diagram instead of the $A - \omega$ plane. Indeed, the $\alpha - \omega$ diagram of Tomita and Kai (1979a) did bear some resemblance to the phase-locking region of the circle mapping (see, e.g., Fig. 4.2), but, unfortunately, the chaotic region was entirely separated from the quasiperiodic regime by periodic motion. Now we know from Fig. 5.20 that the Tomita-Kai section only cut the rind of the "watermelon", and we have to dig into the pulp.

Figure 5.22 shows another $\alpha - \omega$ phase diagram, perpendicular to the $A = 0.48 - 0.2\omega$ slanting line in Fig. 5.20, i.e., parallel to the line where we studied the U-sequence in detail. We see, first of all, a competition between period-doubling and quasiperiodic motion: period-doubling cannot become fully fledged, when a quasiperiodic regime overwhelms. Many strips of chaotic behaviour open up their way into the quasiperiodic sea. If we go upstream from there, some kind of transition must be encountered. In Fig. 5.23, we show four stroboscopic portraits, obtained along the $\omega = 0.85$ vertical in Fig. 5.22. The first figure (a) gives an outline of the free Brusselator's limit cycle, because the coupling α is rather small. (There were 256 points

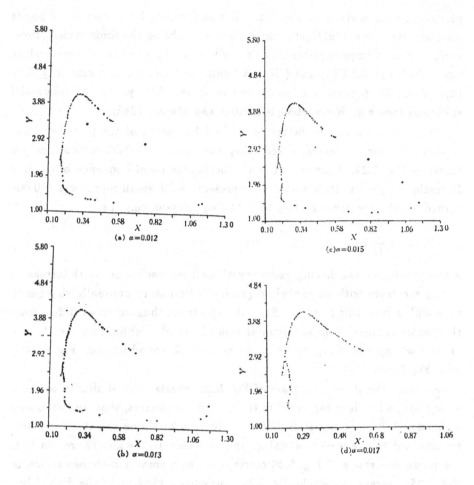

Figure 5.23: Transition to chaos from quasiperiodic regime along the $\omega = 0.85$ vertical line in Fig. 5.22. (a) $\alpha = 0.012$; (b) $\alpha = 0.013$; (c) $\alpha = 0.015$; (d) $\alpha = 0.023$.

plotted in each portrait' of Fig. 5.23. If a sufficiently large amount of points were sampled, one would get a close curve resembling the limit cycle.) However, an almost imperceptible kink on the left-hand part of the curve develops into a fold [Fig. 5.23 (b) and (c)] and results in a chaotic attractor [Fig. 5.23 (d)]. A similar process has been found in some other systems of differential equations (see, e.g., Riela, 1982; Schreiber and Marek, 1982).

This transition can be better recognized by means of the power spectra. A series of Fourier spectra, taken along the same $\omega = 0.85$ vertical line are shown in Fig. 5.24. When $\alpha = 0$, only the fundamental frequency of the free Brusselator ω_{FB} and its harmonics are present. With small coupling $\alpha = 0.008$ turned on, the spectrum is close to 3/14 frequency-locking, i.e.,

$$\omega_{FB} = \frac{3}{14}\omega,$$

where $\omega = 0.85$ is the driving frequency of the linear oscillator. With increasing α, the spectrum with an explicitly quasiperiodic nature gradually changes to that with a broad-band noise. Still, the spectrum thus obtained differs from that which resulted from an accumulation of period-doublings by the absence of regularly spaced sharp peaks in the noise background [cf.,e.g., Fig. 5.24(f) with Fig. 5.15].

Actually, the $A - \omega$ diagram of Fig. 5.20 reveals more similarity to the $\alpha - \omega$ plane, when it is expressed in the $\gamma - \beta$ parameters, that we introduced in Section 5.7.1 [see Equation (5.28)]. Figure 5.25 shows the $A - \omega$ plane, transformed to the $\gamma - \omega$ variables and extended to a wider range. In fact, the parameter range of Fig. 5.20 corresponds to a small fan-shaped region in Fig. 5.25, whereas the whole Fig. 5.25 represents a blow-up of the dashed box in Fig. 5.26, which extends to a much wider parameter range.

Figure 5.25 shows clearly the global feature which is important for an understanding of the transition from quasiperiodicity to chaos. In one region of the parameter space, one sees chaotic regimes separated by periodic zones described by the U-sequence (not necessarily made up of two letters — in principle, symbolic dynamics of more letters could be involved as well). While we have in some other region the mode-locking tongues, ordered according to the Farey sequences and embedded in the quasiperiodic sea. Since these two regimes coexist in one and the same model, they must be somehow interrelated.

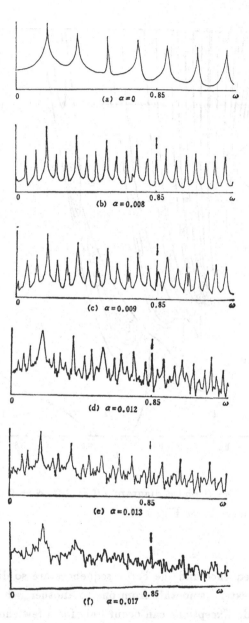

Figure 5.24: The power spectra at the transition from quasiperiodicity to chaos.

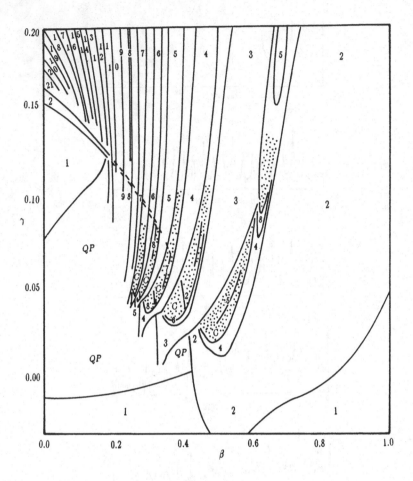

Figure 5.25: The $\gamma - \beta$ phase diagram $(B = 1.2,\ \alpha = 0.05)$. C stands for Chaos. For other legends see Fig. 5.20.

However, the U-sequence and the Farey sequences are so different in nature, that they cannot extend smoothly from one to another, when the parameters are suitably varied. Exceptions can occur only for a few short periods, where uniqueness guarantees a one-to-one correspondence between them. For example, in the U-sequence, the only period 2 and period 3 are described by the words R and RL, respectively, whereas, in the Farey sequences, there exists a unique period 2 (described by winding number $1/2$) and a unique period 3 (the winding numbers $1/3$ and $2/3$ differ only in the sense of rotation). Therefore, these periods can develop smoothly from mode-locking tongues in the quasiperiodic sea into periodic windows interspersed in the chaotic bands, as is actually shown by Fig. 5.25. It is clear that this cannot be the general rule for higher periods. For example, for period 11 there are 93 different primitive words in the U-sequence, but only 5 different rational fractions in the Farey sequence. Most of the U-sequence periods must close in themselves or terminate somewhere. What has just been said provides the physical reason for the folding and bending of the periodic zones in Figs. 5.20 and 5.25. However, the real connection between the U-sequence and the Farey sequence remains to be clarified.

On the other hand, Figure 5.25 can be viewed as a distorted picture of the $A - K$ phase diagram for the circle mapping (4.2), studied in detail, for example, by Belair and Glass (1985). It is interesting to note that, if one keeps staying within the winding number $1/2$ region, one encounters the main period-doubling sequence, passing successfully the winding numbers $2/4$, $4/8$, $8/16$, etc., but one can never form rational fractions with odd denominators, e.g., for periods 3, 5, 7, ... This is why the complete U-sequence, as seen in Fig. 5.25, is formed by invoking bent periodic sequences which originate from other winding numbers.

We continue with more observations relating to Fig. 5.26. In the upper half of this figure, one observes only mode-locked regimes, ordered strictly according to the Farey tree (see Section 4.4); all attempts to find quasiperiodic motion have failed. In fact, we have measured the rotation numbers for all the observed periods (not shown in the figure). The most clearly seen periods are the $1/n$ members along the left shoulder of the Farey tree. All values of n from 1 to 6 have been found without omissions. These periods are capable to enter smoothly the periodic windows, embedded in the chaotic sea, since they correspond to symbolic sequences of RL^{n-2} type, admissible in the symbolic dynamics of two letters.

The structure of the upper part of Fig. 5.26 is expected on the basis of quite general consideration. Since the original system of differential equations does

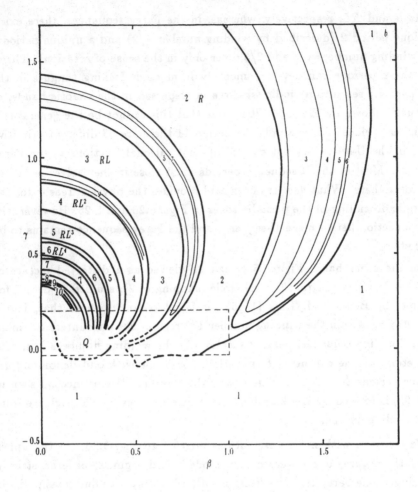

Figure 5.26: The $\gamma - \beta$ phase diagram ($B = 1.2$, $\alpha = 0.05$), extended to a wider parameter range. The dashed box corresponds to Fig. 5.25.

not change upon changing the sign of ω (i.e., β in the new representation), the whole figure should possess mirror symmetry with respect to the $\beta = 0$ vertical line (the γ axis). If the boundaries between different mode-locked regimes go straight upwards or outwards, some mode-locking zones would extend to very large values of γ, which is impossible, because the external forcing should take over. Therefore, these curves must close up at the γ axis, forming a disk-shaped region, surrounded by a 0/1 or 1/1 regime.

Figures 5.22 to 5.24 are taken from Wang and Hao (1984).

5.7.5 Homoclinic and Heteroclinic Intersections

We have learnt in Chapter 4 that the transverse homoclinic and heteroclinic intersections of stable and unstable manifolds play the role of organizing centers for chaotic behaviour and lead to the formation of Cantor set structure of the attractors. Considered already by H. Poincaré (B1899, §397, cf. Section 4.6.3) to be something too complicated to describe, these notions have been interrelated by Smale's theorem (1967). Beautiful and powerful as it may be, this theorem cannot serve as a substitute for a detailed numerical study of "real" models, where these homoclinic and heteroclinic features may be much richer and more intricate. This kind of investigations has been carried out mainly for iterated maps of the plane, e.g., Hénon's map (see Section 4.6) and its piecewise linear counterpart (Lozi, 1978; Tél, 1983), by many authors (see, e.g., Simó, 1979; Franceschini and Russo, 1981).

Although it has been predicted by the general theory, real homoclinic and heteroclinic behaviour has rarely been explored for differential equations in detail. Sparrow (B1982) studied the autonomous Lorenz system of ordinary differential equations from this point of view, however, mainly in a qualitative manner, without constructing the stable and unstable manifolds explicitly. In the case of non-autonomous systems, Holmes (1979) has investigated a system which possesses a fixed point of the saddle type, when the driven force is neglected, and has found that there exist "horseshoes" and that the parameter value of the appearance of a homoclinic orbit is close to that when chaotic behaviour is due to emerge. The existence of a saddle point in the "free" oscillator makes it easier to follow the occurrence of homoclinic intersections.

There exist, however, many forced systems, where the free limit cycle oscillator possesses no fixed point of the saddle type, but has an unstable focus somewhere inside the limit cycle. Both inside and outside the limit cycle, the solutions of the ordinary differential equations approach the cycle. In this case, Holmes' method is of little help, since one cannot start the construction of stable and unstable manifolds from the saddle point of the free oscillator.

One has to follow the formation of homoclinic and heteroclinic crossings in the Poincaré sections of the extended phase space, i.e., with the time axis included. To the best of our knowledge, a detailed study of homoclinic and heteroclinic behaviour in Poincaré sections by direct integration of such non-autonomous systems of ordinary differential equations has not yet been carried out. Perhaps, the reason is that it was believed that such a calculation might be too time-consuming. Nevertheless, owing to the simplicity and efficiency of the periodic orbit following algorithm for driven systems (see Section 5.4.1), this kind of study can be carried out even on medium-size computers. Since one can reach a deeper understanding of chaotic motion only by a combination of geometric (homoclinic and heteroclinic intersections), algebraic (symbolic dynamics), and statistical (Lyapunov exponents, dimensions and entropies) methods, we will now present the result of such a study (Ni, Tong and Hao, 1988).

In our calculation, we fix three of the parameters: $A = 0.4$, $B = 1.2$ and $\omega = 0.85$, but change α. In fact, this is the lower part of the $\omega = 0.85$ vertical line in Fig. 5.17, now studied from a geometric point of view. In order to visualize the "bifurcation diagram", we have integrated (5.26) and recorded one value of y at every multiple of the period $T = 2\pi/\omega$, throwing away a few hundred periods as transients. The y versus α plot has been shown in Fig. 1.8 (page 19). It looks much like a bifurcation diagram of the logistic map, except for the crisis that precedes the merging point into the one-band region. We summarize the observed regimes for the parameter range of $\alpha = 0.0394$ to 0.0628 in Table 5.4. In this table, the abbreviation nP stands for "Period n orbit", and nI — for "n Islands" or "n-piece chaotic attractor".

After the unstable periodic point has been located, the eigenvalues of the matrix A_{ij} are calculated

$$\lambda_{\pm} = \frac{A_{11} + A_{22} \pm \sqrt{(A_{11} - A_{22})^2 + 4A_{12}A_{21}}}{2}.$$

In fact, the stability of the orbit has been monitored by testing whether both $|\lambda_{\pm}|$ are less than unity. Numerical results reveal that the unstable periodic points $2P$ and $4P$ are indeed saddle points, and that their $\lambda_{\pm}$ are negative in the parameter range of interest. Hence the tracing of unstable periodic points can be greatly simplified. Since both the values λ_{+} and λ_{-} are negative, the

α	Attractors
0.0394 - 0.0434	Period 2 orbit (2P)
0.0436 - 0.0508	Period 4 orbit (4P)
0.0510 - 0.0532	Period 8 orbit (8P)
0.0534 - 0.0536	Period 16 orbit (16P)
0.0538	Close to transition to chaos
0.0544 - 0.0566	4-piece chaotic attractor (4I)
0.0568 - 0.0613	2-piece chaotic attractor (2I)
0.0613	A crisis
0.0613 - 0.0628	1-piece chaotic attractor (1I)

Table 5.4: Bifurcations along the $\omega = 0.85$ vertical line.

successive iterates X_1, X_2,...from the Newton or the chord method will fall alternately on two sides of the true periodic point X_0. Therefore, if the process converges, we must have $| X_2 - X_0 | < | X_1 - X_0 |$, whence $| (X_2 + X_1)/2 - X_0 | < | X_1 - X_0 |$. One can take $X_1' = (X_2 + X_1)/2$ as a new initial point in the next step, and get a new X_2' in the same way. The combined use of this interpolation scheme and the original periodic orbit tracing algorithm has led to very quick convergence. This furnishes us with a third method of calculation.

By means of the eigenvalues of the matrix A, we determine the unstable and and stable characteristic directions u and s at the saddle point in the Poincaré section. Let

$$\left| \begin{matrix} s \\ u \end{matrix} \right\rangle = \mathbf{L} \left| \begin{matrix} x \\ y \end{matrix} \right\rangle, \quad \left| \begin{matrix} x \\ y \end{matrix} \right\rangle = \mathbf{L}^{-1} \left| \begin{matrix} s \\ u \end{matrix} \right\rangle,$$

where $\mathbf{L}$ is the matrix that diagonalizes A, i.e.,

$$\mathbf{L}A\mathbf{L}^{-1} = \begin{pmatrix} \lambda_+ & 0 \\ 0 & \lambda_- \end{pmatrix},$$

$$\mathbf{L}\mathbf{L}^{-1} = \mathbf{I},$$

and

$$\mathbf{L} = \begin{pmatrix} \frac{A_{21}}{\lambda_+ - A_{11}} & 1 \\ \frac{A_{21}}{\lambda_- - A_{11}} & 1 \end{pmatrix}.$$

A point located at distance ϵ from the saddle point along the stable characteristic direction s has the coordinates

$$\left| \begin{array}{c} x_1 \\ x_2 \end{array} \right\rangle = \mathbf{L}^{-1} \left| \begin{array}{c} \epsilon \\ 0 \end{array} \right\rangle,$$

that along the unstable characteristic direction u the coordinates

$$\left| \begin{array}{c} x_1 \\ x_2 \end{array} \right\rangle = \mathbf{L}^{-1} \left| \begin{array}{c} 0 \\ \epsilon \end{array} \right\rangle.$$

In order to construct the unstable manifold M^u of an unstable periodic point O of period nP, we should integrate (5.26), starting from a certain number of initial points, spread along the unstable characteristic direction u within a small segment ϵ for time t equal to multiples of nT. However, in view of the alternate nature of the iterates due to negative values of the $\lambda_\pm$, it is better to integrate up to multiples of $2nT$, in order to to pick up one of the two branches of M^u. This situation resembles the Hénon map (Simó, 1979).

In practice, due to the closeness of X_2, obtained from the initial X_1 by integration over $2nT$, we adopt the segment $\overline{X_1X_2}$ as a part of the unstable manifold and divide it into 500 evenly spaced points, to be used as initial points for the construction of a smooth curve M^u. The unstable manifold happens to be highly compressed in one direction, so that its folding and bending show off as rapid oscillations along the stretched direction. This can be clearly seen, when one watches the movement of the plotter pen. With 500 initial points, the calculated M^u can maintain a smooth shape only for up to 10 integrations, i.e., for a total of 5000 points.

The plotting of the stable manifold M^s goes much the same way, except that one has to integrate in $-t$ instead of t; a smaller interval $\overline{X_1X_2}$ must be taken and it must be divided not evenly, but exponentially. This is the result of the exponential escape from the saddle during the backwards integration.

All the figures of the homoclinic and heteroclinic intersections, shown in this subsection, are actual numerical results, not just schematic drawings, of which one sees in plenty of publications nowadays. We emphasize this fact by indicating the exact coordinates of the four corners for each figure.

Homoclinic Intersections

Before proceeding to the presentation of the results, we explain the notation used in the figures and the text. A stable manifold of the ith point of a period n (nP) orbit is denoted by M^s_{np-i}, the unstable manifold of the jth point of a mP orbit by M^u_{mp-j}. Note that the subscripts i or j relate to the point in a cycle, not the branches of the same manifold (usually an invariant manifold has two branches emerging from an unstable fixed or periodic point). In particular, when $n = 1$, there is no need to attach an index i. One distinguishes the following possibilities:

1. Homoclinic intersections, when $M^s_{np-i} \cap M^u_{np-i} \neq \emptyset$.

2. Heteroclinic intersections, originating from different points of the same unstable periodic orbit, i.e., $M^s_{np-i} \cap M^u_{np-j} \neq \emptyset$, where $i \neq j$. We call them heteroclinic intersections of the first type.

3. Heteroclinic intersections between different unstable periodic orbits, i.e., $M^s_{np-i} \cap M^u_{mp-j} \neq \emptyset$, where $n \neq m$ (heteroclinic intersections of the second type).

We will discuss these cases in turn, starting from the homoclinic intersections.

Figure 5.27 corresponds to $\alpha = 0.056$, which is in the parameter range of $4I$, i.e., a four-piece chaotic attractor. Point O is one of the unstable periodic points of $2P$. The two branches of the unstable manifold M^u_{2p} of O oscillate rapidly along AB and CD. The folds are so flat that they look like the widening of a single curve. The manifolds M^s_{2p} and M^u_{2p} do not intersect each other, although the system has already entered a chaotic regime.

The four-piece attractors ($4I$) bands merge into two pieces ($2I$) at $\alpha = 0.0568$. Figure 5.28 shows the situation at $\alpha = 0.057$, i.e., slightly beyond the band-merging point. M^s_{2p} and M^u_{2p} have already intersected near points O, G and H, giving birth to infinitely many homoclinic points. This can be more clearly seen in the blow-up of the vicinity of the point G (Fig. 5.29). The ordinate has been enlarged approximately 200 times, the abscissa 500 times, so that the segment of M^s_{2p} looks like a straight line. Still one could resolve only a few of the folding-backs. Figure 5.30 shows the now obvious intersections of M^s_{2p} and M^u_{2p} at $\alpha = 0.058$. If one keeps on integrating the system long

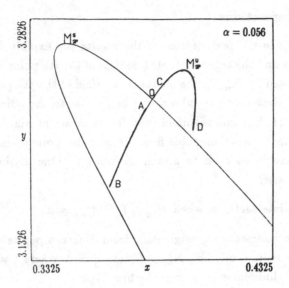

Figure 5.27: M_{2p}^s and M_{2p}^u do not intersect yet. $\alpha = 0.056$, $4I$ region.
Point O denotes one of the unstable $2P$.

enough, the two branches of M_{2p}^u will interweave with each other to outline one
of the two pieces of the $2I$ chaotic attractor.

We have found that M_{4p}^s and M_{4p}^u of the unstable $4P$ cycle begin to intersect each other and to give birth to homoclinic points, only when the system
has entered the parameter range of $4I$. This reminds us of the Hénon map,
where Simó (1979) discovered that only after $a \geq a_{c1} = 1.05$, i.e., when the
system has entered the regime of a one-piece chaotic attractor, the homoclinic
points appear. Note that the fixed point discussed by Simó is just the unstable periodic point of $1P$. From the calculations of Simó and our own, one
can draw the conclusion that in the case of a period-doubling cascade, both in
maps and in ordinary differential equations, the stable and unstable manifolds
of unstable periodic nP can intersect each other only after the system enters
the regime of nI, i.e., when there appears an n-piece chaotic attractor. This
can be explained as follows.

Suppose one of the n unstable periodic points of nP, say, the point O, is
located inside a certain piece of a chaotic attractor; then the two branches of

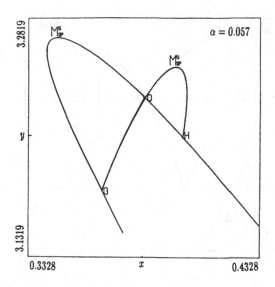

Figure 5.28: Homoclinic intersections near G, H and O. $\alpha = 0.057$, $2I$ region. O denotes one of the unstable $2P$.

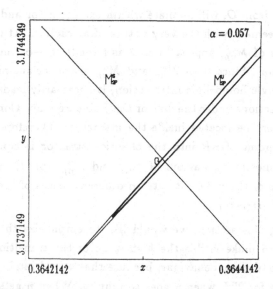

Figure 5.29: Blow-up of the neighbourhood of G in Fig. 5.28.

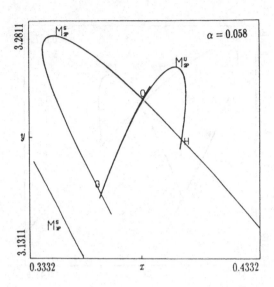

Figure 5.30: M_{2p}^s and M_{2p}^u intersect clearly at $\alpha = 0.058$ ($2I$ region).

its M_{np}^u, starting from O, will oscillate within the attractor and can never get outside it, in agreement with the very nature of attractors. At the same time, the two branches of M_{np}^s, approaching O as t tends to $-\infty$, must eventually enter the attractor. Since both M_{np}^s and M_{np}^u are invariant manifolds, once there occurs a single homoclinic intersection, it necessarily leads to an infinite number of intersections under the flow or the Poincaré map. Consequently, the point O itself must be located inside the attractor. Therefore, the entrance of one periodic point of nP into the chaotic attractor is equivalent to the occurrence of homoclinic behaviour of M_{np}^s and M_{np}^u from the same point. If n different points of the nP orbit enter n different pieces of the attractor, we must have an nI attractor.

In concluding this section, we would like to emphasize that, in practice, it is impossible to make visible the first stage of the transition to chaos by looking at the homoclinic behaviour, because that would imply a study of the Poincaré sections for $2^n I$, when n goes to infinity. When n gets large enough, the instability caused by the integration of the M_{np}^s in $-t$ becomes too severe to yield any sensible result within reasonable computing time. This explains why we have confined ourselves to $1I$, $2I$ and $4I$ bands.

Heteroclinic Intersections

We first look at the heteroclinic intersections between M^s_{np-i} and M^u_{np-j} with $i \neq j$, i.e., intersections of invariant manifolds of different points of the same cycle.

Figure 5.31 corresponds to $\alpha = 0.0608$, in the parameter range of $2I$. O_1 and O_2 denote two unstable periodic points of $2P$. We see homoclinic intersections, formed by their own invariant manifolds, i.e., M^s_{2p-1} with M^u_{2p-1} and M^s_{2p-2} with M^u_{2p-2}, separated by M^s_{1p}, the stable manifold of the unstable $1P$ point O. In order not to blur the picture, we did not draw M^u_{1p} in this figure. In Fig. 5.32, taken at the same parameter value $\alpha = 0.0608$, we omitted M^u_{2p-i} ($i = 1, 2$), but added M^u_{1p}. The latter intersects with both M^s_{1p} (homoclinic point) and M^s_{2p-i} (heteroclinic point of the second type). In fact, M^u_{2p-1} and M^u_{2p-2} are hardly distinguishable from the corresponding segments of M^u_{1p}, since they all approach the $1I$ attractor asymptotically.

At $\alpha = 0.0613$, the system enters the $1I$ regime. In Fig. 5.33, the curve M^u_{2p-1} reaches M^s_{2p-2}, and M^u_{2p-2} reaches M^s_{2p-1}, giving rise to heteroclinic intersections of the first type. The four branches of M^u_{2p-i}, emerging from O_1 and O_2, the two unstable periodic points of $2P$, interweave with each other into one piece of a chaotic attractor. In order to obtain Fig. 5.33, one has to integrate along the different branches of the unstable manifolds many times, so that the resulting M^u_{2p} becomes quite disconnected.

Now we look at heteroclinic intersections of the second type, i.e., those between M^u_{mp} and M^s_{np}, with $m \neq n$. In Fig. 5.34, only one branch of M^u_{2p}, starting from one of the unstable 2-cycle O_1, and the M^s_{4p}, emerging from one of the $4P$ unstable periodic points O, have been drawn. The parameter $\alpha = 0.055$ coreresponds to a $4I$ regime. The unstable manifold M^u_{2p} intersects with M^s_{4p} infinitely many times near E and O, and interweaves with the branches of M^u_{4p} (not shown in the figure). It tends to one of the four pieces of the $4I$ attractor $\overline{AB}$ and oscillates back and forth in between A and B. In the parameter range of $4I$, the stable manifold M^s_{2p} (also not shown) does not go through any piece of the chaotic attractor.

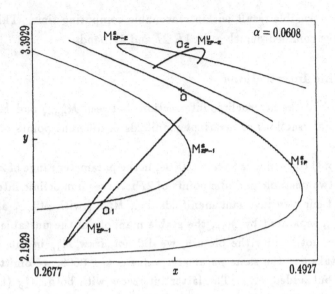

Figure 5.31: Homoclinic intersections of M^s_{2p-i} and M^u_{2p-i} at $\alpha = 0.0608$ (2I region).

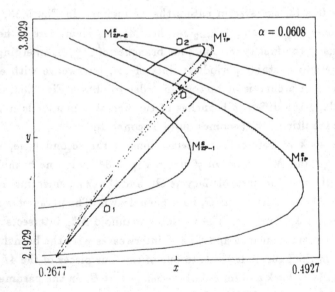

Figure 5.32: The same parameter as the previous figure, but showing M^u_{1p} instead of M^u_{2p-i}.

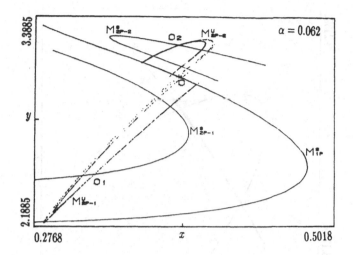

Figure 5.33: Heteroclinic intersections of $\mathcal{M}^{s,u}_{2p-1}$ and $\mathcal{M}^{u,s}_{2p-2}$ at $\alpha = 0.062$ (1I region).

Figure 5.35 is drawn at $\alpha = 0.054$, in the parameter range of 8I. This figure may be quite misleading, if one mistakes transient behaviour for long time asymptotic behaviour. The branch of $\mathcal{M}^u_{2p}$, shown as coming out from the unstable 2P point O_1, seems to interweave with the two branches of $\mathcal{M}^u_{4p}$ from O and to oscillate rapidly along two of the eight pieces of the 8I attractor, labelled $\overline{AB}$ and $\overline{CD}$ in the figure. However, the intersection of $\mathcal{M}^u_{2p}$ and $\mathcal{M}^u_{4p}$ near O, is fictitious and is due to transients. This is better seen from Fig. 5.36, drawn for the same parameter as Fig. 5.35, with the difference that the first 1600 points obtained during the construction of $\mathcal{M}^u_{2p}$ from O_1, were omitted. Now the unstable manifold $\mathcal{M}^u_{2p}$ has eventually been attracted into $\overline{AB}$ and $\overline{CD}$, dropping only a few scattered transient points near $\mathcal{M}^s_{4p}$. If a sufficiently long transient time were admitted, no points would fall outside the 8I attractor. Under a mapping from t to $t+2T$, points in $\mathcal{M}^u_{2p}$ would be translated alternately in between $\overline{AB}$ and $\overline{CD}$.

In general, two manifolds $\mathcal{M}^s_{2^k p}$ and $\mathcal{M}^u_{2^n p}$ ($k > n$) can intersect each other an infinite number of times and give rise to heteroclinic intersections only

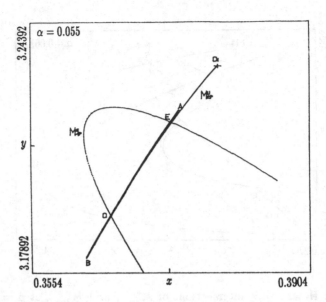

Figure 5.34: Heteroclinic intersections between M_{2p}^u (only one branch shown) and M_{4p}^s near E and O at $\alpha = 0.055$ ($4I$ region). O is one of the unstable $4P$, O_1 is that for $2P$.

when the intersections take place in a chaotic attractor, e.g., in a $2^m I$ band with $k \geq m > n$.

Periodic Solutions

We complete the discussion with a remark on the stable and unstable manifolds in the periodic regime. Take, for example, $\alpha = 0.0534$ in the parameter range of $16P$. Here the $2P$ and $4P$ periodic points have all lost stability and become saddle nodes in the Poincaré sections. We can construct M_{2p}^s, M_{2p}^u, M_{4p}^s, M_{4p}^u, etc. If one draws M_{2p}^u in a simple-minded way, it would intersect with M_{4p}^s a finite number of times. However, this does not mean that it is a heteroclinic intersection. The point is that after the transients die away each branch of M_{2p-i}^u will eventually be attracted to stable attractors of the system, namely, to four of the sixteen periodic points of the $16P$ orbit. M_{4p}^s does not meet these points at all; thus it cannot intersect M_{2p}^u an infinite number of times.

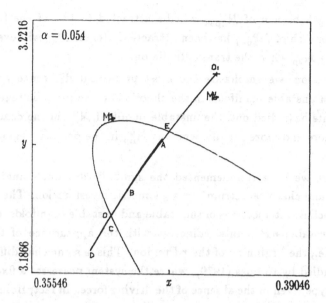

Figure 5.35: Fictitious intersections of $\mathcal{M}^u_{2p}$ and $\mathcal{M}^s_{4p}$ near E, outside the chaotic attractors $\overline{AB}$ and $\overline{CD}$, due to transients, $\alpha = 0.054$, $8I$ region.

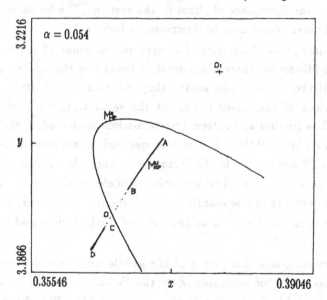

Figure 5.36: The same as the previous figure, drawn after throwing away 1600 transient points. The absence of heteroclinic intersections is clear.

Similarly, each branch of M_{4p-i}^u will be attracted to two of the four points, where one branch of M_{2p-i}^u has been attracted. Hence, M_{4p}^u will eventually be contained in M_{2p}^u when the transients die out.

In conclusion, we emphasize the most prominent difference between the behaviour of unstable manifolds in the chaotic and the periodic regimes. After the transients have died out, the unstable manifold M_{np}^u in the chaotic regime oscillates more and more rapidly, whereas M_{np}^u in the periodic region oscillates periodically.

Therefore we have complemented the symbolic dynamics analysis of the bifurcation and chaos "spectrum" by a geometric investigation. The emergence of the homoclinic intersections of the stable and unstable manifolds, originating from the unstable nP points, coincides with the appearance of the n-piece attractor, i.e., the beginning of the nI region. This is somewhat different from the case studied by Holmes (1979), where the system possesses a fixed point of the saddle type, even in the absence of the driving force. In fact, Holmes studied the nature of the first homoclinic intersection in the phase space (i.e., not in the Poincaré sections), and showed that the appearance of the intersection is equivalent to the emergence of chaos in the system. The intuitive pictures of these two different cases may be described as follows.

In the phase plane of a system of differential equations, that possess centers as well as saddle nodes, there is a separatrix formed by the stable and unstable manifolds of the saddle. The nearer they are to the separatrix, the longer are the periods of the closed orbits; at the separatrix, the period becomes infinite. When friction and external perturbation are included, the separatrix ceases to exist, but still the orbits with longer periods are located nearer to the saddle point. When a cascade of bifurcations occurs, the orbit is drawn nearer and nearer to the saddle. The unstable manifold of the saddle is attracted to the orbit, but the stable manifold stays outside. In the end, a homoclinic intersection occurs in the close neighbourhood of the saddle and chaos comes into being.

In our system, unstable points of the saddle type only appear in the extended phase space, or equivalently, in the Poincaré sections. Take, for example, the unstable fixed point O (i.e., the unstable $1P$). When a cascade of bifurcations occurs, the $2P$, $4P$, $8P$, ... points will appear like a tree on both sides of O. The unstable manifold of O is attracted to these points, but the stable manifold of O stays outside. When the system develops a 2^n-piece chaotic attractor, chaotic behaviour only appears in the neighbourhood of the $2^n P$ points. All $M^s_{2^k p}$, with $k < n$, stay outside, but all $M^u_{2^k p}$ are attracted into the $2^n I$ pieces. At the same time, all $2^m P$ points, with $m \geq n$, together with their unstable manifolds are located inside the corresponding pieces of the attractor. Homoclinic and heteroclinic intersections may occur only at these scales. When the 2^n-piece attractors merge successively into 2^{n-1}, 2^{n-21}, $\ldots$, 4, 2-pieces, the chaotic regions become larger and closer to the O point. At last, the two pieces merge to form a single piece attractor. Only at this stage, the O point with its unstable manifold enters the attractor and leads to its own homoclinic intersections.

Roughly speaking, in the first case, the homoclinic structure starts from the vicinity of the saddle, while in the second case, it culminates at the unstable fixed point. The latter situation resembles the Hénon map and the Lozi map, where the envelope of the unstable manifolds approaches the strange attractor and the stable manifolds (integrated backwards) give the outline of the basin of attraction (see, e.g., Tél, 1983).

Regarding the heteroclinic intersections of invariant manifolds, they occur only in chaotic attractors. In order to be precise, heteroclinic intersections of the first type, i.e., those between $M^u_{2^m p}$ and $M^s_{2^m p}$, can occur only in $2^n I$ bands, with $n < m$. Heteroclinic intersections of the second type, i.e., those between $M^u_{2^n p}$ and $M^s_{2^k p}$, can take place only in $2^m I$ bands, with $k \geq m > n$. M^s and M^u may intersect in the transient process a finite number of times. In order to exclude the latter in practical calculations, one has to wait for the transient to pass away.

It should be pointed out that we have studied only the most regular part of the parameter space, associated with a transition to chaos via an accumulation of period-doubling bifurcations. It would be interesting to look at more intricate parts of the parameter space, where different routes to chaos coexist. Even in the explored region, we have put aside the invariant manifolds originating from tangent bifurcations of higher orders. The nature of the crisis, which distinguishes the bifurcation diagram Fig. 1.8 from that of one-dimensional mappings, requires also further study.

5.8 Case Study 2: the Lorenz Model

In this section, we apply the symbolic dynamics approach to an autonomous system of differential equations, namely, the much-studied Lorenz model, introduced in Section 1.4. We write again this system of three ordinary differential equations (Lorenz, 1963):

$$\dot{x} = \sigma(y - x),$$
$$\dot{y} = rx - y - xz, \tag{5.32}$$
$$\dot{z} = xy - bz,$$

where σ, b and r are control parameters. The best known directions in the parameter space are the r-axis at $\sigma = 10$, $b = 8/3$ (Lorenz, 1963; Sparrow, B1982) and $\sigma = 16$, $b = 4$ (Tomita and Tsuda, 1980). Most of the early investigations of the Lorenz model have been concentrated on the segment from $r \leq 1$ to $r = 28$, where Lorenz discovered the chaotic attractor that now carries his name (Kaplan and Yorke, 1979 a and b; Yorke and Yorke, 1979). Many periodic windows have been observed for $r > 50$ (see, e.g., Sparrow, B1982; Frøyland and Alfsen, 1984). We shall continue our study for the region $r > 50$ and try to find the global systematics of stable periodic orbits.

One could have introduced a symbolic description of the orbits by means of partition in the two-dimensional $x - y$ section of the phase space, as Sparrow did in his book. However, we shall see that, when the discrete symmetry of the system 5.32 is taken into account, a one-dimensional map, i.e., the anti-symmetric cubic map (1.10) with its triples of letters will do a better job.

5.8.1 Nomenclature of Periods for the Lorenz Model

The first problem encountered, when attempting to apply elementary symbolic dynamics to the Lorenz system (5.32) is the lack of an absolute nomenclature for the observed periods. One cannot tell whether this is a period 5 and that is a period 7 orbit, because the fundamental frequency of the system drifts as the parameter varies. This is in sharp contrast to periodically driven systems, such as the forced Brusselator (5.26), where the external period is under our control and serves as a reference for the measure of all other periodic components of the motion. In addition, the presence of a controllable period opens up the

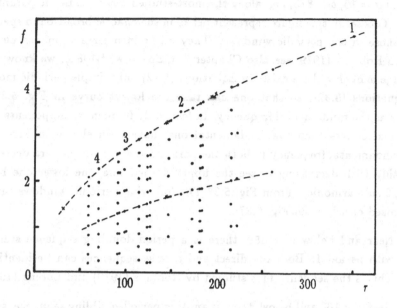

Figure 5.37: Frequency components in the power spectra vs. the parameter r

possibility of using the subharmonic stroboscopic sampling technique to attain a very high frequency resolution, comparable to that obtained for discrete mappings.

On the other hand, power spectrum analysis has been proved to be a very effective tool in the study of bifurcation and chaos in nonlinear systems. It is a little surprising that very few Fourier spectra have been published for the Lorenz model (besides, e.g., two spectra given by Lorenz himself, 1980). Even in the monograph of Sparrow (B1982), devoted entirely to the Lorenz equations, one cannot find a single spectrum. However, extensive power spectrum analysis, combined with known analytical results (Robbins, 1979) valid for large r, will provide us with a calibration curve for the fundamental frequency.

We skip here the technicalities related to power spectrum analysis (see Section 5.6) and summarize the main results in Fig. 5.37. where the location of many peaks, seen in the Fourier spectra, are plotted against the parameter r

at fixed $\sigma = 10$, $b = 8/3$, i.e., along the most-studied direction in the parameter space. Crosses in this figure represent peaks in the chaotic broad-band spectra, dots those in the periodic windows. They all fit into certain smooth curves. Due to Robbins (1979; see also Chapter 7 of Sparrow, B1982), we know that for large enough r there exists in Equations (5.32) only simple periodic motion in Equations (5.32), so that one can take the lowest curve in Fig. 5.37 at large r as the fundamental frequency. In fact, only frequency components that are equal or less than the fundamental one, have been shown in Fig. 5.37. The fundamental frequency in both the periodic and chaotic spectra decreases smoothly with decreasing r (see the upper dashed line, the lower one being the 1/2 subharmonic). From Fig. 5.37 The following periodic windows can be recognized clearly from Fig. 5.37:

1. Near and below $r = 350$, there is a period-doubling sequence starting with period 1. Both the direct and reverse sequences can be identified. This is the sequence first studied by Robbins (1979) and Lorenz (1980).

2. Near $r = 160$ and below, there is another period-doubling sequence, starting with period 2. This is the Manneville-Pomeau (1979) window.

3. Next, one has the Franceschini (1980) window near $r = 100$. It is clearly a period 3 window.

4. Near $r = 71.5$, a period-doubling sequence starts with period 4. Another sequence, starting with period 5, can scarcely be resolved near $r = 59.25$. These windows can be understood better after we clarify their systematics in the next section.

Equipped with the calibration curve of Fig. 5.37, we can assign an absolute period to each numerically observed orbit. For example, the return time, measured from a Poincaré map, corresponds to the lowest peak in the spectrum. One can refer to the curve in Fig. 5.37 to tell which subharmonic it represents.

It is remarkable that the periods measured in this way coincide with the periods that one can read off from the "bifurcation diagram" obtained directly by integration of the differential equations (5.32). We have shown this plot of the stationary x values versus the parameter r in Fig. 1.5 (see page 14). Actually, it was a long diagram cut into three pieces. Note that the abscissa has different scales in different sections, in order to expose details. Although it is quite time-consuming to get this type of bifurcation diagrams for differential equations, they are useful in the visualization of the systematics of the periodic windows. The resemblance of Fig. 1.4 to that of one-dimensional mappings is striking. This observation has motivated us to look for some map which would yield similar systematics of periods.

5.8.2 Systematics of Periods in the Lorenz Model

In order to understand the order of occurrence of different periodic windows in the Lorenz model, one has to devise some symbolic system, or more precisely, to label each periodic window with a suitable word, made up of a certain number of letters, and then to study the underlying symbolic dynamics. This is the way adopted for the study of one-dimensional mappings. However, in higher dimensional systems, the reasonable assignment of words does not seem to be so apparent. In his monograph, Sparrow has introduced a two-symbol system, using the letters x and y. An orbit is assigned a letter x, whenever it encircles the fixed point C_+, while a letter y indicates an orbit around another fixed point C_-. Thus, the Franceschini window, mentioned in Section 5.8.1, acquires the word $x^2y = xxy$.

As we have pointed out already in Chapter 3, an arbitrary partition of the phase space may help us to reduce the description of a dynamical system and to retain some of the essential features of the motion. However, a more thoughtful partition, in accordance with the "physics", may lead to better systematics of the periods. The discrete symmetry of (5.32), as well as the symmetry breaking and symmetry restoration, seen in the bifurcation diagram Fig. 1.5, suggest that the antisymmetric property may play an essential role in the systematics of the periods. Therefore, we have tried to assign a word, made up of three letters L, M and R, to each numerically observed orbit. Our method works as follows.

We start from the Poincaré maps of the Lorenz equations. The $z = r - 1$ plane has been used by many authors as the Poincaré section, since all the interesting trajectories intersect this plane and it contains both fixed points C_- and C_+. In Fig. 5.38, we show a typical Poincaré map for $r = 169.902$. Drawing the successive x_{n+1} versus x_n map will give an outline of the un-

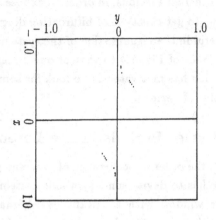

Figure 5.38: A Poincaré map at $r = 169.902$.

derlying one-dimensional mapping, if it exists. This is shown in Fig. 5.39 for the same orbit as in Fig. 5.38. It is remarkable that the resulting figure looks much like a symmetric period 16 orbit for the antisymmetric cubic map (1.10). Similar plots have been obtained for most of the windows studied by us. This observation is crucial and suggests a reasonable approach to the naming of the Lorenz periodic windows, which follows the same rule as for the cubic map in Section 3.7. According to Section 3.7, we identify the largest numerical output x_i as the rightmost R point and attribute x_{i-1} to the left critical point x_c^-. In this way, all the other x_j acquire a unique assignment of letters L, M or R. We demonstrate this procedure in Table 5.5 for two different period 5 orbits at $r = 114.00$ and $r = 83.39$, respectively.

-0.26525	0.69690	0.53309	-0.45523	0.58495
x^-	R	R	L	R
-0.04111	0.74371	0.63140	0.35347	-0.56941
x^-	R	R	R	L

Table 5.5: Assignment of letters to two different period 5 orbits.

It is hard to believe that the x-axis is the only favorable direction, in which

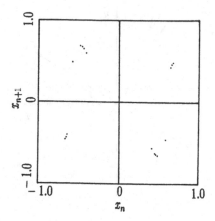

Figure 5.39: The $x_{n+1} - x_n$ map at $r = 169.902$.

the projection of the periodic orbit reveals its "cubic' nature. So we have taken a more flexible projection by introduction of an angle θ, as shown in Fig. 5.40. We found a few cases which do not look cubic in their x-projection, but restore the cubic shape at some $\theta \neq 0$. In general, one has to introduce a one parameter parametrization for the $x_{n+1} - x_n$ map, in order to make it conform with some one-dimensional map. We have made a search for periodic windows along the r parameter axis and have added a few new ones to those already known. It would be too naive to expect that all the periodic windows in a complicated system of differential equations, like that of (5.32), could be put into a simple correspondence to those of a one-dimensional map. Furthermore, it is a highly non-trivial fact that 47 out of 53 primitive (i.e., excluding those due to period-doublings) periodic windows fit into the cubic scheme. In Table 5.6, we have summarized our attempts to assign words to all the known and newly discovered primitive windows. In comparison with Table 3.10 for the cubic map, many windows were lacking, perhaps, due to their narrowness or illegality. Nevertheless, above $r = 166.01$, we have the R-region. Between $r = 92.6$ and $r = 166.01$, the RR-region can be identified. In particular, the periods which we mentioned at the end of Section 5.8.1 correspond to the words RRR, $RRRR$, $RRRRR$, etc. This ordering is characteristic for the antisymmetric cubic map

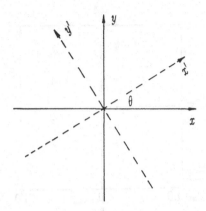

Figure 5.40: $\theta \neq 0$ axes for the projection

(see Section 3.7.3).

It is readily verified that all the words shown in Table 5.6 belong to the symbolic sequence, defined in Section 3.7. The most striking observation arising from Table 5.6 is that the order of all the words below $r = 189.549$, except the one at $r = 76.315$, is exactly the same as that of their one-dimensional counterparts along the increasing A direction. For example, consider $P_1 = RRLR$ at $r = 114.00$ and $P_2 = RRLRRL$ at $r = 107.613$. According to the ordering introduced in Section 3.7, it follows from $P_2 > P_1$ that $r_1 > r_2$, as indeed is the case. We emphasize that the parameter r in the Lorenz model is in a sense opposite to the parameter A in the cubic map (1.10). Therefore, a descending r corresponds to an increasing A. The perfect ordering below $r = 189.549$ reflects the fact that the cubic dynamics, underlying Lorenz periodic windows, depend monotonically on the parameter r, with only one exceptional orbit.

On the other hand, the stable periods in between $r = 209.45$ and $r = 191.99$ are ordered oppositely to the cubic, while the period-doubling cascades within each window still develop towards the descending r direction. The reason for this anomalous ordering is not yet clear. Asymmetric orbits and their symmetry-restored partners are frequently encountered in the table. This is

i	Period	Word	r Range
1	24		215.08-215.06
2	12		214.06-213.95
3	6	$RMRLR(\theta = 5.0)$	209.45-209.31
4	6	$RMRLR$	208.81-208.78
5	10		207.10-207.096
6	8	$RMMLRLR$	205.47-205.46
7	10	$RMMLRLRLR$	204.12-204.11
8	14*	$RMMLRLRLMMRLR(x_c^- RMMLRLx_c^+)$	200.66
9	10*	$RMMLRLMMR(x_c^- RMMLx_c^+)$	198.97-198.95
10	10*	$RMRMRLMLM(x_c^- RMRMx_c^+)$	191.99
11	5	$RMMR$	190.81-190.80
12	7	$RMMRLR$	189.549
13	16*	$RMLRLRLMLMRLRLR(x_c^- RMLRLRLx_c^+)$	187.25
14	12*	$RMLRLMLMRLR(x_c^- RMLRLx_c^+)$	185.80-185.74
15	8*	$RMLMLMR(x_c^- RMLx_c^+)$	180.97-180.93
16	10	$RMLMLMRLR$	180.127
17	10	$RMLMLMLMR$	178.075
18	12*	$RMLMLRLMRMR(x_c^- RMLMLx_c^+)$	177.81-177.78
19	6	$RMLMR$	172.712
20	16*	$RMRMRMLMLMLMLMR$ $(x_c^- RMRMRMLx_c^+)$	169.902
21	10	$RMLMRMLMR$	168.58
22	4*	$RRL(x_c^- Rx_c^+)$	166.01-146.2
23	12		145.99-145.93
24	20		144.38-144.35
25	12*	$RRLMRMLLRML(x_c^- RRLMRx_c^+)$	143.35-143.32
26	6	$RRLML$	136.79
27	10	$RRLMLLRRL$	136.21
28	16*	$RRLMLLRMLLRMRRL(x_c^- RRLMLLRx_c^+)$	135.48
29	8*	$RRLRLLR(x_c^- RRLx_c^+)$	132.04-132.03
30	16*	$RRLRLLRRLLRLRRL(x_c^- RRLRLLRx_c^+)$	129.134
31	6	$RRLRL$	126.44-126.41
32	12*	$RRLRLRLLRLR(x_c^- RRLRLx_c^+)$	123.57-123.56
33	8	$RRLRLRL$	121.687
34	5	$RRLR$	114.00-113.91
35	10*	$RRLRRLLRL(x_c^- RRLRx_c^+)$	110.506
36	7	$RRLRRL$	107.613

i	Period	Word	r Range
37	14*	$RRLRRLRLLRLLR(x_c^- RRLRRLx_c^+)$	106.74
38	8	$RRLRRLR$	104.19-104.18
39	16*	$RRLRRLRRLLRLLRL(x_c^- RRLRRLRx_c^+)$	103.637
40	3	RR	100.07-99.93
41	9		99.28-99.26
42	12*	$RRMLLMLLMRR(x_c^- RRMLLx_c^+)$	94.55
43	6*	$RRRLL(x_c^- RRx_c^+)$	92.60-92.17
44	12*	$RRRLLRLLLRR(x_c^- RRRLLx_c^+)$	90.197
45	5	$RRRL$	83.39-83.36
46	10*	$RRRLRLLLR(x_c^- RRRLx_c^+)$	82.01
47	12*	$RRRMMRLLLMM(x_c^- RRRMMx_c^+)$	76.315
48	4	RRR	71.52-71.41
49	8*	$RRRRLLL(x_c^- RRRx_c^+)$	69.718
50	5	$RRRR$	59.25-59.24
51	10*	$RRRRRLLLL(x_c^- RRRRx_c^+)$	58.71-58.70
52	6	$RRRRR$	52.456
53	12*	$RRRRRLLLLL(x_c^- RRRRRx_c^+)$	52.246

Table 5.6: Periodic windows for the Lorenz model. For the asterisk see text.

also a typical phenomenon for the cubic map (1.10), owing to its antisymmetry. At $r = 209.45$ and $r = 208.81$, there are two period 6 orbits described by the same word $RMRLR$. This can be understood, by looking at their trajectories in the phase space. If we change the sight angle θ slightly from the value given in Table 5.6 for the $r = 209.45$ orbit, then the resulting figure will look topologically the same as the $r = 208.81$ orbit.

There are a few stable periods, which cannot be put into the scheme of one-dimensional cubic map, no matter how one adjusts the projection angle θ. For the interested reader we list all these "non-cubic" words in Table 5.7. Now we are in a position to say that, apart from a few exceptions, the skeleton of the systematics for the periodic windows obeys the "cubic" law. Or, put differently, the symbolic dynamics, underlying the global periodic structure of the Lorenz model below $r = 189.549$, appears to be the same as in the antisymmetric cubic

Number as in Table 5.6	Period	Word
1	24	$(RM)^{11}R$
2	12	$(RM)^{5}R$
5	10	$(RM)^{3}MLR$
23	12	$(R^{2}LM)^{2}R^{2}L$
24	20	$x_c^- P x_c^+ \overline{P} x_c^-, \quad P = R^{2}LMR^{2}L^{2}R$
41	9	$(R^{2}M)^{2}R^{2}$

Table 5.7: "Non-cubic" words in the Lorenz model.

map. Above $r = 189.549$, most of the periods can still be described by words belonging to the U-sequence, but the parameter dependence of the windows becomes more complicated.

Next, we compare briefly our approach with that of Sparrow (B1982). The first "homoclinic explosion" at $r_0 = 13.926$ creates an invariant set, which includes an infinite number of periodic orbits. We call it the original set, in accordance with Sparrow. A lot of the periodic orbits will be liquidated by subsequent homoclinic explosions, beginning at $r_A = 24.06$. Each homoclinic explosion generates a pair of homoclinic orbits, which will leave the origin as r increases. After $r_1 = 30.1$, some of the periodic orbits may become stable and form periodic windows, and then annihilate with other orbits. Windows involving orbits coming from the original set are called normal windows, while the others are called extra windows. Using Sparrow's notation of two letters, the systematics of normal windows is that windows with more consecutive x (or y) occur at smaller r values. For example, the symmetric $x^{n+1}y^{n+1}$ window comes earlier than the $x^n y^n$ window, as r increases. In simple cases, the correspondence between the two symbolic systems is straightforward. For instance, the $x^n y^n$ symmetric orbit corresponds to the $x_c^- R^{n-1} x_c^+ L^{n-1} x_c^-$ orbit in our language, without the use of the third letter M. Therefore, the ordered R^n-regions in Table 5.6 agree with the systematics of the normal windows. The systematics of the extra windows is as yet unclear. Our symbolic system has some advantage, compared to that of Sparrow. With one additional symbol, it helps to reveal a more subtle structure of the periodic windows and makes a comparison with one-dimensional mappings possible. It cannot be excluded that one may need more letters, say 5, to explore finer details of the systematics.

In conclusion of this section, we recall that one of the reasons why Hénon introduced the now well-known two-dimensional map (Hénon, 1976) was to mimic the Lorenz attractor. However, the Hénon map does not share the same discrete symmetry as the Lorenz model. Consequently, it has become something on its own and has not brought about much understanding to the Lorenz system. A two-dimensional mapping, incorporating the antisymmetric property, would be closer to the Lorenz model.

We have described an empirical approach to the application of symbolic dynamics to numerical studies of bifurcation and chaos in autonomous ordinary differential equtions. Combined with a reconstruction of the stable and unstable manifolds and their homoclinic and heteroclinic intersections, as well as with various methods of characterization of the attractors, it may bring about a deeper understanding of chaotic dynamics in nonlinear systems. Our results have demonstrated the importance of the discrete symmetry of the model for the determination of the global systematics of the periodic solutions. There are many more unclear problems or "experimental" facts which certainly will provide more grist for the mathematician's mills.

Chapter 6

Characterization of Chaotic Attractors

Our present understanding of chaos in nonlinear systems would be unthinkable without using computers. However, there might be some serious objections to the use of computers. Anyway, on a computer with a finite word length and in a finite run time, one cannot realize a truly quasiperiodic motion, let alone chaotic trajectories. In principle, it is impossible to distinguish a very long periodic orbit from a quasiperiodic or chaotic one by merely looking at the trajectories. In addition, inevitable round-off errors place doubts on any claim that the observed erratic motion is deterministic chaos. Nevertheless, the situation is far from being hopeless, as it might appear to be at a first glance. The reliability of our knowledge on chaos gained, from computer studies, is theoretically guaranteed by some rigorous mathematical results, and, in practice, by a pursuit of a correct research strategy.

Theoretically, several theorems have been proved that guarantee the existence of true chaotic orbits close to the numerically observed ones, at least within a reasonable number of iterations (see, e.g., Hammel, Yorke and Grebogi, 1987). Although one generates only periodic "pseudo-orbits" on a computer, they are "shadowed" by true chaotic ones. We put aside these theoretical sophistications and turn to the practical aspects of the characterization of

337

chaotic motion.

A correct strategy for computer experiments and, to some extent, also for laboratory experiments, involves two consecutive steps: identification of periodic solutions and characterization of chaotic attractors.

Since periodic orbits of not very long periods are the only kind of motion that one can determine on computers with confidence, one should at first identify these periods and try one's best to explore the global systematics of these periodic solutions. As we have learnt from the previous chapters, symbolic dynamics may be of great help in this endeavour. The systematics of periodic orbits alone tells much about the location and nature of chaotic attractors.

The characterization of chaotic attractors employs such notions as Lyapunov exponents, various dimensions and entropies. Recently, this insight has brought about great success in the understanding of "real" chaos in laboratories and in mathematical models. Now, one has, at least in principle, some quantitative criteria by means of which one can distinguish between chaotic behaviour and random motion and compare different strange attractors. The present chapter will provide a minimal theoretical background and describe a few practical techniques.

We shall treat, in the first place, various notions of dimensions, placed in a more general setting than is needed for chaotic dynamics only. Next, we turn to the calculation of Lyapunov exponents and take the opportunity to show, that the secret of success in the characterization of attractors by means of digital computers lies in the invocation of tangent and phase space together. The chapter concludes with a discussion of the use of various entropies and the relationships between dimension, Lyapunov exponents and entropy.

6.1   Various Definitions of Dimensions

Intuitively, the dimension of a space is the minimal number of coordinates needed to specify in it the location of a point. Geometrical objects with integer dimensions, such as zero-dimensional points, one-dimensional lines, two-dimensional surfaces, three-dimensional bodies and four-dimensional space-time, are familiar notions. These are the topological dimensions that do not change under continuous deformation of the objects. The characterization of chaotic attractors calls for fractal, i.e., not necessarily integer, dimensions, introduced by mathematicians at the beginning of this century, but made a regular tool of science only recently. Credit must be given to B. B. Mandelbrot who coined the term fractal and did a good job of popularizing the idea of fractal geometry.[1] Originally, Mandelbrot defined a fractal as an object, the fractal dimension of which appears to be larger than the topological dimension d.

Many objects are now known to have fractal dimensions. The irregular growth pattern of aggregates or percolation backbones, the nested eddies of turbulent flow, the trajectory of a wandering Brownian particle, the fluctuating spin patterns in a magnet close to the Curie point, and, of course, our favourite chaotic attractors, all come under the notion of *strange sets* that will be characterized by means of various fractal dimensions. Generally speaking, there are three classes of geometric objects:

1. Ordinary geometrical objects, the topological and fractal dimensions of which coincide, when there is no need for the latter.

2. Regular, infinitely nested, self-similar objects, such as simple and multi-scale Cantor sets (see Figures 6.2 and 6.4 below), the fractal dimension of which exceeds their topological dimension.

3. Irregular objects with self-similarity that manifests itself in certain statistical descriptions. As a rule, the fractal dimension of these objects happens to be even larger than the corresponding topological dimension. A typical example is the trajectory of a Brownian particle, which is a continuous, but nowhere differentiable, curve of integer "fractal" dimension 2.

In chaotic dynamics, one frequently deals with objects of the last two classes.

6.1.1 The Fractal Dimension D_0

Consider a square in the plane. Let us increase its linear size to l times its original size in each direction. We get a square which is l^2 times as large. The same increase of linear sizes, when applied to a cube, leads to a cube which is

---

[1]B. B. Mandelbrot, *Fractals. Form, Chance, and Dimension*, 1977; *The Fractal Geometry of Nature*, 1982, W. H. Freeman and Co., San Francisco.

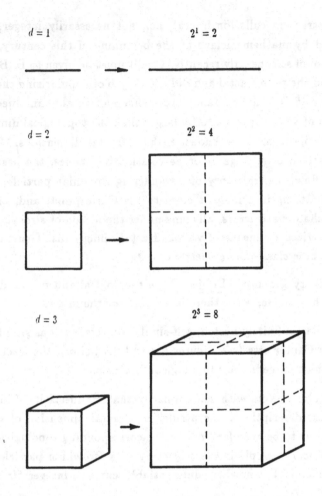

Figure 6.1: Scale change of regular geometric objects.

l^3 times as large (Fig. 6.1). In general, if we have a geometrical object A and we get an object N times larger by increasing its linear sizes to l times their originals, these numbers are simply related by

$$N = l^d, \tag{6.1}$$

where d is the dimension of the object. Taking the logarithm of both sides of (6.1), we get

$$D_0 = \frac{\log N}{\log l} \tag{6.2}$$

For reasons, which will become clear later on, we have changed the notation, attaching a subscript 0 to the capital letter D and reserving the lower case letter d for the integer topological dimension. Equation (6.2), taken as a new definition of the dimension D_0, releases us from the restriction of D_0 being an integer. D_0 is called the *fractal dimension*. To be more precise, D_0 determines the capacity of the strange set under study, and in most cases of interest for chaotic dynamics it coincides with the Hausdorff dimension. There do exist special examples in which the capacity differs from the Hausdorff dimension. It is customary to call D_0 the fractal dimension, when the capacity and the Hausdorff dimension are equal.

A classical example of a geometric object with a noninteger fractal dimension is the Cantor set (Fig. 6.2), obtained by dividing the unit interval $(0, 1)$ into thirds, discarding the middle third, and repeating the operation for the remaining intervals, and so on, *ad infinitum*. In contrast to the original continuous interval, a Cantor set is sparse everywhere. The fractal dimension is a quantitative measure of the capacity of this strange set. In order to calculate the fractal dimension of the Cantor set, take either the right or the left half of Fig. 6.2 as a unit and increase the linear size by a factor $l = 3$; thus we get $N = 2$ of the original units. Therefore, the Cantor set has a dimension

$$D_0 = \frac{\log 2}{\log 3} = 0.6309\ldots.$$

The definition (6.2) works well, whenever there is a way to count the numbers l and N, as is the case with a regular object like the Cantor set. However, when one deals with more intricate objects, such as a strange attractor, straightforward application of (6.2) may encounter difficulties. At this stage, the method of *box-counting* comes to our aid.

Any strange set S lives in an underlying space. For instance, a strange attractor exists in the corresponding phase space, and more often is seen in a two-dimensional plane of the Poincaré section. We divide the underlying space into small boxes of linear size ϵ and count the number of boxes $N(\epsilon)$ which contain at least one point of the set S. Replacing the number N in (6.2) by $N(\epsilon)$, l by $1/\epsilon$, and taking the limit $\epsilon \to 0$, we define the fractal dimension by

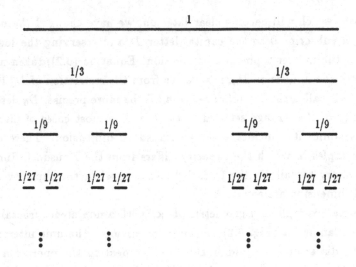

Figure 6.2: The Cantor set.

$$D_0 = - \lim_{\epsilon \to 0} \frac{\log N(\epsilon)}{\log \epsilon}. \tag{6.3}$$

In numerical practice, one has to be content with a finite number of different values of ϵ. The dimension D_0 is found from the slope of a double logarithmic plot: $\log N(\epsilon)$ versus $\log \epsilon$. Usually, one counts $N(\epsilon)$ only once for the smallest possible ϵ, dictated by the available memory, and then makes a recount, when grouping small boxes into biger ones.

Obviously, the fractal dimension D_0 does not reflect the inhomogeneity of the strange set S, since boxes containing few points of S are counted on an equal footing with those entirely filled by points belonging to S. The information dimension D_1, defined next, is designed to remedy this defect.

6.1.2   The Information Dimension D_1

The box-counting procedure allows to extract more detailed information on the structure of the strange set S. It is sufficient to keep a record of the number of times the i-th box is visited by a trajectory, speaking in terms of the measurement of strange attractors, or of the number of points of S it contains, in the general case of strange sets. This number $N_i(\epsilon)$, divided by the total number $N(\epsilon)$ of non-empty boxes, gives the probability or weight $P_i(\epsilon)$ for the i-th box:

$$P_i(\epsilon) = \frac{N_i(\epsilon)}{N(\epsilon)}.$$

Using the logarithmic measure for information

$$I(\epsilon) = -\sum_{i=1}^{N(\epsilon)} P_i(\epsilon) \log P_i(\epsilon)$$

[see Equation (6.19) in Section 6.1.5], we define the information dimension by

$$D_1 = \lim_{\epsilon \to 0} \frac{I(\epsilon)}{\log \epsilon}. \tag{6.4}$$

If every box is visited with the same probability, i.e.,

$$P_i(\epsilon) = \frac{1}{N(\epsilon)} \qquad \forall \ i,$$

then $I(\epsilon) = -\log N(\epsilon)$, i.e., the information dimension D_1 reduces to the fractal dimension D_0, or, in other words, there is no need to introduce D_1.

Although the box-counting algorithm, used above to define both D_0 and D_1, is conceptually simple and clear, it does not furnish an effective method for practical calculations (Greenside, Wolf, Swift, and Pignataro, 1982; Wang, Chen, and Hao, 1984b). The number of boxes grows exponentially with the dimension d of the underlying space, and the convergence becomes worse for higher d. Moreover, one may encounter situations, when d is not known *a priori*. In practice, this method is restricted to $d = 1$ to 3. Anyway, one needs a more effective algorithm in order to estimate the dimension of strange sets for higher values of d.

6.1.3 The Correlation Dimension D_2

The notion of correlation dimension, introduced by Grassberger and Procaccia (1983b and c), suits well experimental situations, when only a single time series is available; it is now being used widely in many branches of the physical sciences. Therefore, we shall discuss it in more detail.

Suppose we have acquired a sequence of data from a computer or laboratory experiment

$$x_1, x_2, x_3, \ldots, x_N, \tag{6.5}$$

where N is a big enough number. These numbers are usually sampled at an equal time interval $\Delta\tau$, and may be viewed as having been projected from a certain trajectory, touring an attractor of unknown dimension on to a single dimension, corresponding to the time series being measured.

Does the sequence (6.5) contain any information on the dimension of the attractor? If so, how can we extract it from these numbers? In fact, the first question was answered in the 1950's, during the early development of the theory of information. In particular, the work of A. Rényi, to be referred to in Section 6.1.5, remains nowadays a source of inspiration for those thinking of dimensions. For an answer to the second question, we require a construction known as the time-difference method to recover the phase space behind the experiments.

Since, in general, both the dimensions of the phase space and of the attractor are unknown, we try to construct with the aid of the data a space of high enough dimension m (the *embedding space*), which would allow the attractor to lie there unconstrained, using the technique suggested by Packard, Crutchfield, Farmer and Shaw (1980), as well as by Takens (1981). First of all, make a guess regarding the dimension m of the embedding space and choose a time delay $\tau = p\Delta\tau$, where p is an integer; then form a set of m-dimensional vectors with components taken from the sequence (6.5)

$$\mathbf{y}_i = (x_i, x_{i+p}, x_{i+2p}, \ldots, x_{i+(m-1)p}), \quad i = 1, 2 \ldots . \tag{6.6}$$

In this way we get M vectors. M is smaller than, but of the same order as N, if N is large enough. There are two parameters m and p in this construction. Since this phase space reconstruction technique is widely used for the calculation of other characteristics, in addition to the dimension, we shall devote a separate section (Section 6.3) to the proper choice of the parameters m and p. For the time being we simply assume that we have somehow chosen the right parameters.

Having constructed explicitly the m-dimensional vectors $\mathbf{y}_i$, one can measure the separation $|\mathbf{y}_i - \mathbf{y}_j|$ of the two vectors by means of any reasonable definition of distance. In numerical practice, the Euclidian distance is not always a good choice. A better definition of distance happens to be

$$|\mathbf{y}_i - \mathbf{y}_j| = \max_\alpha |y_{i\alpha} - y_{j\alpha}|,$$

where α probes all the components from 1 to m. In other words, the maximal difference between the corresponding components determines the distance between two vectors. This definition, combined with a preliminary sorting of the sequence (6.6), leads to a more effective numerical algorithm.

If the distance is less than a preset number ϵ, we say that these two vectors are correlated; otherwise, they are uncorrelated. Now count the number of correlated pairs among these M vectors. It is better to normalize this number by the total number of possible pairs M^2 to arrive at the *correlation integral*

$$C(\epsilon) = \frac{1}{M^2} \sum_{i,j=1}^{M} \theta(\epsilon - |\mathbf{y}_i - \mathbf{y}_j|), \tag{6.7}$$

where the unit-step, or Heaviside, function

$$\theta(x) = \begin{cases} 1 & \text{if } x > 0, \\ 0 & \text{if } x \leq 0, \end{cases}$$

simply counts the correlated pairs.

What happens, if we choose ϵ too large so that it is larger than the separation between any pair of vectors? All vectors $\mathbf{y}_i$ would be "correlated" and one would have

$$C(\epsilon) = 1,$$

or, equivalently,

$$\log C(\epsilon) = 0.$$

For a proper choice of m and not too big a value of ϵ, it has been shown by Grassberger and Procaccia (1983c) that the correlation integral $C(\epsilon)$ behaves like

$$C(\epsilon) \propto \epsilon^\nu. \tag{6.8}$$

Thus, one can define a dimension in analogy with the information dimension (6.4)

$$D_2 = \lim_{\epsilon \to 0} \frac{\log C(\epsilon)}{\log \epsilon}. \tag{6.9}$$

This is the correlation dimension. The meaning of the subscript 2 will be explained later on in Section 6.1.4.

The other extreme of too small a value of ϵ is more interesting from a physicist's point of view. Generally speaking, two mechanisms contribute to the separation between vectors: firstly, the intrinsic dynamics under study; secondly, the random noise of the ambience and the measuring apparatus. The latter factor acts on each of the m components of the vectors. Therefore, if ϵ is chosen to be smaller than the noise floor, Relation (6.8) would become

$$C(\epsilon) \propto \epsilon^m.$$

This is, of course, an artifact which, nevertheless, may be useful in telling the precision limit of our measurement.

We summarize schematically what has just been said in Fig. 6.3. There are three linear segments in this idealized figure. The horizontal segment, with zero slope, corresponds to values of ϵ that are too big. On the steeper segment with slope m, the dimension of the embedding space corresponds to too small values of ϵ. Only the intermediate segment tells about the geometrical scaling property of the strange set. It is sometimes called the *scaling region*. In practice, several different values of m must be tried, while the scaling region should remain approximately the same. The presence or absence of such a region is an objective fact, which is independent of one's desire. If it is absent, there is nothing to do but give up the attempt to calculate the dimension. Perhaps, this is an appropriate place to draw attention to the difference of how physicists and mathematicians deal with scaling properties. In idealized mathematical models, e.g., the Cantor set of Fig. 6.2, one can change the length scale infinitely many times, towards either or both zero and infinity. In physics, we are always confined to a scaling region, which is limited in both directions, where other characteristic lengths come into play. It is extremely important to be aware of the range of the scaling region, and not to push the scaling arguments beyond the limits.

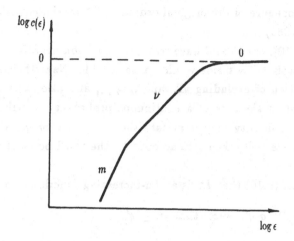

Figure 6.3: The scaling region in the determination of D_2 (schematic).

6.1.4   High Order Information Dimensions D_q

The fractal dimension D_0, the information dimension D_1, and the correlation
dimension D_2, introduced in the last few sections, turn out to be particular
cases of the q-th order information dimension D_q for $q = 0$, 1, 2, respectively.
There are at least three different points of view from which to look at D_q.

The first one is the box-counting point of view, familiar to us. Suppose we
cover the strange set S, using boxes of size ϵ, and there are $N(\epsilon)$ non-empty
boxes, the i-th box being visited with a probability p_i. Then D_q is defined by
(Hentschel and Procaccia, 1983; Grassberger and Procaccia, 1984)

$$D_q = \frac{1}{q-1} \lim_{\epsilon \to 0} \frac{\log(\sum_{i=1}^{N(\epsilon)} p_i{}^q)}{\log(\epsilon)}. \tag{6.10}$$

Although box-counting seems to be the most straightforward way for the calculation of D_q in computer experiments, the use of (6.10) is impractical, when the
dimension d of the underlying phase space gets higher, say, $d > 2$ (Greenside,
Wolf, Swift and Pignataro, 1982). The difficulty lies not only in the growth
of the box number and the memory requirement, but also in that the inhomogeneous filling up of the boxes often makes it numerically very inefficient. In
passing, we mention that the box-counting algorithm, when applied to differ-

ential equations, can be improved by taking into account corrections due to the discretization procedure of the original ordinary differential equations (Wang, Chen and Hao, 1983).

In writing (6.10), one should have restricted q to non-negative values, since otherwise zero probability terms would cause trouble. Nevertheless, we introduce the convention of excluding all vanishing p_i, and also extend q to $-\infty$. We shall see that in the case of a continuous probability distribution, zeros or infinities in the density may dominate the $q \rightarrow -\infty$ or $q \rightarrow +\infty$ limit. These situations are well taken into account by the third point of view on D_q (Section 6.2).

It follows from (6.10) that D_q is a non-increasing function of q, i.e.,

$$D_q \leq D'_q \quad \forall \; q, q', \text{ such that } q \geq q'.$$

For $q > 1$, this is a consequence of $p_i \leq 1$. For $q < 1$, an additional minus sign comes to our aid in order to keep the order unchanged. Adding the topological dimension d, we have the relation

$$d \leq D_2 \leq D_1 \leq D_0 \leq \ldots .$$

The second approach may be called a subset point of view. It requires some knowledge about the structure of the strange set S. Suppose that at the top level the set S has a length scale $l = 1$ and is being visited with probability $p = 1$. At the next level of resolution, one sees that S consists of n pieces, each having a length scale l_i with a probability p_i, where $\sum_{i=1}^{n} l_i < 1$ due to holes of different sizes within the set S, but still $\sum_{i=1}^{n} p_i = 1$, because the "physics" takes place only on the set S, not in the holes. At the third level, each of the previous pieces decomposes into n smaller pieces, each scaled down in the same manner and being visited with the same relative probability, and so on and so forth, *ad infinitum*. Figure 6.4 shows the described hierarchical structure schematically. This is an example of so-called multiscale Cantor sets. They are still quite regular objects, characterized by a triplet (n, l_i, p_i). In what follows, we shall, for the sake of brevity, call n "the number of pieces". More examples of strange sets will be encountered in our further study of chaos.

It has been argued recently (Halsey, Jensen, Kadanoff, Procaccia, and Shraiman, 1986) that D_q may be determined from the following sum rule

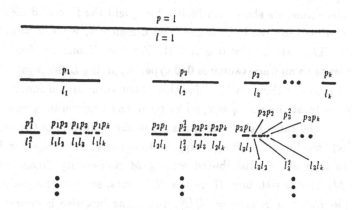

Figure 6.4: A multiscale Cantor set.

$$\sum_{j=1}^{n} \frac{p_j{}^q}{l_j{}^{(q-1)D_q}} = 1. \tag{6.11}$$

We shall give a straightforward derivation of this formula in Section 6.1.5.

The third approach may be called a singularity point of view. It employs two parameters α and f, which characterize the singularity of the measure on S: α determines the strength of the singularity and f describes how densely these singularities are populated (see Halsey *et al.*, cited above). In fact, α and f are related. Each q picks up an α from the spectrum $f(\alpha)$ and determines the dimension in the following manner:

$$D_q = \frac{1}{q-1}(\alpha q - f(\alpha)). \tag{6.12}$$

This approach has led to an important recent development of the so-called thermodynamic formalism of dynamical objects, which, in general, may have wide implications for the study of complex systems. We shall return to this new development in Section 6.2.

It is far from being obvious that the D_q, defined by Equations (6.10), (6.11) and (6.12), are one and the same quantity. Our derivation of (6.11) in the next section will demonstrate the consistency of the first two definitions. Their relation to (6.12) will be discussed in Section 6.2.1.

In the meantime, we show that (6.10) does yield the fractal dimension, the information dimension, and the correlation dimension, when $q = 0, 1$ and 2, respectively. The case of $q = 0$ is trivial. For $q = 1$, one has $\log \sum_i p_i = 0$ and (6.10) leads to an uncertainty of $0/0$ type. Applying L'Hospital's rule with respect to q recovers Formula (6.4) for the information dimension.

In order to treat the case $q = 2$, let us recall the box-counting concept with boxes of size ϵ, the same ϵ that has appeared in the definition of the correlation integral $C(\epsilon)$ [see Equation (6.7)]. Suppose that there are altogether N points from the strange set S, distributed among M non-empty boxes, numbered from 1 to M. In the i-th box, there are N_i points, so that the probability of finding a piece of S in it is $p_i = N_i/N$. Since the box size is chosen to be ϵ, all the N_i points in the box are correlated and there are N_i^2 correlated pairs. Next, one should count the number of correlated pairs, formed by the points from the neighbouring boxes, etc. Therefore, we have

$$
\begin{aligned}
C(\epsilon) \;&=\; \frac{1}{N^2} \sum_{i,j=1}^{N} \theta(\epsilon - |\mathbf{y}_i - \mathbf{y}_j|) \\
&=\; \frac{1}{N^2} \sum_{i}^{M} N_i^2 + \text{contribution from neighbours} \\
&\propto\; \sum_{i}^{M} p_i^2 + \cdots
\end{aligned}
$$

Hence, to this approximation, the quantity $C(\epsilon)$ indeed leads to D_2.

6.1.5   Derivation of the Sum Rule (6.11)

We shall derive the sum rule (6.11) by incorporating an old idea of A. Rényi[2] with some simple scaling consideration.

In order to warm up, let us look at the particular case of (6.11), taken at $q = 0$,

$$
\sum_{j=1}^{n} l_j^{D} = 1. \tag{6.13}
$$

[2] A. Rényi, *Selected Papers of Alfred Rényi*, vol. 2, p.526, Akademiai Kiodo, Budapest, 1976.

If we cover the set S by using boxes of size ϵ, the number $N(\epsilon)$ comes from the numbers of each piece, i.e.,

$$N(\epsilon) = \sum_{j=1}^{n} N_j(\epsilon). \tag{6.14}$$

If one covers the j-th piece by using smaller boxes of size $l_j\epsilon$, one would get the same number $N(\epsilon)$ of non-empty boxes, due to the self-similar structure of S:

$$N_j(l_j\epsilon) = N(\epsilon). \tag{6.15}$$

On the other hand, from the Definition (6.3) of the fractal dimension, we have asymptotically

$$N(\epsilon) \sim \epsilon^{-D},$$

whence

$$N_j(\epsilon) \sim \left(\frac{\epsilon}{l_j}\right)^{-D},$$

Inserting these relations into (6.14), we get the sum rule (6.11).

Note that in the period-doubling case $n = 2$, $l_1 = \alpha^{-1}$, and $l_2 = \alpha^{-2}$, where $\alpha = 2.5029\ldots$ is the Feigenbaum scaling factor. Equation (6.13) reduces to the golden mean equation and yields

$$D_0 = \frac{\log\left(\sqrt{5}+1\right)/2}{\log(\alpha)} = 0.524\ldots, \tag{6.16}$$

which may be compared with $D_0 = 0.538\ldots$, obtained from a more precise and elaborate calculation by Grassberger (1981), the difference being caused by "corrections to scaling", speaking in the language of the theory of critical phenomena.

We will now derive the sum rule (6.11). The derivation itself is a good review of the development of the quantitative measure of information. For the sake of clarity, we proceed in four steps.

Step 1. As early as 1928, Hartley[3] introduced the notion of the logarithmic measure of information gained when picking up an element from a set of N elements by letting

[3]R. V. Hartley (1928), *Bell System Technical J.*, **7**, 535.

$$I = \log_2 N = \log_2(1/p), \quad p = 1/N. \tag{6.17}$$

Step 2. Twenty years later, Shannon[4] considered a set S of events, consisting of n subsets S_j, to each of which the simple formula (6.17) applies, i.e.,

$$I_j = \log_2(1/p_j), \quad p_j = N_j/N, \quad N = \sum_j N_j. \tag{6.18}$$

Shannon took the weighted average of I_j to be the total information for S:

$$I = \sum_{j=1}^{n} p_j I_j = -\sum_{j=1}^{n} p_j \log_2 p_j, \tag{6.19}$$

a formula known to schoolboys nowadays.

Step 3. Rényi made a further step by applying the so-called general mean with a nonlinear law $h(x)$ to define the average, i.e., by writing

$$I = h^{-1}[\sum_{j=1}^{n} p_j h(I_j)]. \tag{6.20}$$

The general mean is a common notion in the theory of probability, e.g., we may generalize the arithmetic mean to

$$\bar{x} = \left[\frac{x_1^n + x_2^n + \cdots + x_k^n}{k}\right]^{1/n},$$

$n = 1$ being the arithmetic mean. Formula (6.20) happens to be our old friend, if we write down the canonical ensemble definition for the free energy F in terms of the microscopic energy spectrum E_j of the system as

$$F = h^{-1}[\sum_{j=1}^{n} g_j h(E_j)]$$

by taking h to be the Boltzmann factor $h(x) = e^{-x/kT}$ and g_j to be the degeneracy of the j-th level. It is not incidental that h happens to be an exponential function; Rényi has shown that, in order to retain the additivity of information for independent events, the function h has to be either exponential or linear. When it is linear, we are led back to Shannon's definition of information (6.19). Rényi took

[4]C. E. Shannon (1948), *Bell System Technical J.*, **27**, 379, 623.

$$h(x) = e^{(1-q)x} \tag{6.21}$$

to derive from (6.20)

$$I = \frac{1}{1-q} \log \sum_{j=1}^{n} p_j{}^q \tag{6.22}$$

by substituting I_j from (6.18). (We have changed to the radix e, while Rényi worked with 2. This is, of course, unessential.) In order to apply (6.22) to the box-counting situation, we change n to $N(\epsilon)$ and write $I(\epsilon)$ instead of I. Then the q-th order information dimension is defined by

$$D_q = -\lim_{\epsilon \to 0} \frac{I(\epsilon)}{\log(\epsilon)}, \tag{6.23}$$

which is nothing but the box-counting formula (6.10).

Step 4. Now consider a strange set S, characterized by a triple (n, l_j, p_j), where n is "the number of pieces", l_j is the length scaling factor, and p_j the probability of some events, e.g., visits, to occur on the j-th piece. In order to measure this set at a certain level of resolution by box-counting, one would get

$$I(\epsilon) = \frac{1}{1-q} \log \sum_{i=1}^{N(\epsilon)} t_i{}^q,$$

where we have written the probability of visits as t_i, in order to avoid confusion. Note that the index i runs from 1 to $N(\epsilon)$, the total number of non-empty boxes, and the probabilities t_i normalize to unity:

$$\sum_{i=1}^{N(\epsilon)} t_i = 1.$$

Now let us change to smaller boxes of size $l_j\epsilon$ and concentrate on the j-th piece of the set S. Clearly, the probability of visiting the i-th box becomes $s_i = p_j t_i$, and these s_i normalize to p_j, instead of 1, where p_j is the probability of the "physics" (visits) to occur on the j-th piece of the strange set. The index j runs from 1 to n, the number of pieces. To view the j-th piece as an independent object, we ought to ensure the correct normalization. Therefore, we have[5]

---

[5]An inexactitude in our riginal derivation has been corrected here.

$$I_j(l_j;\epsilon) = \frac{1}{1-q}\log\sum_{i=1}^{N(\epsilon)} t_i e^{(q-1)\log(p_j t_j)} = \frac{1}{1-q}\log\sum_{i=1}^{N(\epsilon)}\frac{(p_j t_i)^q}{p_j},$$

where we have used the scaling relation (6.15). Taking out the factor $p_j{}^q$ from the sum and changing ϵ to ϵ/l_j, we get

$$I_j(\epsilon) = -\log p_j + I(\frac{\epsilon}{l_j}). \tag{6.24}$$

Now invoke Rényi's general mean (6.20) once more over the subsets, with p_j being the weight of the j-th piece, to define the total information

$$I(\epsilon) = \frac{1}{1-q}\log\sum_{j=1}^{n} p_j e^{(1-q)I_j(\epsilon)}. \tag{6.25}$$

Note that this is where the sum over subsets comes into play. Substituting (6.24) into (6.25), we are led to

$$(1-q)I(\epsilon) = \log\sum_{j=1}^{n} p_j{}^q e^{(1-q)I(\epsilon/l_j)}. \tag{6.26}$$

Using the asymptotic form of (6.23), i.e.,

$$I(\epsilon) \sim -D(q,\epsilon)\log(\epsilon),$$

to rewrite (6.26) and assuming the existence of the limit

$$D_q = \lim_{\epsilon\to 0} D(q,\epsilon),$$

we get, after cancellation of identical terms on both sides of the equation, the equality

$$\log\sum_{j=1}^{n} p_j{}^q l_j{}^{(1-q)D_q} = 0,$$

which yields (6.11).

6.1.6 D_q for the Limiting Sets of Period-n-tupling Sequences

For simple fractals, such as the Cantor set, when there exists only one length scale l and one probability p for the "physics" taking place on it, there is no need to introduce so many dimensions. In fact, all the q-th order information dimensions D_q as well as the exponents α and f reduce to one and the same quantity. For instance, for the one-third Cantor set, we have

$$\alpha = f = D_q = \log 2/\log 3 \qquad \forall \ q.$$

If a strange set happens to be inhomogeneous, either in its geometric structure, i.e., having different l_j, or in its "physics", i.e., having different p_j, or in both, then the notion of D_q and the $f(\alpha)$ spectrum may be of some help in providing a more detailed description.

As an application of the sum rule (6.11), we calculate the q-th order information dimension D_q for the limiting sets of the period-n-tupling sequences (i.e., sequences of periods n^k, $n = 2, 3, 4, 5,\ldots$, and $k = 1, 2, \ldots$, see Section 3.4) of the logistic map (Zeng and Hao, 1986). Asymptotically, i.e., when the order k of bifurcations rises high enough, this case fits well into the scheme of a multiscale Cantor set. Namely, at every bifurcation, each of the n-piece attractor, seen at the previous level of resolution, splits into n smaller pieces. These pieces are visited with the same relative probability $p_j = 1/n$ for all j. The scaling factor l_j of the j-th piece, with respect to its "parent", can be calculated either numerically. These factors are listed in Table 6.1 for each sequence in descending order.

Now we have all the ingredients (n, p_j, l_j) for the calculation of D_q, using the sum rule (6.11) as an implicit equation. The numerical results are drawn as D_q versus q curves for various words in Fig. 6.5. For clarity only one of the $n = 6$ curves, namely, that for RL^4, is drawn in Fig. 6.5. Since D_1 cannot be calculated directly from (6.11) by putting $q = 1$, we multiply both sides of (6.11) by the product of l_j, raised to the power $(1-q)D_q$, and then take the derivative with respect to q. Now let $q = 1$ and get

$$D_1 = \frac{\sum_{j=1}^{n} p_j \log p_j}{\sum_{j=1}^{n} p_j \log l_j}. \tag{6.27}$$

This is a slight generalization of the well-known formula for the information dimension D_1 [see Equation (6.4)]; it reduces to the latter when $l_j = 1/\epsilon$ for all j. The value of D_1, calculated from (6.27), fits into the D_q curve smoothly.

n	Word	$l_j, j = 1$ to n
2	R	0.40573, 0.16929
3	RL	0.10820, 0.04214, 0.01187
4	RL^2	0.025807, 0.009369, 0.002606, 0.0006678
5	RLR^2	0.04965, 0.01939, 0.01640, 0.008021, 0.002461
5	RL^2R	0.02183, 0.008016, 0.005664, 0.001775, 0.000479
5	RL^3	0.006157, 0.00219, 0.000597, 0.000153, 0.0000385
6	RLR^3	0.04780, 0.02108, 0.00628, 0.0199, 0.008534, 0.002524
6	RL^2RL	0.00868, 0.00314, 0.00216, 0.000996, 0.000288, 0.0000755
6	RL^2R^2	0.047928, 0.019835, 0.018822, 0.009594, 0.006814, 0.002313
6	RL^3R	0.0048165, 0.0017175, 0.001143, 0.000347, 0.0000913, 0.0000232
6	RL^4	0.0015439, 0.0005468, 0.0001482, 0.0000378, 0.00000951
		0.00000238

Table 6.1: Scaling factors l_j for the period-n-tupling sequences.

All the $D_q - q$ curves, shown in Fig. 6.5, look alike. Indeed, if one rescales them by D_0, then all $D_q/D_0 \sim q$ points fall on one and the same curve (Figure 6.6).[6] This observation suggests that there exists a super-universal scaling property that goes beyond the word-dependent "universal" exponents. In fact, it would be enough to show that the combination $D_q \log \alpha_W / \log N_W$ is a function of q only, independent of the word W, which has a length N_W and a scaling factor α_W. The last statement may be proved by considering scaling property of the limiting sets of three sequences P^{*n}, Q^{*n} and $(P * Q)^{*n}$, related by the *-composition[7].

The fractal dimension D_0, calculated in this way may be compared with the results of computations, using the $q = 0$ sum rule (6.13) (Chang and McCown, 1984). In order to check the quality of the l_j, we list in the last column of Table 6.2 the sum rule values.

In connection with the l_j values, listed in Table 6.1, we make the following remark. For a given word, the smallest of the scaling factors appears to be the

[6]Cao Ke-fei, Master Thesis, Department of Physics, Yunnan University, 1988; Peng Shou-li, unpublished.

[7]Yang Wei-ming, and Zheng Wei-mou, in preparation.

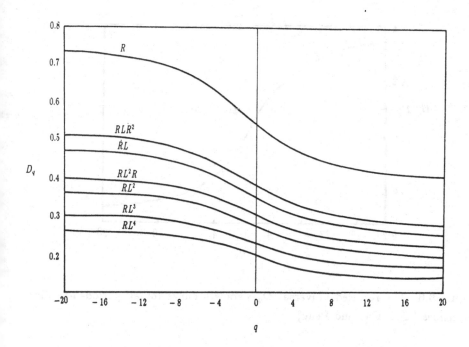

Figure 6.5: The D_q versus q curves for the limiting sets of the period-n-tupling sequences.

square of the largest one, i.e., the following equality holds

$$l_n = l_1{}^2. \tag{6.28}$$

Equation (6.28) can be derived from the renormalization group equation

$$\alpha^{-1}g(\alpha x) = g^{(n)}(x) \tag{6.29}$$

with the usual boundary conditions $g(0) = 1$, $g'(0) = 0$. We will need two relations, which follow from (6.29):

$$g^{(n)}(0) = g^{(n-1)}(1) = 1/\alpha, \tag{6.30}$$

and

$$\frac{d}{dx}g^{(n-1)}(x)\bigg|_{x=1} = \alpha. \tag{6.31}$$

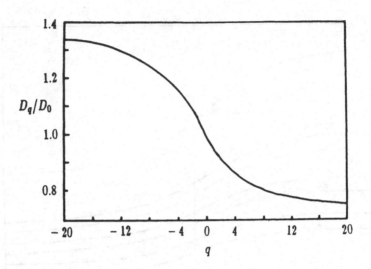

Figure 6.6: The rescaled universal D_q versus q curve for all period-n-tupling sequences (after Cao and Peng).

In order to obtain the last formula, one differentiates both sides of (6.29) and then applies L'Hospital's rule to the result

$$g'(g^{(n-1)}(x))g'(g^{(n-2)}(x))\ldots g'(g(x)) = \frac{g'(\alpha x)}{g'(x)}$$

for $x \to 0$.

The iterates $x_k \equiv g^{(k)}(0)$, $k = 0, 1, \ldots$ ($g^{(0)}(x) \equiv x$), do not repeat themselves exactly, but show a certain recurrent feature with period n. The points

$$\ldots, x_{(-n)}, x_0, x_n, \ldots$$

all lie in the largest central island, while the points

$$\ldots, x_{-(n-1)}, x_1, x_{n+1}, \ldots$$

lie in the rightmost and smallest island. To choose a characteristic length for each island, we may take the difference of some points within one and the same island. For example, we define

$$l_1 = x_n - x_0 = 1/\alpha,$$
$$l_n = x_1 - x_{-(n-1)} = 1 - g^{-(n-1)}(0).$$

When writing the last line, we have in mind the rightmost preimage of 0, which is close to 1. Now, rewriting the last line in the form

$$g^{(n-1)}(1 - l_n) = 0,$$

expanding it to the first order in l_n, and making use of (6.30) and (6.31), we arrive at (6.28). For the time being, we do not know any explicit formula connecting the intermediate l_j.

For $q \to +\infty$, the only term that dominates the sum (6.11) is associated with $l_{min} = l_n$, which yields

$$D_{+\infty} = -\frac{\log n}{\log l_{min}}.$$

In the other extreme, $q \to -\infty$, the only dominant term is $l_{max} = l_1$ and we have

$$D_{-\infty} = -\frac{\log n}{\log l_{max}} = 2 D_{+\infty}.$$

n	Word	Zeng & Hao	Chang & McCown	$\sum_j^n l_j^{D_0}$
2	R	0.53784	0.5379	1.0003
3	RL	0.35038	0.350	0.999997
4	RL^2	0.26906	0.2689	1.0002
5	RLR^2	0.38352	0.3835	1.0002
5	RL^2R	0.30290	0.3029	0.99998
5	RL^3	0.22480	0.2253	0.9999
6	RLR^3	0.42118		0.9999
6	RL^2RL	0.27344		0.9999
6	RL^2R^2	0.42089		1.0002
6	RL^3R	0.24336		1.00001
6	RL^4	0.19679		0.9999

Table 6.2: Fractal dimensions for the limiting sets of the period-n-tupling sequences.

Word	R	RL	RL^2	RLR^2	RL^2R
$D_{+\infty}$	0.3839	0.2470	0.1895	0.2860	0.2104

Word	RL^3	RLR^3	RL^2R^2	RL^3R	RL^4
$D_{+\infty}$	0.1581	0.2946	0.1887	0.1679	0.1384

Table 6.3: $D_{+\infty}$ for the period-n-tupling sequences.

The last equality in the above line comes from (6.28). Using the tabulated values of l_n in Table 6.1, we calculate $D_{+\infty}$ for various words, as shown in Table 6.3.

6.1.7   Calculation of D_q from Time Series

The idea, described in Section 6.1.3, of recovering the correlation dimension D_2 from an experimental time series, may be extended to calculate the whole D_q curve (Pawelzik and Schuster, 1987). We explain the key points.

We start from the box-counting definition (6.10) for D_q. First move the $1/(q-1)$ factor into the logarithm and write it in the form

$$D_q = \lim_{\epsilon \to 0} \frac{\log C_q(\epsilon)}{\log \epsilon},$$

where

$$C_q(\epsilon) = \left[\sum_{i=1}^{N(\epsilon)} p_i p_i^{(q-1)} \right]^{\frac{1}{q-1}} = h^{-1}\left[\sum_{i=1}^{N(\epsilon)} p_i h(p_i) \right] \qquad (6.32)$$

is a generalization of the correlation integral $C(\epsilon)$, defined by (6.7). In writing the second equality in the expression above, we have employed the general mean (6.20) with the exponential $h(x) = x^{(q-1)}$.

Recall that we have used in the box-counting formula homogeneously distributed boxes and counted the number of non-empty boxes $N(\epsilon)$. Since each of these non-empty boxes contains at least one point of the trajectory, we can concentrate on the attractor and recount them by using a natural measure on the attractor. The natural measure $\mu(x)$ may be thought of as a density function at the point x on the attractor and be normalized to unit volume, i.e.,

$$\int d\mu(x) = 1, \tag{6.33}$$

corresponding to the normalization condition

$$\sum_{i=1}^{N(\epsilon)} p_i = 1.$$

Now we can compare p_i to the measure of a small ball $B_\epsilon(x)$ of radius ϵ, centered at a point x *on* the trajectory, i.e., let p_i correspond to $\mu(B_\epsilon(x))$. Therefore, in order to give the two normalization conditions a similar appearance, it is better to present (6.33) in the form

$$\int \mu(B_\epsilon(x)) \frac{d\mu(x)}{\mu(B_\epsilon(x))} = 1.$$

Consequently, we have the correspondence

$$\sum_{i=1}^{N(\epsilon)} p_i^q \Rightarrow \int [\mu(B_\epsilon(x))]^{q-1} d\mu(x).$$

Note the change of the power from q to $q-1$, due to the change from counting homogeneneously distributed boxes to counting boxes, inhomogeneously distributed along the trajectory. Now cover the attractor by small boxes, each centered at a point $\mathbf{X}_j$ on a trajectory. Suppose there are altogether M such boxes and we can replace the integral by a sum

$$\frac{1}{M} \sum_{j=1}^{M} \tilde{p}_j^{q-1}(\epsilon),$$

where the pobability $\tilde{p}_j$ is given by the normalized number N_j of "correlated pairs" along the trajectory and seen from the given point:

$$\tilde{p}_j(\epsilon) = \frac{N_j}{M} = \frac{1}{M} \sum_{i=1}^{M} \theta(\epsilon - |\mathbf{X}_i - \mathbf{X}_j|),$$

[cf. Equation (6.7) for the correlation integral].

The generalized correlation integral (6.32) now reads

$$C_q(\epsilon) = h^{-1}[\sum_{j=1}^{M} g_j h(\tilde{p}_j(\epsilon))],$$

where the weight $g_j = 1/M$ is the same for all j. Written explicitly, we have

$$C_q(\epsilon) = \left\{ \frac{1}{M} \sum_{j=1}^{M} \left[\frac{1}{M} \sum_{i=1}^{M} \theta(\epsilon - |\mathbf{X}_i - \mathbf{X}_j|) \right]^{q-1} \right\}^{\frac{1}{q-1}}. \tag{6.34}$$

Replacing the $\mathbf{X}_i$ by the vectors $\mathbf{y}_j$ in the embedding space, reconstructed from the time series [see Equation (6.6)], one calculates D_q from (6.34).

Another method for the determination of D_q from a time series, based on calculation of unstable periodic orbits, has been suggested by Grebogi, Ott and Yorke (1987).

6.2 Thermodynamic Formalism for Multifractals

Now we turn to the third approach to high order information dimensions, namely, the singularity point of view.

6.2.1 The Singularity Point of View

The box-counting formula (6.10) can be thought of as a time average along a trajectory, because the boxes are visited successively, as the trajectory develops in time. One can transform it into a kind of "ensemble" average by considering the strange set as being made of points of different colours, where each colour contributes to the singularity in its own way.

As a parallel to the phase transition theory, we recall the block-spin picture of a ferromagnet[8]. Close to the critical temperature T_c, when the correlation length $\xi(T)$ goes to infinity, one can divide the magnet into blocks of linear size ϵ, when the limit $\xi \to \infty$ is equivalent to looking at the limit $\epsilon \to 0$ (cf. the beginning of Section 2.7.1). A block's total magnetization M scales with the block size as

$$M \propto \epsilon^x, \tag{6.35}$$

[8]L. P. Kadanoff, *Physica*, 2(1966), 263.

where the exponent x describes the singularity, which appears when ϵ goes to zero. For the averaging of M over a block, one has to divide it by the volume ϵ^d, where d is the spatial dimension of the magnet. x and d are critical exponents, the former being anomalous and the latter, normal. Compared to strange sets of more general types, phase transitions represent quite degenerate cases, where only a finite number of exponents are required for the description of the singular behaviour near the transition point.

When a strange set is composed of points of different colours, we can label these colours by an index α. Instead of the magnetization M, we consider the probability p for some "physics" to take place on the set. Again divide the set into cells of linear size ϵ. Points of colour α contribute to the singularity of p, in analogy with (6.35):

$$p \propto \epsilon^\alpha, \tag{6.36}$$

using the same notation α for the exponent of the colour α. In order to average over the "volume", we permit points of different colours to be packed into subsets of different dimensions, i.e., instead of a normal d, we use a $f(\alpha)$ for each colour α and write the corresponding volume element as $\epsilon^{f(\alpha)}$.

Now we can replace the sum along the trajectory by an integral over colours, when the index α varies continuously:

$$\sum_i^{N(\epsilon)} p_i^q \implies \int d\alpha' \rho(\alpha') \frac{\epsilon^{q\alpha}}{\epsilon^{f(\alpha)}}, \tag{6.37}$$

where we have placed primes on the integration variable and introduced the density of colours $\rho(\alpha')$, which normalizes to one and drops out from the final result. The integral (6.37) must be estimated in the limit $\epsilon \to 0$. The most significant contribution comes from the α' that makes the exponent of

$$\epsilon^{q\alpha - f(\alpha)}$$

into a minimum, i.e., that α which is determined by the solution of the problem

$$\frac{d}{d\alpha'}[q\alpha' - f(\alpha')] = 0,$$

$$\frac{d^2}{d\alpha'^2}[\alpha'q - f(\alpha')] > 0,$$

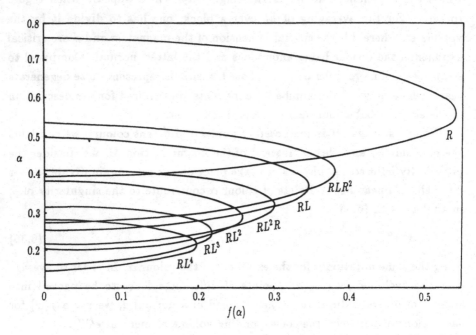

Figure 6.7: The $f(\alpha)$ spectra for the period-n-tupling sequences.

i.e.,

$$q - \frac{d}{d\alpha} f(\alpha(q)) = 0,$$

$$\frac{d^2}{d\alpha^2} f(\alpha(q)) < 0. \tag{6.38}$$

We denote it simply by α. Taking the integral in (6.37) only at the maximum of the integrand, we have

$$\sum_i^{N(\epsilon)} p_i^q \Longrightarrow \epsilon^{\alpha(q)q - f(\alpha(q))}.$$

Substituting this result back into the box-counting formula (6.10), we find

$$D_q = \frac{1}{q-1}[q\alpha(q) - f(\alpha(q))], \tag{6.39}$$

i.e., Equation (6.12). Note also that whenever $\alpha = f(\alpha)$, one always has

$$D_q = \alpha(q) \quad \text{for} \quad \alpha(q) = f(\alpha(q)).\tag{6.40}$$

Another way of writing (6.39) introduces the new notation τ_q

$$\tau_q \equiv (q-1)D_q = q\alpha(q) - f(\alpha(q)).\tag{6.41}$$

Differentiating (6.41) with respect to q

$$\frac{d}{dq}[(q-1)D_q] = \alpha(q) + [q - \frac{df(\alpha(q))}{d\alpha}]\frac{d\alpha(q)}{dq},$$

we see that the last term vanishes due to the first conditions (6.38), and we are led to

$$\alpha(q) = \frac{d\tau_q}{dq} = \frac{d}{dq}[(q-1)D_q].\tag{6.42}$$

Equations (6.39) and (6.42) may be used in two ways. Either the "spectra" $f(\alpha)$ and $\alpha(q)$ are known, then one calculates D_q from (6.39), or when the D_q — q dependence is known, one determines first of all $\alpha(q)$ from (6.42) and then gets $f(\alpha(q))$ from (6.39).

As an example, we show the $f(\alpha)$ versus α curves for the limiting sets of the period-n-tupling sequences, calculated from the known D_q versus q curves shown in Section 6.1.6 (Figure 6.5). These $f(\alpha)$ spectra are drawn in Fig. 6.7 (Zeng and Hao, 1986). Note the seemingly monotonic dependence of the location of these curves on the order of words in the symbolic dynamics of two letters. Furthermore, all these curves may be brought into one curve by rescaling them as

$$\frac{\alpha(q)}{\alpha(0)} \sim \frac{f(\alpha(q))}{f(\alpha(0))}$$

curves, as shown in Fig. 6.8 (see the footnote on page 356).

The $f(\alpha)$ — α dependence has been measured from the forced Rayleigh-Bénard convection experiment (Jensen, Kadanoff, Libchaber, Procaccia and Stavans, 1985; Glazier, Jensen, Libchaber and Stavans, 1986). For a model of the one-dimensional quasicrystal, the $f(\alpha)$ spectrum has been calculated analytically (Zheng, 1987a).

In most cases, the $\alpha(q)$ versus q, or conversely, the $q(\alpha)$ versus α, as well as the D_q versus q dependences are smooth functions. Figures 6.5 and 6.7 are

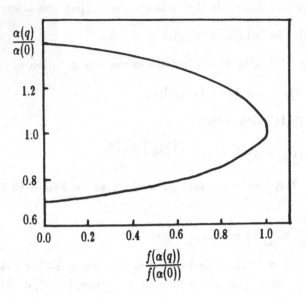

Figure 6.8: The rescaled universal $f(\alpha)$ spectrum for the period-n-tupling sequences (after Cao and Peng).

typical in this sense. It has become customary in the literature to draw the $f(\alpha)$ versus α relationship, instead of the $\alpha(f)$ versus f dependence, as we did in Figs. 6.7 and 6.8. A typical $f(\alpha)$ curve is shown in Fig. 6.9, where we have labelled the ordinate by y. The auxiliary construction, involving a dash-dot line, will be explained later on.

A few general properties of this curve follow from the earlier formulae:

1. f is a convex function, as $f'' < 0$.

2. Since $df/d\alpha = q$, $f(\alpha)$ is increasing for $q > 0$ and decreasing for $q < 0$.

3. $D_0 = f(\alpha(q))\,|_{q=0}$ occurs at the maximum point A of the $f(\alpha)$ curve.

4. The $f(\alpha)$ versus α curve contacts the bisector $f = \alpha$ at $q = 1$ (point B in Fig. 6.9). At this point, we have the equality of the three characteristics: $D_1 = \alpha(1) = f(\alpha(1))$.

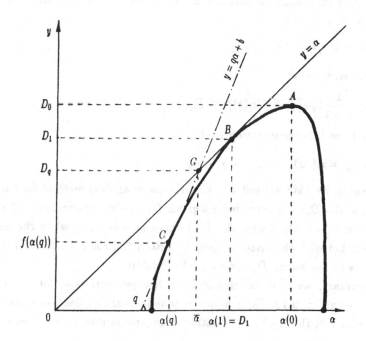

Figure 6.9: A typical $f(\alpha)$ curve and graphical calculation of D_q.

5. Since $df/d\alpha = \pm\infty$ at $q = \pm\infty$, if the $f(\alpha)$ curve intersects the α axis, then it intersects vertically at $\alpha(\infty)$ and $\alpha(-\infty)$.

We use Fig. 6.9 to show how to determine D_q by graphical construction. Draw a straight line of slope q

$$y = q\alpha + b$$

(the dash-dot line) and move it to contact the $f(\alpha)$ curve at the point C. This point determines a particular $\alpha(q)$ and the corresponding $f(\alpha(q))$ shown in the figure, thus fixing the value of b to be $f(\alpha(q)) - q\alpha(q)$. Consequently, the equation of the dash-dot line reads

$$y = f(\alpha(q)) + q(\alpha - \alpha(q)).$$

Now this line intersects the bisector at the point G, where y and α have the same value $\overline{\alpha}$, i.e.,

$$\overline{\alpha} = f(\alpha(q)) + q(\overline{\alpha} - \alpha(q)).$$

Solving for $\overline{\alpha}$, we find

$$\overline{\alpha} = \frac{1}{q-1}(q\alpha(q) - f(\alpha(q))) \equiv D_q.$$

It follows from this construction that

$$D_{+\infty} = \alpha(\infty), \quad D_{-\infty} = \alpha(-\infty) \ .$$

Therefore, we have arrived at a very simple graphical method for the determination of the $D_q - q$ curve from a given $f(\alpha) - \alpha$ dependence: take a ruler and bring it into contact with the $f(\alpha)$ curve at a slope q, when the height of its intersection with the bisector gives D_q. Rotating the ruler around the $f(\alpha)$ curve leads to the entire D_q versus q relationship.

In summary, we show schematically a few possible cases of the $D_q - q$ dependence in Fig. 6.10. For a homogeneous fractal set, like the classical one-third Cantor set, the $f(\alpha) - \alpha$ relation degenerates into a single point on the bisector. Therefore,

$$D_q = \alpha = f(\alpha) \quad \forall \ q,$$

and there is no need to introduce so many "dimensions". This is shown in Fig. 6.10(a). Figure 6.10(b) shows a general smooth $D_q - q$ curve constructed by the aforementioned method from a smooth $f(\alpha) - \alpha$, like the one shown in Fig. 6.9.

When the smooth dependence of D_q on q (or $f(\alpha)$ on α) ceases to exist at some point, a "phase transition" is said to have taken place, in analogy with the theory of phase transitions and critical phenomena [Figure 6.10(c)]. We shall show a few simple examples in the next section.

6.2.2 Phase Transitions in the Thermodynamic Formalism

In order to obtain a deeper feeling for the $f(\alpha) - \alpha$ and $D_q i - q$ relationship, let us calculate the D_q dependence for the fully developed chaotic regime at the

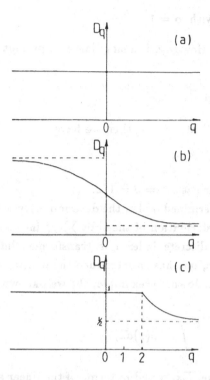

Figure 6.10: Different possibilities for D_q (schematic): (a). A single-scale homogeneous Cantor set. (b). A multiscale Cantor set. (c). A "phase transition".

RL^∞ point, namely, at $\mu = 2$ for the logistic map. In this case, the invariant density is known explicitly to be

$$\rho(x) = \frac{1}{\pi\sqrt{1 - x^2}}. \tag{6.43}$$

[see Equation (2.57) in Section 2.5.4]. Take a small segment $(x_0 - \epsilon, x_0 + \epsilon)$ in the vicinity of x_0; the probability that a point of the orbit falls within this segment is

$$p(x_0) = \int_{x_0-\epsilon}^{x_0+\epsilon} \rho(x)dx.$$

If $\rho(x)$ does not have any singularity in this segment, we simply get

$$p(x_0) \propto \epsilon^\alpha \quad \text{with} \quad \alpha = 1, \tag{6.44}$$

[cf. Equation 6.36)]. However, if a singularity is present at a particular point x_0, e.g.,

$$\rho(x) \propto |x - x_0|^{-\beta}, \tag{6.45}$$

as is the case in (6.43) at $x = \pm 1$, then we have

$$p(x_0) \propto \epsilon^{1-\beta}, \tag{6.46}$$

i.e., $\alpha = 1 - \beta$. In our case, $\alpha = \beta = 1/2$.

So far we have determined α. For the determination of the other exponent f, we consider the transformation of the sum $\sum_i p_i^q$ into an integral. In general, when a sum over a discrete index i is transformed into an integral over a continuous variable x, one has to introduce the density or measure $\rho(x)$ with respect to x and write down, for example, the well-known relation of solid state physics:

$$\frac{1}{(2\pi)^3} \sum_i \ldots = \frac{1}{V} \int \ldots \rho(x) dx, \tag{6.47}$$

where V is the volume. Expressed in terms of the linear size l, one has $V = l^d$, where d is the dimension. If the density $\rho(x)$ does not contain a singularity, this is all that one needs. However, when the density distribution has singularities, concentrated in low dimensional regions, say, at some points in $\rho(x)$ or along some lines in a two-dimensional density $\rho(x, y)$, one must be more cautious. Although these regions have zero volume compared to V, their contribution to the sum may not be neglected because of the presence of singularities. Therefore it is better to separate them from the sum. For point-wise regions, the corresponding terms in the sum may be left on the right-hand side of (6.47), i.e., we have $d = 0$ at these points. For lines of singularities, we should write down a one-dimensional integral with $d = 1$ on the right-hand side of (6.47), and so on. These d values play the role of f in the thermodynamic formalism.

For the logistic map, the underlying space is an interval and the density is continuous on this interval, except for the two end points, where a singularity

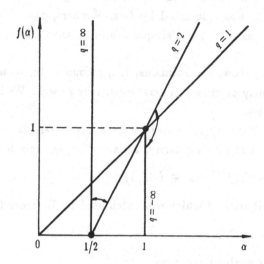

Figure 6.11: The $f(\alpha)$ — α dependence for the surjective logistic map reduces to two isolated points.

with $\beta = 1/2$ emerges. Therefore we have $f = 1$ and $\alpha = 1$ [see Equation (6.44)] at all regular points. At the two end points, however, we have $f = 0$ and $\alpha = 1/2$ [see Equation (6.46)]. Therefore, the $f(\alpha)$ versus α curve shrinks into two isolated points in the f — α plane, as shown in Fig. 6.11.

We use the graphical construction for the calculation of D_q. Since the $f(\alpha)$ curve has now degenerated into two points, there are only two possibilities:

1. For $q = \infty$ to 2, all the straight lines touch the point $(\alpha, f) = (1/2, 0)$, when Equation (6.39) leads to

$$D_q = \frac{q}{2(q-1)}.$$

2. For $q = 2$ to $-\infty$, all lines touch the point $f = \alpha = 1$, consequently, Equation (6.40) yields $D_q = \alpha = 1$.

We thus have

$$D_q = \begin{cases} 1 & q = -\infty \text{ to } 2, \\ \frac{q}{2(q-1)} & q = 2 \text{ to } \infty. \end{cases} \tag{6.48}$$

In fact, this is the D_q versus q curve, drawn in Fig. 6.10(c).

Formula (6.48) was calculated by Ott, Withers, and Yorke (1984). The above derivation, as well as the simple examples that follow, are due to Yang Wei-ming[9].

Figure 6.10(c) shows a continuous, i.e., second order phase transition. In principle, there may be discontinuous transitions as well. We look at a number of simple examples.

Example 1. If β in (6.45) is negative, we have a zero in the density instead of an infinity. It is also a singularity. As an example, consider the map

$$x_{n+1} = 1 - 2|x_n|^{1/2} \quad x \in (-1, 1),$$

the density distribution of which was calculated by Hemmer (1984) to be

$$\rho(x) = (1 - x)/2.$$

According to our analysis, we have

$$\alpha = f = 1$$

for regular points, and

$$\beta = -1, \quad \alpha = 2, \quad f = 0,$$

at the end point $x = 1$. The graphical construction yields

$$D_q = \begin{cases} \dfrac{2q}{q-1} & -\infty < q \leq -1/2, \\ 1 & -1/2 < q < \infty. \end{cases}$$

Example 2. If both infinity and zero are present in the density $\rho(x)$, one may get two consecutive transitions. For instance, if, in addition to the $\beta = 1/2$ endpoint singularities, there is a zero

$$\rho \propto |x - x_0|^{\tilde{\alpha}}$$

near x_0, then we have

$$D_q = \begin{cases} \dfrac{q(1+\tilde{\alpha})}{q-1} & -\infty < q \leq -\tilde{\alpha}^{-1}, \\ 1 & -\tilde{\alpha}^{-1} \leq q \leq 2, \\ \dfrac{q}{2(q-1)} & 2 \leq q < \infty. \end{cases}$$

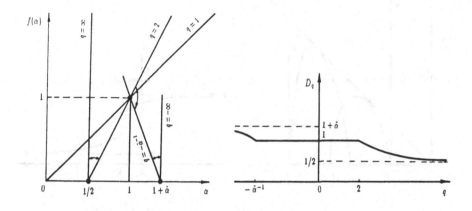

Figure 6.12: Two consecutive phase transitions in the D_q versus q dependence (schematic).

The graphical construction is shown in Fig. 6.12. Several exactly solvable mappings, listed by Katsura and Fukuda (1985), may lead to this kind of transitions in their D_q dependence.

Example 3. For the generation of a discontinuous transition, we take a "composite" strange set, represented by two shifted $f(\alpha)$ curves, shown in the left figure of Fig. 6.13. The graphical construction leads to the $D_q - q$ curve shown on the right. There is one point worth mentioning. It follows from the graphical construction that, when more than two $f(\alpha)$ curves are present, only the peripheral segments determine the form of the D_q curve. The information dimension D_1, however, is determined by all the $f(\alpha)$ curves. Suppose three curves $f^{(a)}$, $f^{(b)}$ and $f^{(c)}$ with their own $D_1^{(a)}$, $D_1^{(b)}$ and $D_1^{(c)}$ comprise the composite set; it is readily shown that D_1 may be calculated from the weighted reciprocals:

$$\frac{1}{D_1} = g^{(a)} \frac{1}{D_1^{(a)}} + g^{(b)} \frac{1}{D_1^{(b)}} + g^{(c)} \frac{1}{D_1^{(c)}},$$

where

[9]Yang Wei-ming, unpublished.

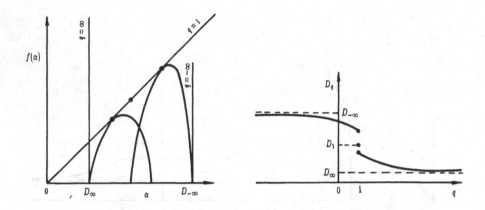

Figure 6.13: A "composite" strange set (schematic).

$$g^{(a)} = \frac{\sum p_i^{(a)} \log p_i^{(a)}}{\sum p_i^{(a)} \log p_i^{(a)} + \sum p_i^{(b)} \log p_i^{(b)} + \sum p_i^{(c)} \log p_i^{(c)}} ,$$

etc. This relation generalizes naturally to more components. For its derivation, we have made use of (6.27).

Our remark concerning D_1 has some experimental consequence. If a time series has covered too long a period, so that the data actually come from several different attractors, and one tries to calculate D_q, what will be the result? All D_q, with $q < 1$, will be determined by the rightmost piece (or pieces) of the $f(\alpha)$ curves, while all D_q with $q > 1$, determined by the leftmost piece (pieces); only D_1 will be given by the weighted average of all D_1, as shown above[10].

What is the moral which we can learn from these simple examples? A smooth $f(\alpha)$ emerges, when there is a multiscale fractal structure in the attractor, e.g., in the case of the accumulation points of period-doubling or period-n-tupling sequences, or, in other words, at the borderline to chaos. As regards fully chaotic motion, corresponding to RL^∞ or $\rho\lambda^\infty$ in the sense of the generalized composition rule (Section 3.6.6), the $f(\alpha)$ dependence degenerates into

---

[10]Ling Fu-hua has shown that D_0 tends to the largest value and D_2 to the smallest one, but D_1 is undetermined — private communication via Xu Jing-hua.

a few isolated points in the $f - \alpha$ plane, giving rise to discontinuities in the D_q curve.

6.3 Time-difference Method for the Reconstruction of Phase Space Dynamics

In Section 6.1.3, we have used the time-difference method to calculate the correlation dimension D_2. In fact, the reconstruction of phase space dynamics from a time series is a crucial step in the characterization of the "experimental" attractors and serves as a common start for the calculation of dimensions, Lyapunov exponents and entropies. Recently, much progress has been achieved in the implementation of this technique. Therefore we devote this section to a more detailed discussion of the subtleties involved.

We have already known from the discussion on power spectra in Section 5.6 that a proper choice of the sampling interval $\Delta \tau$ and of the length of the series N are two decisive factors for the success of an analysis. In order to span the embedding space, we must choose an additional parameter — the embedding dimension m. Usually, a time series

$$x_1, x_2, \ldots, x_i, \ldots, x_N$$

is sampled at equal time intervals $\Delta \tau$. For the formation of the components of the m-dimensional vectors in the embedding space, we select points separated from each other by a time difference $p\Delta \tau$:

$$\mathbf{y}_i = (x_i, x_{i+p}, x_{i+2p}, \ldots, x_{i+(m-1)p}), \quad i = 1, 2 \ldots, M, \qquad (6.49)$$

where $M < N$, but of the same order as N, provided N is big enough. Prior to a discussion of the choice of the parameters m and p in the following two sections, we emphasize that the very possibility of a recovery of phase space dynamics has its root in the presence of the strong nonlinearity that mixes all degrees of freedom. Otherwise, a single measured signal cannot contain information on other "modes".

6.3.1   On the Choice of the Embedding Dimension

The suggestion by Packard *et al.* (1980) and Takens (1981) that the time-difference method be used for the reconstruction of the phase space was based on the topological embedding theorem of Whitney[11]. The general idea of this embedding theorem is easily explained.

Take an arbitrary one-dimensional curve and confine it to a two-dimensional surface. Generally speaking, the curve may intersect itself at some points of the surface. These intersections cannot be removed by means of small deformations. Consequently, a general curve, confined to a surface, does not feel "free", being forced to touch itself somewhere. In contrast, if a curve is placed in a three-dimensional space, then all self-intersections may be removed by infinitesimal perturbations. Hence these intersections may be considered to be incidental. Now let us try to generalize this observation to geometrical objects of an arbitrary dimension.

Suppose we have two objects A and B with dimensions d_A and d_B, respectively. They both live in an underlying space of dimension d. How do we find the dimension of the intersection $A \cap B$ of these two objects? For this purpose, we define the "codimensions"

$$cod_A = d - d_A,$$
$$cod_B = d - d_B,$$
$$cod_{A \cap B} = d - d_{A \cap B}.$$

(Note that the usage of the term codimension here seems to be different from the application in other contexts, e.g., at the end of Section 4.2. Nevertheless, they are essentially the same, being the number of one's degrees of freedom to extend the dimensionality of an object or a parameter space.) The codimensions of intersections have the nice property of additivity:

$$cod_{A \cap B} = cod_A + cod_B, \tag{6.50}$$

which is a particular case of the general formula

$$cod_{\cap_i A_i} = \sum_i cod_{A_i}.$$

[11]H. Whitney, *Ann. Math.*, **37**(1936) 645.

It follows from (6.50) that

$$d_{A \cap B} = d_A + d_B - d. \tag{6.51}$$

The correctness of (6.51) may be checked by a few examples. Put two curves $(d_A = d_B = 1)$ in a plane $(d = 2)$; generally speaking, they will intersect at points, since according to (6.51) $d_{A \cap B} = 0$. However, they miss each other in a three-dimensional space, as $d = 3$ leads to $d_{A \cap B} = -1$, whereas a surface $(d_A = 2)$ and a curve $(d_B = 1)$ may intersect at points, when $d = 3$, etc. All these observations must be understood in a generic sense, excluding specially posed cases, e.g., two parallel lines A and B.

Now apply (6.51) to one and the same object $A = B$. We wish it to live in the d-dimensional space "comfortably", i.e., any part of A may stretch itself without being forced to touch other parts of A somewhere else. Since $d_{A \cap A}$ may be a misleading notation, we simply require the left-hand side of (6.51) to assume the value -1. Therefore, the embedding space must have the dimension

$$d = 2\,d_A + 1. \tag{6.52}$$

This is usually more than enough. The embedding dimension (denoted by m in Section 6.1.3 and at the beginning of Section 6.3) must be at least twice the dimension of the attractor, in order to allow for an unbiased reconstruction. In practice, m should be chosen by trial and error. When m is large enough, the scaling region in Fig. 6.1.4 will remain the same, regardless of the particular m.

At present, the phase space reconstruction technique is being used to calculate other characteristics of chaotic motion. Generally speaking, the minimal dimension of the embedding space depends on what kind of quantity one tries to recover from the time series. The trajectory (or the phase portrait, as it is sometimes called) seems to be the most elaborate object, one might really require $d \geq 2d_A$ to achieve it. If one is concerned with Lyapunov exponents or fractal dimensions — something averaged along the trajectory, not so elaborate as the trajectory itself — one may need $d_A < d < 2\,d_A$.

6.3.2   On the Choice of the Time Difference

For the choice of the other parameter (the integer p or rather the delay time $\tau = p\Delta\tau$, see the beginning of Section 6.3) in the phase space reconstruction scheme, Fraser and Swinney (1986), at the suggestion of R. Shaw, have worked out a criterion which is based on a search for the time τ that produces the first local minimum of so-called mutual information. Although the application of this criterion involves some computational details, the basic idea can be explained in a few words.

If we have chosen p too small, then x_i and x_{i+p} will depend strongly on each other. In order to make the chosen data points "independent" components of a vector y_i, it would be better to look for a p that causes the autocorrelation function $\langle x_i x_{i+p} \rangle$ to vanish. However, as has been shown by Fraser and Swinney, the minimum of the autocorrelation function does not always provide a good τ. In order to minimize more general, functional dependence, let us form the two series

$$S = \{s_i\} = \{x_i\},$$

and

$$Q = \{q_i\} = \{x_{i+p}\}.$$

By box-counting in the $S - Q$ plane, one may find the joint probability distribution $P_{sq}(s_i, q_j)$ in addition to the probability distributions $P_s(s_i)$ and $P_q(q_j)$. The joint distribution yields the "joint" information [see, e.g., Equation (6.19)]:

$$H(S, Q) = -\sum_{i,j} P_{sq}(s_i, q_j) \log P_{sq}(s_i, q_j). \tag{6.53}$$

For an emphasis of the "mutual" part of the information (6.53), one must subtract the uncorrelated components

$$H(S) = -\sum_i P_s(s_i) \log P_s(s_i),$$

and

$$H(Q) = -\sum_j P_q(q_j) \log P_q(q_j).$$

This defines the *mutual information*

$$I(S, Q) = H(Q) + H(S) - H(S, Q).$$

Obviously, $I(Q,S) = I(S,Q)$. In principle, the dependence of $I(S,Q)$ on p must be calculated, in order to find the value that produces the first local minimum. Moreover, one must compromise the refinement of the boxes with the number of points in each box, when calculating the joint probability distribution. Sending the interested reader for details to the original paper (Fraser and Swinney, 1986), we note only that the mutual information criterion has furnished a good idea, but further tests on more realistic data are needed.

There has been an indication that the time difference, determined from minimal mutual information, does not always guarantee a good reconstruction (Conte and Dubois, 1988). Another physical consideration may be of some help. Since usually there exists a certain characteristic orbital time in the system (the fundamental period at nearby parameter values, or the time corresponding to the most clearly seen peak in the power spectrum, etc.), it is reasonable to use such a portion of the time series, that is close to once or twice the characteristic time, for the reconstruction of a vector in the embedding space. In other words, the time $(m-1)\tau$ [called also the *embedding time*, see Equation (6.5)] must be of the order of the characteristic time of the system. In practice, the zero autocorrelation time, the minimal mutual information time and the orbital time do not differ in orders of magnitude. The determination of the best τ is still a matter of trial and error.

6.4 The Lyapunov Exponents

As we have said at the beginning of this chapter, it is, in principle, impossible to realize a truly quasiperiodic or chaotic orbit on a digital computer with a finite word length and within a finite run time. However, things are not so hopeless, if one pursues a correct strategy in carrying out the numerical work. One concentrates first on the systematics of the periodic orbits, that can be identified with confidence. The transition to chaos or quasiperiodicity takes place as one or another limit of the periodic regimes which are described by certain symbolic dynamics or Farey sequences. In this way, we can locate regions in the parameter space, where chaos is to be expected. The rule is simple: look at your neighbours and guess who are you. In doing so, one can deal at any time with a single orbit and there is no need to study simultaneously nearby trajectories.

This is no longer true when we set ourselves the task to characterize a chaotic attractor or to compare the "strangeness" of two attractors. One should recall that in dissipative systems strange attractors are formed from the compromise of two opposite trends: dissipation plays a global stabilizing role, by contracting the phase space volume, while local orbital instability forces initially neighbouring orbits to separate exponentially. Phase volume contraction does not imply shrinking of nearby trajectories in all directions. There may be other directions, along which the distances between these orbits are being stretched all the time. Therefore, a more detailed study of the strange attractors inevitably involves the behaviour of a bunch of orbits instead of a single orbit.

We speak of exponential separation of orbits, because this is the typical symptom of an instability. Just consider a linear system of ordinary differential equations

$$\dot{\mathbf{x}} = \mathbf{A}\mathbf{x}, \tag{6.54}$$

where $\mathbf{x}$ is a vector function and $\mathbf{A}$ a constant $n \times n$ matrix. Multiply (6.54) from the left by a constant matrix $\mathbf{P}$ that diagonalizes $\mathbf{A}$

$$\mathbf{P}\dot{\mathbf{x}} = \mathbf{P}\mathbf{A}\mathbf{P}^{-1}\mathbf{P}\mathbf{x};$$

then, in terms of the new vector $\mathbf{y} = \mathbf{P}\mathbf{x}$, we shall have n separated first order differential equations of the form

$$\dot{y}_i = \lambda_i y_i, \tag{6.55}$$

where λ_i is the i-th eigenvalue of $\mathbf{A}$. The solution of (6.55) is simply

$$y_i(t) = e^{\lambda_i t} y_i(0);$$

it will grow or decay exponentially, depending on the sign of λ_i, while the marginal value $\lambda_i = 0$ will keep it constant. A system of nonlinear differential equations can be linearized at each and every point along its trajectory, when systems of the type (6.54) will arise [see Equation (6.62) below]. These eigenvalues may be real or complex; they characterize the stability of the system at one point only. One would like to thread all the points along a trajectory and to get a set of global characteristics, in terms of real numbers for the entire orbit. If this is done properly, these numbers should characterize the stretching and contracting behaviour for a bunch of trajectories with the same destiny, i.e., going to the same attractor. This will be accomplished by calculation of the Lyapunov exponents.

In order to explain this idea, we have avoided any complications, say, related to multiple eigenvalues of $\mathbf{A}$, which would lead to a Jordan normal form instead of a simple diagonal matrix. There are standard procedures for dealing with these complications; by the way, they are not needed in the calculation of the Lyapunov exponents, as will be explained in Section 6.4.3.[12]

6.4.1 Invoking the Tangent Space — the Key to Success

We have just mentioned in passing that a system of nonlinear differential equations may be linearized at every point of the trajectory in order to produce systems of the type (6.54). By this statement, we have inadvertently disclosed the secret to success in the characterization of the attractors, since the act of linearization has taken us out of the phase (or configuration) space into the tangent space. The systematics of periods can be accomplished in the phase space only, whereas the faithful characterization of attractors must invoke the tangent space.

We explain the idea by the example of one-dimensional mappings. First of all, the definition of the Lyapunov exponent (called also the Lyapunov number in this case) is very simple. Take a map

$$x_{n+1} = f(\mu, x_n)$$

and iterate it from two close initial points x_0 and x_0'. After N iterations, the separation between the results will be

$$
\begin{aligned}
x_N - x_N' &= f^{(N)}(x_0) - f^{(N)}(x_0') \\
&\approx \frac{d}{dx} f^{(N)}(x_0)(x_0 - x_0').
\end{aligned}
$$

[12]The presentation of this section, especially 6.4.3 through 6.4.6, has benefitted much from discussions with Gu Yan and R. Conte.

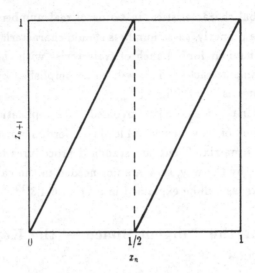

Figure 6.14: The shift map (6.57).

The reasonable assumption of exponential growth or decay of the "dimension-less" separation with the discrete time N

$$\frac{x_N - x_N'}{x_0 - x_0'} \approx \frac{d}{dx} f^{(N)}(x_0) \propto e^{\lambda N}$$

leads to the definition of the Lyapunov number λ:

$$\lambda = \lim_{N \to \infty} \frac{1}{N} \log \frac{d}{dx} f^{(N)}(x_0). \qquad (6.56)$$

Now apply this definition to a simple piecewise linear map, i.e., the *shift map*

$$x_{n+1} = f(x_n) \equiv 2x_n \pmod{1}, \quad x_n \in [0,1] \qquad (6.57)$$

(see Fig. 6.14). However, let us first look at the result, when we confine our-selves to the phase space, i.e., to the set $\{x_i\}$. Since $f'(x) = 2 > 1$ everywhere, all orbits will be unstable, and we expect to see chaotic iterates. However, on a computer of finite word length, this map realizes a left shift by one bit, with, at each iteration, a zero complemented at the rightmost bit. This means that, irrespectively of the initial point

$$x_0 = 0.\alpha_1 \alpha_2 \ldots \alpha_{n-1} \alpha_n,$$

where $\alpha_i = 0$ or 1, from which we start, we shall always come to a stable fixed point:

$$x_1 = 0.\alpha_2\alpha_3 \ldots \alpha_n 0$$
$$x_2 = 0.\alpha_3\alpha_4 \ldots 0\,0$$
$$\ldots \ldots$$
$$x_n = 0.00 \ldots 0\,0$$

i.e., 0.0. Clearly, it would be nonsensical to try to characterize the motion by means of phase space data $\{x_i\}$ only. However, if one calculates the Lyapunov number, according to (6.56),

$$\lambda = \lim_{n \to \infty} \frac{1}{n} \log \frac{df^{(n)}}{dx} = \lim_{n \to \infty} \frac{1}{n} \log \prod_{i=1}^{n} f'(x_i), \qquad (6.58)$$

we get the correct result $\lambda = \log 2 > 0$. In this example, the number λ is also the entropy (see Section 6.5). What can we learn from this trivial example? Taking the derivatives in (6.58) implies transition to the tangent space. For a reliable characterization of the attractors, one has to invoke the tangent as well as the phase space. In higher dimensions, this must be accomplished numerically. We shall see this clearly, when we calculate the Lyapunov exponents in the subsequent sections.

6.4.2 General Considerations

Before going to the definition and practical calculation of the Lyapunov exponents, we first summarize briefly the basic facts about these exponents and fix notation.

An n-th order system of ordinary differential equations has n Lyapunov exponents. They are real numbers, ordered by descending magnitude and denoted by

$$\lambda_1 \geq \lambda_2 \geq \ldots \geq \lambda_{n-1} \geq \lambda_n. \qquad (6.59)$$

Therefore, when writing λ_1, we always have in mind the largest Lyapunov exponent (positive for chaotic motion), while, as a rule, λ_n is the smallest, most negative exponent. Among these numbers, there may be m_+ positive ones, m_0 zeros, and m_- negative ones, representing the number of stretching, marginal and contracting directions in the phase space:

$$m_+ + m_0 + m_- = n.$$

For an autonomous system, there must be at least one zero Lyapunov exponent, if the solution is bounded and does not end up at a fixed point. The reason is simple: as time elapses, the trajectory evolves without going to an end, and no contraction will be felt along the orbit itself (for a proof see, e.g., Haken, 1983). This is even more true for a non-autonomous system, when no contraction occurs along the direction of the time t.

The sum of all Lyapunov exponents measures the rate of the phase volume contraction, i.e., the same quantity as the divergence of the vector field $\mathbf{F}$ which defines the differential equations [see Equations (6.61) in the next subsection]:

$$\sum_{i=1}^{n} \lambda_i = \mathrm{div}\mathbf{F}. \tag{6.60}$$

Equation (6.60) holds, if $\mathrm{div}\mathbf{F}$ is a constant, as in the Lorenz model. If it varies with the coordinates, as in the forced Brusselator, then the divergence must be averaged along the trajectory (Andrey, 1986):

$$\sum_{i=1}^{n} \lambda_i = \lim_{t \to \infty} \frac{1}{t} \int_0^t \mathrm{div}\mathbf{F}(\mathbf{x}(s))ds.$$

Consequently, for conservative systems, all Lyapunov exponents sum up to zero, i.e.,

$$\sum_{i=1}^{n} \lambda_i = 0,$$

(in fact, they cancel in pairs, if the system is Hamiltonian); the sum is negative for dissipative systems, and thus reflects the overall phase space contraction.

The appearance of the first positive Lyapunov exponent signals the transition to chaos. The signature of the Lyapunov exponents provides a classification for attractors, e.g.,

$(-, \ -, \ -, \ -)$ fixed point,
$(\ 0, \ -, \ -, \ -)$ periodic motion,
$(\ 0, \ 0, \ 0, \ -)$ quasiperiodic motion on a torus,
$(+, \ 0, \ -, \ -)$ chaos,
$(+, \ +, \ 0, \ -)$ superchaos à la Roessler.

Consequently, the fact that a Lyapunov exponent passes through zero when control parameters are varied indicates bifurcation. In this way, we can have a phenomenological classification of attractors and transitions between them. Even the sole knowledge of the largest Lyapunov exponent would tell us much about the attractor. In fact, for an autonomous system of three variables, λ_1 is the only exponent which must be calculated, since $\lambda_2 = 0$ and λ_3 may be obtained from (6.60).

6.4.3 Calculation of Lyapunov Exponents from the Evolution Equations

We first treat the case when the dynamical equations are known explicitly, using ordinary differential equations as an example. The extraction of Lyapunov exponents from experimental data, when neither the equation nor the dimension of the phase space are known, will be discussed in Section 6.4.6.

For the definition of the Lyapunov exponents for a system of autonomous ordinary differential equations

$$\frac{dx}{dt} = F(x), \tag{6.61}$$

(x and F are n-dimensional vectors, we avoid writing their components explicitly as long as possible), we adopt a purely geometric approach that will yield only real numbers. For an initial value $x(t = 0) = x_0$, Equation (6.61) leads to an integral curve or solution curve

$$x(t) = \Phi^t(x_0).$$

The nonlinear system (6.61) may be linearized at any point $x(t)$ of the trajectory, in order to obtain a linearized system of equations for a small variation $\delta x(t)$ from $x(t)$

$$\frac{d\delta x(t)}{dt} = J(x(t))\,\delta x(t), \tag{6.62}$$

where

$$J(x(t)) = \frac{\partial F(x)}{\partial x}\Big|_{x=x(t)}$$

is an $n \times n$ matrix, which is real if $\mathbf{F}$ is real. Note that Equations (6.62) are non-autonomous, even though Equations (6.61) are autonomous. The solution of the linear equations (6.62) may be expressed by means of an evolution operator $\mathbf{U}(t)$ which satisfies the same linear equations with the initial value $\mathbf{U}(0) = \mathbf{I}$, $\mathbf{I}$ being the unit matrix. In fact, we have seen all these equations in Chapter 5 (Sections 5.3.2 and 5.4.1). We present them again for the sake of convenience.

Normalizing the variation vector $\delta\mathbf{x}(t)$ by its length at $t = 0$ and taking the limit $|\delta\mathbf{x}(0)| \to 0$, we arrive at the tangent vector

$$\mathbf{W}(t) = \lim_{\delta\mathbf{x}(0) \to 0} \frac{C\,\delta\mathbf{x}(t)}{|\delta\mathbf{x}(0)|},$$

where C is an arbitrary constant. Clearly, the vector $\mathbf{W}(t)$ obeys the same linear equations (6.62). If we follow the evolution of the tangent vector $\mathbf{W}(t)$ along the trajectory, its length may grow or contract at an exponential rate; this rate determines the Lyapunov exponent (denoted for the time being by λ)

$$\lambda = \lim_{t \to \infty} \frac{1}{t} \log |\mathbf{W}(t)|. \tag{6.63}$$

The value of λ certainly depends on the direction of the chosen $\mathbf{W}$. It would appear that n independent choices of the direction would lead to n different values of λ. However, due to the stretching effect, almost all tangent vectors $\mathbf{W}$'s will align with the direction of the strongest stretch and lead to the largest Lyapunov exponent λ_1. The value of λ might depend on the initial point $\mathbf{x}(0)$, but this was excluded by a theorem due to Oseledec (1968), which says that for almost all choices of $\mathbf{x}(0)$ one would get the same exponent. We skip the mathematical subtleties, such as what does "almost" mean, and note that as yet Equation (6.63) does not provide a practical working formula. Anyway, the tangent vector must be calculated by means of the evolution operator

$$\mathbf{W}(t) = \mathbf{U}(t)\mathbf{W}(0),$$

so that a more rigorous definition of the Lyapunov exponents (see, e.g., Eckmann and Ruelle, 1985) starts with the positive operator

$$\Lambda(\mathbf{x}(0)) = \lim_{t \to \infty} [\mathbf{U}(t)^* \mathbf{U}(t)]^{\frac{1}{2t}} \tag{6.64}$$

(the implicit dependence of $\mathbf{U}$ on $\mathbf{x}(0)$ through $\mathbf{J}$ has not been indicated here). The Lyapunov exponents are defined as the logarithms of the eigenvalues λ_i of

$\Lambda(\mathbf{x}(o))$; the Oseledec theorem guarantees the practical independence of the λ_i from $\mathbf{x}(o)$. Moreover, if one succeeds in choosing an initial vector $\mathbf{x}(o)$ strictly in the eigenspace of Λ, corresponding to an eigenvalue λ_i, but not in the eigenspace of λ_{i+1} [remember the ordering convention (6.59) for the λ_i], then (6.63) would yield the i-th Lyapunov exponent λ_i. However, this is unlikely to happen in practice. A randomly chosen vector tends to take a general orientation in a high dimensional space rather than to lie in a predefined subspace, perpendicular to the direction of greatest stretch. This is why Equation (6.63) always leads to the largest λ_1.

We mention in passing that the above definition may be used directly when dealing with discrete mappings. The evolution operator $\mathbf{U}$ is replaced by the product of the tangent maps (e.g., the map $\mathcal{T}$, defined in (6.68) of Section 6.4.6) at each iteration, and the operator Λ becomes a matrix of finite order, i.e., the same order as the dimension of the underlying mapping.

A practical algorithm for the calculation of all Lyapunov exponents may be derived by an extension of the geometric picture described above (Benettin, Giorgilli, Galgani and Strelcyn, 1978, 1979; Shimada and Nagashima, 1979).

For the computation of the largest Lyapunov exponent, i.e., λ_1, choose at random a small vector $\mathbf{W}$ which sits at the orbit and points to a nearby location. Integrating the starting point one step forward along the exact trajectory, i.e., using the nonlinear equations (6.61), and integrating the end point of the vector $\mathbf{W}$ according to the linearized equations (6.62), one obtains a new vector $\mathbf{U}(t)\mathbf{W}$. Take the length ratio of these two vectors, then repeat the calculation for the next steps. Averaged along the entire trajectory, this ratio would approach the largest exponent λ_1. (In practice, one employs a sufficiently long segment of the trajectory.) We have thus explicitly

$$\lambda_1 = \lim_{t \to \infty} \log \frac{|\mathbf{U}(t)\mathbf{W}|}{|\mathbf{W}|}.$$

We know from Section 5.4.1 that $\mathbf{U}(t)$ should be calculated as the product of vectors $\mathbf{U}(kh, (k-1)h)$, evaluated step by step from $k = 1$ to a big enough value of k.

In order to calculate the next Lyapunov exponent, one should choose two non-colinear vectors which point out from the trajectory, and integrate the nonlinear as well as the linear equations one step forward to get the end points.

Then calculate the ratio of the areas of the parallelograms formed by these vectors before and after the step. Averaged along the trajectory, this ratio will give the sum of the two largest Lyapunov exponents, i.e., $\lambda_1 + \lambda_2$, provided the initial vectors were chosen randomly and transversally to the trajectory. For the calculation of λ_3, proceed similarly with a parallelepiped, and so on, and so forth. This procedure usually converges quite quickly, at least, for the largest λ_1 (Wang, Chen, and Hao, 1985). There is, however, one point worth mentioning: due to the presence of negative exponents, the different vectors, chosen at the beginning, tend to align themselves, causing the area or volume to vanish. In order to avoid numerical difficulties, one should orthogonalize the vectors every few steps. However, if one deals with two-dimensional vectors only, it is unnecessary to use the Schmidt orthogonalization procedure: simply recall the fact that the vector $(-b, a)$ is orthogonal to the vector (a, b).

In concluding of this subsection, we mention a technical problem that often arises in numerical practice. Numerically, one can hardly get a net zero from the procedure described above. How can one identify a true zero exponent from several, small, positive or negative numbers? We know that in some cases the sign of a very small number may be incidental, while the zero exponent is so important for, say, the determination of the dimension (see Section 6.6 below). An idea has been advanced by Ruelle.[13] We know that in a Poincaré section, perpendicular to the flow, the zero exponent, that corresponds to the unconstraint development parallel to the flow, does not appear. Generally speaking, the Lyapunov exponents of a flow and the corresponding map differ by a factor (the orbital time T between two sections). If a flow has the Lyapunov spectrum

$$\lambda_1, \ldots, \lambda_{k-1}, \lambda_k = 0, \lambda_{k+1}, \ldots, \lambda_d,$$

then the corresponding Poincaré map should have a spectrum

$$\underbrace{\lambda'_1, \ldots, \lambda'_{k-1}}_{\text{positive}}, \underbrace{\lambda'_{k+1}, \ldots, \lambda'_d}_{\text{negative}}.$$

Therefore, in principle, the number of positive exponents of a flow may be determined by inspection of the spectrum of its Poincaré map. In practice, this requires additional calculation.

[13] R. Conte, private communication.

6.4.4 Lyapunov Exponents for Periodic Orbits

Lyapunov exponents are useful not only for the characterization of chaotic motion, but also for the tracing of periodic and quasiperiodic orbits. We have mentioned in Section 5.4 that it is desirable to monitor the Lyapunov exponents, whenever the periodic orbit locating procedure is employed, because the method works equally well for both stable and unstable orbits, and one can cross the border of stability without noticing it. Fortunately, it is not necessary to take the limit $t \to \infty$ in this case. The calculation of the Lyapunov exponents can be accomplished by integration of one more period after one has settled down on a periodic orbit. In fact, according to the Floquet theorem[14], the time evolution operator $\mathbf{U}(t)$ for a periodic solution of period T can be decomposed:

$$\mathbf{U}(T) = \mathbf{K}e^{\mathbf{\Lambda}T},$$

where

$$\mathbf{K}^2 = \mathbf{I}.$$

Therefore, to get rid of $\mathbf{K}$, one must calculate it at $2T$. Then the Lyapunov exponents will be given by the real parts of the eigenvalues of the matrix

$$\mathbf{\Lambda} = \frac{1}{2T} \log(\mathbf{U}(2T)). \tag{6.65}$$

6.4.5 The Most Stable Manifold and Destruction of Invariant Tori

In Chapters 4 and 5, we have become acquainted with the importance of invariant manifolds such as the stable and unstable manifolds of a saddle node and their transverse intersections. Up till now, in this chapter, we have been dealing with the largest Lyapunov exponent. It is not true, however, that the smallest (i.e., the most negative) Lyapunov exponent would have least significance. In this section, we shall study a type of generally non-invariant manifolds, related to the most negative Lyapunov exponent, namely, in our notation (6.59), λ_n.

[14]For the Floquet theorem and the notion of Floquet multiplier see, e.g., H. Haken, *Advanced Synergetics*, Springer-Verlag, 1983.

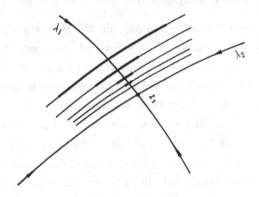

Figure 6.15: The most stable manifolds in the vicinity of a sink (schematic). The emphasized segments show their motion and contraction.

These are the *most stable manifolds* (see Gu, 1987, 1988) figured in the title of this section.

In order to get the idea, let us first look at a sink (a stable node) in a two-dimensional system. As long as it is stable, both eigenvalues of the tangent map are negative: $0 > \lambda_1 > \lambda_2$, λ_2 being the most negative one. All points on nearby orbits will approach the sink with the slower rate λ_1, except those on the two orbits that coincide with the eigen-direction of λ_2. The latter set of points converge to the sink with the faster rate λ_2. These two orbits plus the sink itself form the most stable manifold in the vicinity of the sink. In this particular case, it is also an invariant manifold.

A similar situation arises near a saddle point in the plane. Now, there is one positive and one negative eigenvalue: $\lambda_1 > 0 > \lambda_2$. All orbits will depart from the saddle at the rate λ_1, except for two orbits along the eigen-direction of λ_2 that will converge to the saddle with the rate λ_2. Once again we have an invariant most stable manifold.

In general, we know that the Lyapunov exponents themselves do not depend on the starting point $\mathbf{x}(0)$, from which we perform the average along the

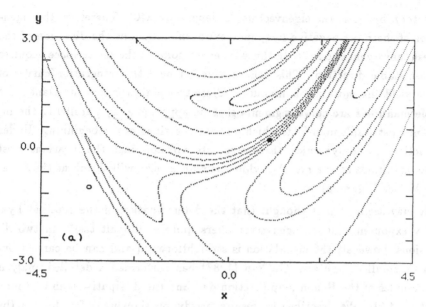

Figure 6.16: The most stable manifolds for the Hénon map at $\mu = b = 0.3$. The solid circle marks the sink, and the open circle — the saddle.

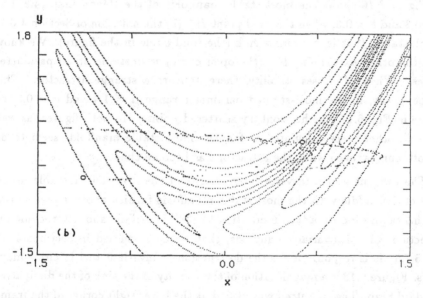

Figure 6.17: The most stable manifolds for the Hénon map at $\mu = 1.4, b = 0.3$, when there is a strange attractor.

trajectory, but that the eigenvectors do depend on $\mathbf{x}(0)$. Therefore, the eigenvector of the most negative Lyapunov exponent picks out the direction of the
fastest convergence to $\mathbf{x}(0)$. In this way, every point in the phase space acquires
a most stable direction. This defines a vector field the integration curves of
which yield the most stable manifold. Close to a sink, the non-invariant most
stable manifolds are curve segments that lie approximately parallel to the invariant most stable manifolds which we have described at the beginning. Under
the action of the dynamics, these segments will approach the invariant most
stable manifolds at the rate λ_1, while their linear size will shrink at the faster
rate λ_2 (see Fig. 6.15).

It may seem at first thought that the determination of the smallest Lyapunov exponent and its eigenvector offers quite a difficult task. In two dimensions, however, the calculation is straightforward and can be carried out
even on small computers. Gu Yan (1987) has performed a detailed analysis
on examples of the Hénon map (Section 4.6) and the dissipative standard map
(Section 4.7). We describe the results briefly by showing a few figures (by
courtesy of the author).

Figure 6.16 shows the most stable manifolds of the Hénon map (4.21) at
$\mu = 0.3$ and $b = 0.3$, when the fixed point H^+ (in the notation of Section 4.6.1)
is still stable, that is, remains a sink (the solid circle in the figure). We know
that the other fixed point H^- (the open circle) is unstable at all parameter
values. When H^+ loses stability, there appears a strange attractor. This
happens, say, in the most-studied parameter range $\mu = 1.4$ and $b = 0.3$, as
shown in Fig. 6.17 (the horizontally scattered points). In both figures, as well
as in all other cases calculated so far, the most stable manifolds seem to be
smooth curves, even when the attractor has become chaotic.

The tangent and transverse intersections of the invariant unstable manifolds of the saddle with the most stable manifolds furnish another mechanism
for chaotic motion to occur, in addition to the homoclinic and heteroclinic intersections of the invariant manifolds, that we have studied in Sections 4.6.3
and 5.7.5. It is responsible for the destruction of invariant tori in circle mappings. Figure 6.18 is a magnification of the vicinity of the sink of the dissipative
standard map. The sink itself was placed at the lower right corner of the frame
as the origin of coordinates. All the local most stable manifolds become par-

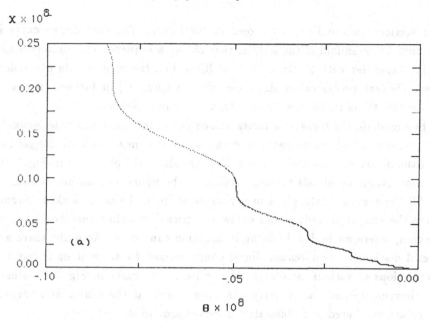

Figure 6.18: The first cubic tangencies of the unstable manifold of the saddle with the non-invariant most stable manifolds (dissipative standard map).

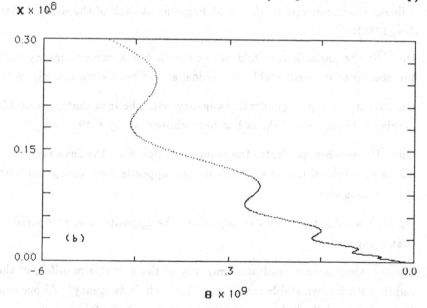

Figure 6.19: The first quardratic tangencies of the unstable manifold of the saddle with the invariant most stable manifold of the sink.

allel vertical lines and are not shown in the figure. The only drawn curve is the unstable manifold of the saddle, that shows a sequence of infinitely many cubic tangencies with the local vertical lines, i.e., the most stable manifolds. These tangent points accumulate towards the sink. Upon further change of parameters these tangencies become transverse intersections.

In Fig. 6.19, the transverse intersections of the invariant unstable manifold of the saddle have been enhanced to such a stage that new quadratic tangencies with the invariant most stable manifold of the sink take place. As in Fig. 6.18, the most stable manifolds become vertical in the figure and are not shown.

All these events take place in a narrow strip in the mode-locked tongue, above the tongue boundary and below the period-doubling threshold, i.e., in a region, where no period-doubling bifurcation can occur. Actually, there are several distinctive demarcation lines, characterized by tangent or transverse intersections of various manifolds. We collect these cases in Fig. 6.20, where the abovementioned narrow strip has been drawn at the center and various stages are numbered and illustrated in boxes around it.

In the boxes of Fig. 6.20, a solid circle denotes the sink and an open circle — the saddle. The direction of the most stable manifolds are indicated by dashed lines. We summarize briefly what happens at each of the demarcation lines (Gu, 1988):

1. On $\overline{CB}$: the unstable manifold of the saddle has a cubic tangency with the non-invariant most stable manifolds, as has been shown in Fig. 6.18.

2. On $\overline{CG}$: it develops a quadratic tangency with the invariant most stable manifold through the sink, as has been shown in Fig. 6.19.

3. On $\overline{CH}$: another quadratic tangency develops with the invariant most stable manifold of the sink, but from the opposite side, compared with the previous one.

4. On $\overline{CF}$: a cubic tangency develops from the opposite side, compared to that along $\overline{CB}$.

5. On $\overline{JG}$: there occurs quadratic tangency of the unstable manifold of the saddle with its own stable manifold (a homoclinic tangency). Above this line, other mode-locked tongues may overlap with the 0/1 tongue.

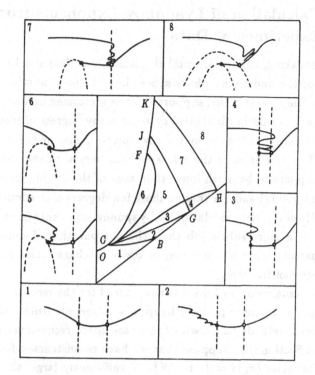

Figure 6.20: Bifurcation structure observed in the 0/1 mode-locked tongue of the dissipative standard map (schematic, for details see text; courtesy of Gu Yan).

6. On $\overline{KH}$: again homoclinic tangency occurs, but with orientation opposite to that along $\overline{JG}$. Upon crossing $\overline{KH}$, both branches of the stable manifold of the saddle extend to negative infinity, and a strange attractor emerges.

The above analysis was based on detailed numerical work and has removed some inexactitude in previous qualitative discussions (see, e.g., Aronson et al., 1982). Having had to confine ourselves to a schematic presentation, we refer the interested reader to the original papers (Gu, 1987, 1988).

6.4.6   Calculation of Lyapunov Exponents from Experimental Data

Generally speaking, in experimental situations, one does not know the actual dimension of the underlying phase space, i.e., the total number of Lyapunov exponents. The present belief, supported by an increasing number of laboratory measurements, is that in dissipative systems many degrees of freedom cease to influence the long time behaviour and asymptotic evolution takes place on some attractor. The dimension of the attractor may vary with the nonlinearity, but usually it happens to be much lower than that of the original phase space, and there exists a good chance for tracing these few degrees of freedom that span the attractor. However, the calculation of Lyapunov exponents from experimental time series is a more subtle job than the computation of dimensions, since one must extract more information, in order to characterize separately each relevant degree of freedom.

At least two algorithms have been suggested for the recovery of the largest and, perhaps, a few more positive Lyapunov exponents from a single time series. All these methods begin with the phase space reconstruction technique, described in Section 6.3. Suppose that we have reconstructed from the experimental time series $\{x_i \,|\, i = 1 \text{ to } N\}$, N sufficiently large, the M (where is of the same order as N) m-dimensional vectors by choosing an appropriate embedding dimension m and time difference $\tau = p\Delta\tau$ (where p is an integer, and $\Delta\tau$ the sampling interval for the original time series). In order to rid the discussion of redundant indices, we consider, first of all, the case $p = 1$, i.e., we assume that a suitable time difference has been used as the sampling interval in the data acquisition system. Thus, we have the vectors [cf. Equation (6.49)]:

$$\mathbf{y}_i = (x_i, x_{i+1}, \ldots, x_{i+m-1}), \quad i = 1 \text{ to } M. \tag{6.66}$$

The method suggested by Wolf, Swift, Swinney and Vastano (1985) is conceptually similar to the calculation in Section 6.4.3, when the evolution equations are known explicitly. Choose several neighbouring vectors $\mathbf{y}_j$ of a reference vector $\mathbf{y}_i$, and calculate the distance of their images $\mathbf{y}_{j+1}$ to the new reference vector $\mathbf{y}_{i+1}$. The ratio of the lengths $\mathbf{y}_{j+1} - \mathbf{y}_{i+1}$ to $\mathbf{y}_j - \mathbf{y}_i$, averaged over a sufficiently long time interval, will lead to the largest Lyapunov exponent λ_1. Similarly, the averaged area ratio of the parallelograms, spanned by the new and old vectors, will yield the sum $\lambda_1 + \lambda_2$, etc. This method has been further discussed by Vastano and Kostelich (1986), and we shall not go into details here.

Another method has been suggested by Eckmann and Ruelle (1985) (see also Sano and Sawada, 1985), and has been elaborated by many researchers recently (see, e.g., Eckmann, Oliffson Kamphorst, Ruelle and Ciliberto, 1986; Conte and Dubois, 1988). It starts from the discrete version of the positive operator Λ [see Equation (6.64) and its comments]. The Lyapunov exponents are defined as the logarithms of the eigenvalues of Λ. The central idea is to reconstruct the tangent maps, step by step, by numerically fitting the experimental data. It seems to us that, when there are enough sampled data, this method may be quite powerful, although its implementation is very time-consuming. We devote the rest of this section to a description of its essential points.

A Few Numerical Algorithms

Since the suggestion of Ruelle and Eckmann involves a few more sophisticated numerical algorithms, we put together the required concepts (for a brief introduction, see, e.g., the book by Nash[15]).

The QR Decomposition of a Matrix. For speeding up of the computation, it is a good numerical practice to decompose matrices before any actual manipulation is performed. A real $m \times n$ (in particular, $m = n$) matrix $\mathbf{T}$ may be decomposed into the product of an orthogonal matrix $\mathbf{Q}$ and a right(upper) triangular matrix $\mathbf{R}$:

$$\mathbf{T} = \mathbf{QR}.$$

Suppose we have to calculate the product of matrices

$$\Pi = \mathbf{T}^{(N)}\mathbf{T}^{(N-1)} \ldots \mathbf{T}^{(2)}\mathbf{T}^{(1)},$$

required in the definition of the positive operator (6.64) for mappings; the correct procedure involves the sequence of decompositions

---

[15] J. C. Nash (1979), *Compact Numerical Methods for Computers: Linear Algebra and Function Minimisation*, Adam Hilger.

$$\mathbf{T}^{(1)} = \mathbf{Q}^{(1)}\mathbf{R}^{(1)},$$
$$\mathbf{T}^{(2)}\mathbf{Q}^{(1)} = \mathbf{Q}^{(2)}\mathbf{R}^{(2)},$$
$$\dots \dots$$
$$\mathbf{T}^{(k)}\mathbf{Q}^{(k-1)} = \mathbf{Q}^{(k)}\mathbf{R}^{(k)},$$
$$\dots \dots$$
$$\mathbf{T}^{(N)}\mathbf{Q}^{(N-1)} = \mathbf{Q}^{(N)}\mathbf{R}^{(N)},$$

leading to

$$\mathbf{\Pi} = \mathbf{Q}^{(N)}\mathbf{R}^{(N)}\mathbf{R}^{(N-1)}\dots\mathbf{R}^{(2)}\mathbf{R}^{(1)}.$$

On the formation of the positive product $\mathbf{\Pi}^T\mathbf{\Pi}$ ($\mathbf{\Pi}^T$ is the transpose of $\mathbf{\Pi}$), the orthogonal matrix $\mathbf{Q}^{(N)}$ disappears, and we are left with the product of N upper triangular matrices. Now we can appreciate the saving in time, because the product of two upper triangular matrices

$$(\mathbf{RR})_{ij} = \sum_{k=1}^{j} \mathbf{R}_{ik}\mathbf{R}_{kj}$$

requires a much smaller number of multiplications and additions, as compared with the product of two general matrices

$$(\mathbf{AA})_{ij} = \sum_{k=1}^{m} \mathbf{A}_{ik}\mathbf{A}_{kj}.$$

Moreover, the upper triangular form of the result automatically simplifies our later task of the diagonalization of the matrices.

The Least Square Solution of an Overdetermined Linear System. Suppose we have a system of linear algebraic equations (in matrix notation)

$$\mathbf{Ax} = \mathbf{b}. \tag{6.67}$$

If $\mathbf{A}$ is an $n \times n$ and non-singular matrix, $\mathbf{b}$ an n-component given vector, we can solve for the unknown $\mathbf{x}$. However, if the right-hand side of (6.67) is required to satisfy several different vectors $\mathbf{b}$, then, in general, there exists no solution, and the system is said to be overdetermined. Nevertheless, one can look for a solution in a different sense. Note that now we must have a rectangular $m \times n$ matrix A, applied to an n-component $\mathbf{x}$ to get the m-component $\mathbf{b}$. For example, if (6.67) is required to satisfy two different vectors $\mathbf{b}$, then we must put $\mathbf{A}$ on top of another $\mathbf{A}$ to get a $2n \times n$ matrix, and $\mathbf{b}_1$ on top of $\mathbf{b}_2$ to get a $2n$-component $\mathbf{b}$. In general, in what follows the quantities $\mathbf{A}$ and $\mathbf{b}$ are such $m \times n$ matrices and such m-component vectors, respectively.

Let us try to adjust $\mathbf{x}$ so as to make the difference vector

$$\mathbf{r} = \mathbf{Ax} - \mathbf{b}$$

as small as possible. We need a norm in order to be able to define the smallness of a vector. Since the components of $\mathbf{r}$ may take either signs, we form the positive square

$$\mathbf{r}^T \mathbf{r} = \mathbf{x}^T \mathbf{A}^T \mathbf{A} \mathbf{x} - \mathbf{x}^T \mathbf{A}^T \mathbf{b} - \mathbf{b}^T \mathbf{A} \mathbf{x} + \mathbf{b}^T \mathbf{b}$$

and require it to be minimal, i.e.,

$$\frac{\partial \mathbf{r}^T \mathbf{r}}{\partial \mathbf{x}^T} = 0,$$

which leads to the so-called normal equations

$$\mathbf{A}^T \mathbf{A} \mathbf{x} = \mathbf{A}^T \mathbf{b}.$$

When the rank of the matrix $\mathbf{A}$ is less than n, the solution is not unique. In order to guarantee uniqueness, one has to impose an additional requirement for the length $\mathbf{x}^T \mathbf{x}$ to be minimal. This is the minimal length, least-square solution of the overdetermined linear system.

The Singular Value Decomposition of a Matrix. An $m \times n$ matrix $\mathbf{A}$ can be represented by

$$\mathbf{A} = \mathbf{USV}^+,$$

where $\mathbf{U}$ is an $m \times n$ column-orthogonal matrix, i.e.,

$$\mathbf{U}^T \mathbf{U} = \mathbf{I},$$

and

$$\mathbf{S} = \text{diag}(s_1, s_2, \ldots, s_r, 0, \ldots, 0)$$

is diagonal with r nonzero elements and $n - r$ zeros, which are called the singular values of $\mathbf{A}$, while $\mathbf{V}$ is an $n \times n$ orthogonal matrix. The singular value decomposition may be used to define the generalized inverse matrix

$$\mathbf{A}^+ = \mathbf{V}\mathbf{S}^+\mathbf{U}^T,$$

where $\mathbf{S}^+ = \mathrm{diag}(p_1, p_2, \ldots, p_r, 0, \ldots, 0)$ with

$$p_i = \begin{cases} 1/s_i, & s_i \neq 0, \\ 0, & s_i = 0. \end{cases}$$

In numerical practice, one can rarely achieve precise zeros for the singular values, so it is more convenient to compare the s_i with a preassigned tolerance. The generalized inverse $\mathbf{A}^+$ is then used to define the least square solution for the overdetermined linear equations; in fact, we have

$$\mathbf{x} = \mathbf{V}\mathbf{S}^+\mathbf{U}^T\mathbf{b}.$$

It is important to note that there is no need to write down the routines for the performance of the above matrix operations. Nowadays, one can find many long-tested subroutines in major programme libraries such as LINPACK, IMSL, NAG or EISPACK.[16]

Recovery of the Tangent Map by Least-Square Fitting

We are now well prepared to continue the calculation of Lyapunov exponents from experimental data. We shall work in the m-dimensional embedding space spanned by the vectors $\{\mathbf{y_i} | i = 1 \text{ to } M\}$. At the start of the process we choose a vector $\mathbf{y}_1$ as the central point and build a ball of radius r around it. From the data file, we find all the vectors $\mathbf{y}_j$ located within this ball and call them neighbours of $\mathbf{y}_1$. When the vector $\mathbf{y}_1$ evolves into $\mathbf{y}_2$, the neighbours evolve into $\mathbf{y}_{j+1}$. (In practice, one allows for $\mathbf{y}_1$ to evolve into $\mathbf{y}_{1+k}$ and look for the new neighbours $\mathbf{y}_{j+k}$. In order to give an idea of the process, we only consider the simple case $k = 1$.)

There are a few factors, which dictate the choice of the radius r. It must be so small that the evolution of the vectors $\mathbf{y}_j - \mathbf{y}_1$ into $\mathbf{y}_{j+1} - \mathbf{y}_2$ may be described by the linearized dynamics, i.e., by the (yet unknown) tangent map $\mathcal{T}$. At the same time, it must be large enough to contain a sufficient number of neighbours, in order to ensure the precision in the later fitting of the tangent

---

[16]Readers who know Chinese may consult the Appendices of Hao Bai-lin, FORTRAN *Programming*, The People's Post and Telecommunication Press, Beijing, 1980, for a singular value decomposition routine and other related sources.

map. Anyway, the radius r should be much smaller than the linear size of the attractor under study. In practice, it is varied at each step by aiming at a compromise between these factors.

Since the small vectors $y_j - y_1$ evolve according to the linearized dynamical equations into $y_{j+1} - y_2$, we can represent the tangent map by a constant matrix T with undetermined elements. How big should this matrix be? Since it describes an evolution of state vectors in the embedding space, an $m \times m$ matrix will do. However, it may appear to be a singular matrix, if the vectors y_j do not span the entire embedding space. This certainly happens, when one has incidentally chosen an embedding dimension m that exceeds the dimension of the actual phase space. In addition, there may not be a large enough number of neighbours in the ball to ensure a good fitting. Therefore, a scheme with a possibly smaller matrix has been suggested (Eckmann, Ollifson Kamphorst, Ruelle and Ciliberto, 1986). We use an $m \times m$ matrix to outline the method, thus writing

$$T_{\alpha\beta}^{(1)}(y_{j\beta} - y_{1\beta}) = y_{j+1\alpha} - y_{2\alpha}, \tag{6.68}$$

where the Greek subscripts α and β label the components of the vectors and run from 1 to m, the dimension of the embedding space, the Latin subscript j runs from 1 to n, the number of neighbours in the ball, and summation over the repeated Greek subscripts β is implied. Note that the unknowns here are the elements of $T^{(1)}$. How many independent elements are there? An $m \times m$ matrix has at most m^2 elements. However, our choice of $p = 1$ in (6.66) and the evolution by one time unit (τ) in (6.68) has reduced us to m unknowns, all others being 1 or 0. This can be seen easily from the construction of the vectors y_j: the components of y_{j+1} are obtained by shifting those of y_j, only the last one has not been contained in y_j. Therefore, the matrix T looks like

$$\begin{bmatrix} 0 & 1 & 0 & \cdots & 0 \\ 0 & 0 & 1 & \cdots & 0 \\ \cdots & \cdots & \cdots & \cdots & \cdots \\ 0 & 0 & 0 & \cdots & 1 \\ t_{m1} & t_{m2} & t_{m3} & \cdots & t_{mm} \end{bmatrix}.$$

Since all but the last of the Equations (6.68) have become identities, we would need m different neighbouring vectors (i.e., j) in order to arrive at enough equations for the determination of all elements of T. However, the results thus obtained would be quite sensitive to the selection of these $\mathbf{y}_j$. To compensate for the scattering of results, we take more vectors in the ball, in order to convert (6.68) into an overdetermined system, and then look for the minimal length least-square solution. In this way, we have left out the possible singularity of T, because the minimal length condition automatically guarantees even in this case a unique solution. In order to avoid any confusion in the comparison of (6.68) with (6.67), one must remember that $T_{\alpha\beta}^{(1)}$ corresponds to the unknown vector $\mathbf{x}$, while $(\mathbf{y}_j - \mathbf{y}_1)_\beta$ plays the role of $\mathbf{A}$.

Having calculated the tangent map $T^{(1)}$ for the first step, we take $\mathbf{y}_2$ as the new center, and repeat the procedure using the new neighbours in order to arrive at the tangent map $T^{(2)}$. All the tangent maps, thus obtained, are multiplied via the QR decomposition. The procedure may be summarized as follows:

1. Preparation: let $Q^{(0)} = I$ and $R = I$, I being the unit matrix.

2. Loop from $i = 1$ to K:

   - Calculate $T^{(i)}$ by least-square fitting the tangent map from step i to step $i + 1$.

   - Perform the QR decomposition $T^{(i)}Q^{(i-1)} = Q^{(i)}R^{(i)}$.

   - Perform the matrix multiplication $R^{(i)}R \rightarrow R$.

3. Closing stage: calculate $R^T R$, i.e., the quantity needed in the calculation of Λ.

One advantage of the Eckmann-Ruelle algorithm is the possibility of a recovery of the eigen-space, corresponding to a calculated Lyapunov exponent.

Compared to the calculation of dimensions from experimental time series, we have more parameters to adjust in the computation of the Lyapunov exponents. Besides the embedding dimension m and the time delay τ (or the integer p), we have the radius of the ball ϵ, the number of neighbours to be chosen in the ball, the time step between two adjacent fittings for the tangent maps, not to mention the tolerance in the numerical algorithm of the singular value decomposition. All these aspects cause the recovery of Lyapunov exponents from time series to be more like an art, and one has to rely on trial and error.

6.5 Information and Entropy

We have learnt so far enough facts to see that chaotic motion appears as a compromise between two opposite trends in a system: local orbital instabilities tend to separate exponentially the initially neighbouring trajectories, while global stabilizing factors, such as conservation laws and, to an even larger extent, dissipation keep the motion in a bounded region of the phase space. The former feature manifests itself by the sensitive dependence on the initial conditions and excludes the possibility of long-term prediction of a motion, whereas the latter guarantees the reliability of the prediction of some averaged characteristics over long periods of time.

The sensitive dependence on the initial conditions may be formulated in terms of information in two seemingly opposite, but essentially identical ways, either as information generated by the dynamics, or as information loss caused by the motion, depending on how one looks at the problem. Both points of view lead us to the notion of entropy.

If two initial points are located so closely together that we are unable to distinguish them at the beginning, then, as the spatial separation increases, they become more and more distinguishable. In this sense, we say that information has been generated by the motion. Roughly speaking, the information is characterized by the number of distinguishable trajectories N, which grows exponentially with the time, i.e.,

$$N \propto e^{Kt}.$$

The quantity K which characterizes the mean rate of generation of information is nothing else but the entropy.

On the other hand, if we have fixed the initial point with very high precision, our initial information still keeps on decreasing with the motion continuing, due to the same orbital instabilities and the finite precision of the measurements. The quantity K^{-1}, with the dimension of time, measures the *predictibility time* (Shaw, 1981), starting from the fixed initial conditions.

The introduction of various entropies, as well as of the dimensions and the Lyapunov exponents, discussed above, has been based on the ergodic theory of dynamical systems which on its own has developed into a well-shaped chapter of mathematics. To this end, we send the reader to the excellent review by Ruelle and Eckmann (1985). To keep to the style of this book, we skip the mathematical details and turn to the practical problems of the calculation of entropies for a given map or from experimental data.

6.5.1   The Metric Entropy

The notion of the metric entropy, denoted hereafter by K, was introduced by Kolmogorov in 1958, and improved later by Sinai. It is also called the Kolmogorov-Sinai entropy or simply the entropy. The original motivation for the introduction of K was to find a global characteristic of mappings (and dynamical systems) that would not change under topological conjugation (see Section 2.5.3 for the topological conjugacy of maps). Nowadays it has become one of the most important criteria for making a distinction between various types of motion: $K = 0$, for a regular trajectory; $K = \infty$, for purely random motion; K acquires a finite positive value for deterministic, but chaotic motion.

The mathematical definition of metric entropy, based on a probability measure of partitions of the phase space and its refinement under the dynamics, involves the estimate of some supremum which is, in practice, not an easy job. We turn to a more physical way of thinking. When dealing with the information dimension D_1 in Section 6.1.2, we were interested in the probability p_i of the i-th box being visited. Now we are concerned with distinguishable trajectories. It is natural to consider the joint probability $p(i_1, i_2, \ldots, i_m)$ that the moving point is in box i_1 at time t, in box i_2 at time $t + \Delta t$, ..., and in box i_m at time $t + (m - 1)\Delta t$. Then the metric entropy is defined, via the information, by (Farmer, 1982b; Grassberger and Procaccia, 1983)

$$K = - \lim_{\Delta t \to 0} \lim_{\epsilon \to 0} \lim_{m \to \infty} \frac{1}{m\Delta t} \sum_{i_1, \ldots, i_m} p(i_1, \ldots, i_m) \log p(i_1, \ldots, i_m), \quad (6.69)$$

where ϵ is the box size. Equation (6.69) may be readily generalized to the q-th moment of the joint probability, just as the definition of D_0 has been extended to D_q:

$$K_q = - \lim_{\Delta t \to 0} \lim_{\epsilon \to 0} \lim_{m \to \infty} \frac{1}{m\Delta t} \frac{1}{q-1} \sum_{i_1, \dots, i_m} p(i_1, \dots, i_m) \log p^q(i_1, \dots, i_m).$$

It follows from this definition that $K_1 = K$ and $K_q < K_{q'}$ for all $q > q'$. These K_q's have the advantage of being easily calculated from a time series. Grassberger and Procaccia (1983) considered the $q = 2$ case and estimated $\sum p^2(i_1, \dots, i_m)$ by counting the number of trajectory pairs that remain close to each other from time t to $t + (m-1)\Delta t$, i.e., by letting

$$C_m(\epsilon) = \lim_{N \to \infty} \frac{1}{N^2} \sum_{i,j} \theta(\epsilon - [\sum_{k=0}^{m-1} (\mathbf{y}_{i+k} - \mathbf{y}_{j+k})^2]^{1/2}).$$

In fact, this counting may be carried out for a general q, by changing to balls centered at points along the trajectory, as we did in Section 6.1.7 with D_q (Pawelzik and Schuster, 1987). The result of this process is

$$K_q = \lim_{\epsilon \to 0} \lim_{m \to \infty} \left[-\frac{1}{m\Delta t} \log C_m^q(\epsilon) \right],$$

where $C_m^q(\epsilon)$ is the further generalization of the generalized correlation integral (6.34):

$$C_m^q(\epsilon) = \left\{ \frac{1}{N} \sum_i \left[\frac{1}{N} \sum_j \theta \left(\epsilon - \left(\sum_{l=0}^{m-1} (\mathbf{y}_{i+l} - \mathbf{y}_{j+l})^2 \right)^{\frac{1}{2}} \right) \right]^{q-1} \right\}^{\frac{1}{q-1}}.$$

In the above formula, the $\mathbf{y}_i$ are vectors in the embedding space, reconstructed from the time series. In practice, the number m can be chosen to be the same as the embedding dimension. The calculation of K_q takes a few percent more computing time than the sole computation of K_2.

To conclude this section, we note that two topologically conjugate maps have the same metric entropy. For instance, the surjective logistic map and the tent map are, in this sense, equivalent. However, the converse is not, in general, true. Therefore, metric entropy is not a complete invariant.

6.5.2 The Topological Entropy

The topological entropy h, also an invariant under topological conjugations, is an even weaker characteristic of chaos, compared with the metric entropy K. Due to the inequalities

$$h \geq K \geq 0,$$

a positive topological entropy does not neccesarily guarantee a positive metric entropy, but that something chaotic does happen in the system: a set of trajectories of measure zero may scramble, there may be a chaotic transient, etc. The situation reminds us of the Li-Yorke statement "Period 3 implies chaos", which does not say anything about the stability or observability of the chaotic orbits. As we shall see, h does have a positive value in the period 3 window (and all other periodic windows, generated by tangent bifurcations) of the logistic map, although for the dominant regime there is certainly the attracting periodic orbit. When the positivity of h is taken as a criterion for chaos, people speak about "topological chaos" (see, e.g., MacKey and Tresser, 1986). Nevertheless, topological entropy has an intimate relationship to symbolic dynamics: for the piecewise linear tent map, one can write down the equation for h by simply looking at the corresponding symbolic sequences. This explains why much attention has been paid to the computation of topological entropy for mappings (see, e.g., Crutchfield and Packard, 1982, 1983; Collet, Crutchfield and Eckmann, 1983; Hsu and Kim, 1985; Chen and Chen, 1986).

The Kneading Determinant

In a mathematical paper, which seems to have remained for ever a preprint[17], Milnor and Thurston suggested a practical method for the calculation of the topological entropy of the tent map by looking at its symbolic sequence

$$\sigma_1 \sigma_2 \ldots \sigma_k \ldots ,$$

where $\sigma_i = R$, L, or C (for the tent map, we still use the letter C to denote the central point, where $f'(C)$ does not exist). One assigns parity -1 to each R and $+1$ to each L, as we did before. In addition, the letter C may appear among the σ_i. If this happens, it is assigned the product of the parities of all the preceding letters. Let us denote these parities by ϵ_i. The *kneading determinant* or the *characteristic function* is defined by

---

[17]J. Milnor and W. Thurston, Princeton preprint, 1977. This paper apparently played an essential role in the understanding of symbolic dynamics, in general, and topological entropy, in particular. Much to our regret, however, it is unavailable in China.

$$P(\tau) = 1 + \sum_{n=1}^{\infty} \prod_{j=1}^{n} \epsilon_j \tau^n. \tag{6.70}$$

It has been proved that the smallest positive root $\tau_{\min}$ of this polynomial determines the topological entropy h, i.e., that

$$h = -\log \tau_{\min}.$$

Consider, for example, the period 3 window (RLR, RLC, RLL). The assignment of parities proceeds as follows:

	R	L	R	R	L	R	R	L	R	$\cdots$
	R	L	C	R	L	C	R	L	C	$\cdots$
ϵ_i	-1	1	-1	-1	1	-1	-1	1	-1	$\cdots$
$\prod \epsilon_i$	-1	-1	1	-1	-1	1	-1	-1	1	$\cdots$
	R	L	L	R	L	L	R	L	L	$\cdots$
ϵ_i	-1	1	1	-1	1	1	-1	1	1	$\cdots$
$\prod \epsilon_i$	-1	-1	-1	1	1	1	-1	-1	-1	$\cdots$

The first two symbolic sequences yield the same polynomial, which can be summed into the rational fraction

$$P(\tau) = \frac{1 - \tau - \tau^2}{1 - \tau^3}.$$

The sequence $(RLL)^\infty$ leads to

$$P(\tau) = \frac{1 - \tau - \tau^2}{1 + \tau^3}.$$

Both functions have the same smallest positive root $(\sqrt{5} - 1)/2 = 0.618\ldots$. Therefore, the topological entropy remains constant in the entire window:

$$h = -\log((\sqrt{5} - 1)/2) = 0.4812118\ldots.$$

It is a general result that the topological entropy does not change in the whole periodic window, including its period-doubled tail.

The kneading determinant has the advantage that it can be written down for periodic as well as chaotic orbits, provided the symbolic sequence is known. Many examples of explicit calculations were given in the paper of Hsu and Kim (1985). If one is only interested in the entropy of periodic orbits, there is a method for obtaining the numerator of the $P(\tau)$ function directly from the symbolic sequence via the Stefan matrix.

The Stefan Matrix

The term Stefan matrix appeared in Derrida, Gervois and Pomeau (1978) with a reference to a private communication from P. Stefan. The topological entropy is obtained as the logarithm of the largest eigenvalue of the Stefan matrix, which can be readily constructed from the word. We describe only the working rule.

First, recall the λ autoexpansion of $1 < \lambda < 2$, when it corresponds to a finite period n (see Section 3.5.2):

$$\lambda = \sum_{i=0}^{n-2} \frac{a_i}{\lambda^i},$$

where $a_0 = a_1 = 1$, $a_i = \pm 1$. Next, try to represent these a_i as a product of $\alpha_j = \pm 1$:

$$a_i = \alpha_0 \alpha_1 \ldots \alpha_{i-1} \alpha_i,$$

with $\alpha_0 = 1$. Then these α_j are related to the word in the following manner:

Word	R	L	$\ldots$	σ_j	$\ldots$	σ_{n-2}	,
j	0	1	$\ldots$	j	$\ldots$	$n-2$	
α_j	1	1	$\ldots$	± 1	$\ldots$	± 1	

(Remember that an admissible word always starts with the letters RL.) The rule is: α_0 is always 1, $\alpha_j = -1$ if $\sigma_j = R$, $\alpha_j = 1$ if $\sigma_j = L$. Then the characteristic determinant of the Stefan matrix is simply

$$\det(\mathbf{S} - \lambda \mathbf{I}) = \begin{vmatrix} 1-\lambda & \alpha_1 & 0 & \cdots & 0 \\ 1 & -\lambda & \alpha_2 & \cdots & 0 \\ 1 & 0 & -\lambda & \cdots & 0 \\ \vdots & \vdots & \vdots & \vdots & \vdots \\ 1 & 0 & 0 & \cdots & \alpha_{n-2} \\ 1 & 0 & 0 & \cdots & -\lambda \end{vmatrix}.$$

A couple of examples may help the reader to master this rule.

Example 1. The period 5 word RL^2RC. The α_j are $1, 1, -1$ for $j = 1$, to 3. We do not list α_0, which is always 1 and does not enter the matrix. The characteristic equation now reads

$$\text{Det}(S - \lambda I) = \begin{vmatrix} 1-\lambda & 1 & 0 & 0 \\ 1 & -\lambda & 1 & 0 \\ 1 & 0 & -\lambda & -1 \\ 1 & 0 & 0 & -\lambda \end{vmatrix} = \lambda^4 - \lambda^3 - \lambda^2 - \lambda + 1 = 0,$$

which looks the same as the corresponding factor in the numerator of the kneading determinant. This is, however, a coincidence, since the particular equation preserves its form, when we make the substitution $\lambda = 1/\tau$.

Example 2. The period 3 word RLC leads to the even simpler determinant

$$\det(S - \lambda I) = \begin{vmatrix} 1-\lambda & 1 \\ 1 & -\lambda \end{vmatrix} = \lambda^2 - \lambda - 1 = 0,$$

the largest root of which is $\lambda_{max} = (\sqrt{5} + 1)/2$. On replacing λ by $1/\tau$, we get the factor $1 - \tau - \tau^2$ in the kneading determinant, the smallest real root of which is $\tau_{min} = 1/\lambda_{max} = (\sqrt{5} - 1)/2$, in agreement with our result in the previous section.

In order to see the origin of the Stefan matrix, we look at the period 3 iterations from a slightly different point of view. The three points C, $f(C)$ and $f^{(2)}(C)$ divide the interval into two segments: $f^{(2)}(C) - C - f(C)$. Count these segments from left to right as X_1 and X_2. Under the map, these segments transform into linear combinations of themselves, namely,

$$\begin{aligned} X_1 &\rightarrow X_2, \\ X_2 &\rightarrow X_1 + X_2, \end{aligned}$$

or, by introducing a transfer matrix,

$$R = \begin{pmatrix} 0 & 1 \\ 1 & 1 \end{pmatrix}, \tag{6.71}$$

$$\begin{pmatrix} X_1' \\ X_2' \end{pmatrix} = R \begin{pmatrix} X_1 \\ X_2 \end{pmatrix}.$$

The transfer matrix (6.71) can be brought into the form of the matrix S by a similarity transformation, corresponding to a regrouping of the segments X_i into new linear cobminations (Derrida, Gervois and Pomeau, 1978). It is clear that similar matrix may be constructed for chaotic orbits of the type $\rho\lambda^\infty$, since the symbolic sequence also divides the interval into a finite number of segments.

When repeated use of the transformation $\mathbf{R}$ leads to a fixed point of $\mathbf{R}^n$, or, equivalently, an n-cycle of $\mathbf{R}$; the trace of the matrix $\mathbf{R}^n$ gives the number of fixed points $N(n)$. For $n \gg 1$, it will be determined by the largest eigenvalue λ_{max} of $\mathbf{R}$. In fact, the topological entropy can be calculated by counting $N(n)$ directly.

The Number of Inverse Paths

For unimodal mappings, the topological entropy may be calculated by counting the number $I(n)$ of inverse paths of n or less steps, starting from the critical point C. For the notion of an inverse path, it is better to recall Fig. 3.7: there were two $f^{-1}(C)$ paths (solid circles) and $I(2) = 3$; four $f^{-2}(C)$ paths (solid squares) make $I(3) = 7$; there were only six $f^{-3}(C)$ paths (triangles), because the inverse paths from the point, labelled RL go beyond the scope of the map, hence $I(4) = 13$, etc. Furthermore, from the construction of Fig. 2.19, it follows that $I(n)$ also gives the number of monotonic branches in the map $f^{(n)}$. Topological entropy is then defined as

$$h(f) = \lim_{n \to \infty} \frac{\log I(n)}{n} \tag{6.72}$$

(see, e.g., Collet, Crutchfield and Eckmann, 1983; Chen and Chen, 1986).

This counting problem looks very simple in a few extreme cases. For the surjective logistic map, all inverse paths are present. Therefore, $I(n) = 2^n - 1$ and $h = \log 2$. We have already seen this figure as the Lyapunov number of the piecewise linear shift map in Section 6.4.1.

In the logistic map, when the parameter μ is less than the Feigenbaum accumulation point μ_∞, $I(n)$ changes like $2n - 1$. Consequently, the topological entropy

$$h = \lim_{n \to \infty} \frac{\log 2n}{n} = 0.$$

In general, the counting process may be cast into a set of recursion relations (Chen and Chen, 1986).

6.6 Relation Between Dimension, Lyapunov Exponents and Entropy

Since the dimension of an attractor measures the number of effectively excited degrees of freedom, it must be related to the number of non-contracting directions that span the attractor, i.e., the number of positive and zero Lyapunov exponents. Highly contracting directions do not show off in the creation of the attractor, but the first few negative exponents may compensate the expanding ones and modify somehow the attractor. These considerations lead us to the Kaplan and Yorke (1979) conjecture on the relation between D_0 and the Lyapunov exponents.

Suppose the Lyapunov exponents have been ordered in the usual way, i.e.,

$$\lambda_1 \geq \lambda_2 \geq \dots.$$

Let us add the exponents one by one, starting from the largest end, and watch the sign of the sum S. The sum of all positive and zero exponents clearly yields a positive number. Now add negative exponents and S diminishes. Suppose the sum S_k of the k exponents, including the positive, zero, and first negative ones, is still positive, but adding the next λ_{k+1} reduces the sum S_{k+1} to less than zero. It is reasonable to think that the dimension of the attractor lies between k and $k + 1$. In order to determine the fractional part Δk of the dimension, we interpolate linearly between S_k and $S_{k+1} = S_k + \lambda_{k+1}$:

$$\Delta k : S_k = (1 - \Delta k) : (|\lambda_{k+1}| - S_k).$$

For the sake of safety, let us call the dimension, thus defined, the *Lyapunov dimension* and denote it by D_Λ. We have

$$D_\Lambda = k + \frac{S_k}{|\lambda_{k+1}|}.$$

In practice, the Lyapunov dimension converges much quicker than the box-counting process, and in most cases $D_\Lambda = D_0$.

Among other relations between these ergodic characteristics we indicate only the Ruelle result that the metric entropy is bounded by the sum of positive Lyapunov exponents:

$$h \leq \sum (\lambda_i > 0).$$

The equality holds often, but not always (see Eckmann and Ruelle, 1985).

6.7   Peculiarity of One-dimensional Mappings

In what follows, we shall confine ourselves to one-dimensional unimodal mappings, because they have some specificities and at the same time are simple enough to be treated "almost analytically". In this case the point is that, we have only one Lyapunov number λ. If $\lambda > 0$, the dimension of the limiting set must be $D_0 = 1$, due to the existence of a continuous invariant measure on it. If $\lambda < 0$, the dimension of the limiting set, being a set of a finite number of periodic points, must be zero. Therefore, the only possibility for a non-integer dimension must be associated with those parameter values, where the Lyapunov number λ approaches zero. For unimodal maps, this happens in three different ways:

1. At every period-doubling bifurcation point μ_k, where λ approaches zero, while remaining negative on both sides of μ_k. The dimension of the limiting set must be zero by arguments of continuity. However, due to critical slowing down near μ_k (see Hao, 1981, and Section 7.1 below), one always sees clusters of points in the plot of the mapping, no matter how long one keeps on calculating. Therefore, if one measures the dimension of the limiting set, using a box-counting algorithm, one would always get a non-zero result. This "operational dimension" happens to be exactly 2/3 at any μ_k, provided k is finite (see Wang and Chen, 1984, and below). Although the introduction of the operational dimension is caused by transient phenomenon, it is appropriate to discuss it in the context of box-counting algorithm for dimensions.

2. At the accumulation point μ_∞ of the period-doubling bifurcation sequences, where the Lyapunov number λ changes from a negative value to a positive one, signaling the onset of chaos. It has been well known that the limiting set at μ_∞ has a Cantor set structure. We have given a simple first approximation [see Equation (6.16) in Section 6.1.5] to the exact value $D_0 = 0.538\ldots$, calculated by Grassberger (1981). In fact,

we have calculated all the D_q versus q curves for the limiting sets of the period-n-tupling sequences (see Section 6.1.6).

3. At the point of an intermittent transition from chaos to periodicity, the Lyapunov number passes through zero, being positive on the chaotic side and negative in the periodic region. The fractal dimension of the limiting sets at these points can be shown to be 1/2 by using mean-field arguments (Wang and Chen, 1984).

If one thinks a little about the two values of dimension cited in the first two points above, a question arises naturally: how does D_0 change from $0.666 \cdots$ at μ_k, k finite, to $0.538 \cdots$ as k goes to infinity? The answer (Hu and Hao, 1983) is that there is a scaling function $D(k, \epsilon)$, which depends both on the order k of the bifurcation and on the size of the boxes ϵ. The two different values are obtained by taking different limits of one and the same function $D(k, \epsilon)$, namely,

$$
\begin{aligned}
\lim_{k \to \infty} \lim_{\epsilon \to 0} D(k, \epsilon) &= 2/3 = 0.666\ldots, \\
\lim_{\epsilon \to 0} \lim_{k \to \infty} D(k, \epsilon) &= 0.538\ldots .
\end{aligned}
\tag{6.73}
$$

The interplay between k and ϵ can be incorporated into a scaling function $D(\theta)$ of a single argument $\theta = \epsilon^{1/k}$. We sketch below the derivation of this scaling function in order to display yet another example of the "almost analytical" discussion for the unimodal maps.

For the sake of simplicity, let us consider a unimodal mapping $x_{n+1} = f(\mu, x_n)$ at its first bifurcation point μ_1 (the discussion carries over to a general μ_k as well). Shifting the origin to the fixed point, one has

$$
x_{n+1} = -x_n + a x_n^2,
$$

where a is determined by the second derivative, i.e., the curvature of the map at the fixed point. Note that now the x_n are small deviations from the fixed point. Neglecting the terms of order x_n^4, we get

$$
x_{n+2} = x_n - 2a^2 x_n^3,
$$

which can readily be approximated by a differential equation, when n gets large enough:

$$\frac{dx}{dn} = -a^2 x^3.$$

It follows from this equation that, if we write $dn = \rho(x)dx$, the density of points $\rho(x)$ near the fixed point must be

$$\rho(x) = -\frac{1}{a^2\,|x|^3}.$$

Owing to critical slowing down close to every bifurcation point, one always gets clusters of points centred at the would-be periodic points in the plot of the mapping. The function $\rho(x)$ describes the distribution of the points within one cluster. In the case of μ_1, let us cover the cluster by boxes (intervals) of size ϵ and count the number $N(\epsilon)$ of boxes, needed to cover all the points. Due to the monotonicity of $\rho(x)$, there must exist an $x_0(\epsilon)$, satisfying

$$\epsilon\,\rho(x_0) = 1, \tag{6.74}$$

such that, when $x < x_0$, there are several points in each box, but for $x > x_0$, each point requires a separate box to cover it. Thus we have

$$N(\epsilon) = \frac{2x_0(\epsilon)}{\epsilon} + 2\int_{x(\epsilon)}^{\infty} \rho(x)dx = 3a^{-2/3}\epsilon^{-2/3}. \tag{6.75}$$

which, by the definition of fractal dimension, leads to the result

$$D_0 = \lim_{\epsilon \to 0} \frac{\log N(\epsilon)}{\log(\frac{1}{\epsilon})} = 2/3.$$

In order to study the behaviour at μ_∞, let us consider μ_k with both k and ϵ fixed. The iterated map $f^{(2)}$ has 2^k fixed points which are the periodic points of f itself at μ_k. Equation (6.75) can be applied to each of these points locally, so the number $N(k, \epsilon)$ of boxes needed to cover all points is

$$N(k, \epsilon) = \sum_{i=1}^{2} 3[a_i(k)]^{-2/3}\epsilon^{-2/3}, \tag{6.76}$$

where a_i is the curvature of the map at the i-th fixed point, and $a_i(k)^{-1}$, being a kind of characteristic length, is scaled by the same universal factor $\alpha = 2.5029\ldots$ as the distance between fixed points. Therefore, it must satisfy the estimate

$$\alpha^{-2k} \; < \; a_i(k)^{-1} \; < \; \alpha^{-k}. \tag{6.77}$$

The dependence of $a_i(k)^{-1}$ on i may be very complicated and non-universal. Nevertheless, one can replace it, on the basis of the estimate (6.77), by an averaged value

$$\overline{a^{-1}} = \alpha^{-\gamma k}. \tag{6.78}$$

For $k \gg 1$, the quantity $\overline{a^{-1}}$ measures the average distance between points in the limiting set, which, in turn, is related simply to the fractal dimension by

$$\lim_{k \to \infty} \frac{\log 2^k}{-\log a(k)^{-1}} = D_0. \tag{6.79}$$

By making use of the universal result (Grassberger, 1981) $D_0 = 0.538\cdots$, the exponent factor γ acquires the universal value

$$\gamma = \frac{\log 2}{D_0 \log \alpha} = 1.4043\ldots.$$

Combining Equations (6.78) and (6.76), we are led to

$$N(k, \epsilon) = C\, 2^k \alpha^{-\gamma k} \epsilon^{-2/3}, \tag{6.80}$$

where C is a finite constant, which does not depend on k and ϵ. Reflected in (6.80) is the interplay between k and ϵ. The larger k is, the higher the rate of convergence towards the limiting set, and the more points lie in the vicinity of the attractor. For a given ϵ, there exists such a k_0 that, when

$$x_0(k_0, \epsilon) = \epsilon, \tag{6.81}$$

the box covering a fixed point will eventually contain all the transients towards that point. From Equations (6.74) and (6.81), the parameter k_0 is defined by

$$k_0 = -\frac{\log \epsilon}{\gamma \log \alpha}. \tag{6.82}$$

At this k_0, the last two factors in (6.80) cancel out and one goes back to (6.79), i.e.,

$$D(k_0, \epsilon) = -\frac{\log N(k_0, \epsilon)}{\log \epsilon} = -\frac{k_0 \log 2}{\log \epsilon}.$$

It is obvious that for $k > k_0$, as more transients will be attracted to the fixed point, one always gets D_0, when measuring the dimension. On the contrary, when $k < k_0$, the number of boxes is given by (6.80), which, on taking into account (6.82), leads to

$$
\begin{aligned}
D(k, \epsilon) &= -\frac{\log N(k, \epsilon)}{\log \epsilon} \\
&= \frac{2}{3} + k\left(\frac{2}{3}\frac{\gamma \log \alpha}{\log \epsilon} - \frac{\log 2}{\log \epsilon}\right) \\
&= \frac{2}{3} + \frac{k}{k_0}(D_0 - \frac{2}{3}) \quad (k < k_0).
\end{aligned}
$$

Putting together the above arguments, we have

$$
D(k, \epsilon) = \begin{cases} \frac{2}{3} + \frac{k}{k}(D_0 - \frac{2}{3}), & k < k_0, \\ D_0, & k \geq k_0. \end{cases}
$$

Now it is a trivial matter to introduce the scaling variable θ and to present our result in the form

$$
D(\theta) = \begin{cases} \frac{2}{3} + \frac{\gamma \log \alpha}{\log \theta}(D_0 - \frac{2}{3}), & \theta > \alpha^\gamma, \\ D_0, & \theta \leq \alpha^\gamma. \end{cases} \tag{6.83}
$$

We note that a similar case of introducing a scaling variable to connect an apparent discontinuity has been treated in Zisook (1981), who studied the relation between the Feigenbaum convergence rate $\alpha = 4.669\ldots$ in dissipative systems and $\alpha = 8.721\ldots$ in conservative systems at the limit of vanishing dissipation.

Chapter 7

Transient Behaviour

So far, we have been studying the asymptotic behaviour of nonlinear systems as the time goes to infinity. In order to settle down at a stationary regime, one always has to throw away a certain number of iterates or periods while carrying out the numerical work. However, how could one be confident that the transients have really faded out? In studies of critical phenomena and phase transitions in physical systems, it has been known since a long time that there exists so-called critical slowing down: it would take a very very long time for a system to reach a new equilibrium state, after having been perturbed from an equilibrium state, if it had been very close to or was right at the critical point. The same happens with the time evolution of nonlinear systems, when they are close to a bifurcation point (see Section 7.1 below). We already knew of some consequence of this slowing down, when introducing the "operational dimension" in Section 6.7.

A nice example of transients can be seen in the system of differential equations[1]

$$\dot{x} = -x - 5y - zy,$$
$$\dot{y} = -y - 85x - xz, \qquad \qquad (7.1)$$
$$\dot{z} = 0.5z + xy + 1.$$

[1]It was discovered by Volta — a student at the Department of Physics, Genova University, in 1984, when doing his thesis with Prof. A. Borsellino and Dr. F. Arcardi.

417

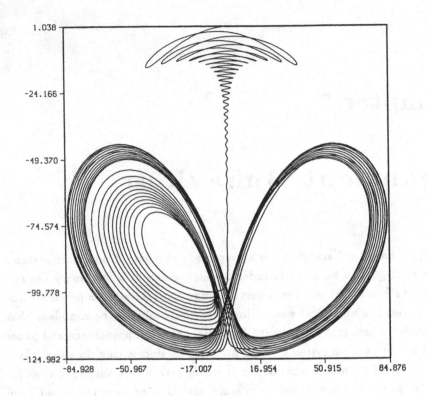

Figure 7.1: A transient trajectory of equations (7.1).

Figure 7.1 is a trajectory, started from $(x, y, z) = (8, 2, 1)$, and projected into a certain plane in the phase space (the initial point was close to the middle of the upper frame of the figure). If one were not patient enough, the lower part of the figure might be mistakenly considered a Lorenz type attractor. Actually, it was merely the fairly regular transient to the simple limit cycle, shown in Fig. 7.2.

As another set of examples, we recall the Poincaré maps of transitions between quasiperiodic and periodic regimes, shown in Figs 5.7 to 5.9. The beautiful patterns of transients indicate that transient behaviour may be much richer than the stationary state.

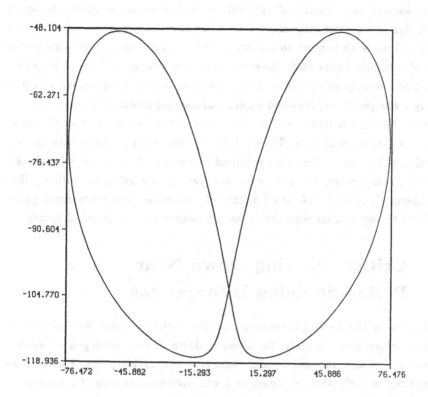

Figure 7.2: The long time limit of the trajectory in Fig. 7.1.

The examples mentioned above are associated with simple attractors. An experimentally observed chaotic attractor may be even more deceitful, as it might well be a chaotic transient, i.e., an erratic escape from a strange repeller towards a much simpler regular attractor (see Section 7.4 below).

In addition to the "negative" sides listed above, transient behaviour certainly plays a more constructive role in our understanding of complex systems. In the phase space of a complex system , besides trivial and strange attractors, there may exist many unstable and metastable objects. Transitions among these objects, before the system settles down at some asymptotic attractor, may give rise to a rich variety of transient phenomena. Moreover, all biological, ecological and economical systems, to which we are tempting to apply chaotic dynamics, exist only in a finite time range. Anyway, we ourselves are only transients with respect to history, so why not pay more attention to the study of transient behaviour? However, transient phenomena are rich and diverse, being mostly subject of current reserach. In this chapter, we will pick up only a few problems, centered on critical slowing down.

From the critical dynamics of phase transitions, we know that transient phenomena are more diverse. The equilibrium universality classes may be subdivided into finer ones. We have explained in Sections 2.7 and 2.9, why chaotic dynamics, being nonequilibrium in nature, reveals a parallelism with equilibrium thermodynamics. We shall again rely on some experience from phase transition theory to deal with the transient behaviour in dynamical systems.

7.1 Critical Slowing Down Near Period-doubling Bifurcations

We start with the basic phenomenon in the "critical dynamics" of period-doubling bifurcations, namely, the slowing down of the convergence, when a map is iterated at a parameter value close to a bifurcation point. For the sake of simplicity of derivation, we consider a one-dimensional map of the form

$$x_{n+1} = \mu f(x_n), \tag{7.2}$$

and denote the p-th iterate of this function by

$$F(p, \mu, x) = \underbrace{\mu f \circ \mu f \circ \cdots \mu f \circ \mu f(x)}_{p \text{ times}},$$

where $p = m 2^k$, for a period-doubling sequence associated with a primitive word of period m and μ_k, is the k-th bifurcation point. Without loss of generality, we assume that bifurcation occurs when μ crosses μ_k from below. Then, given p (or k), the mapping

$$x_{n+1} = F(p, \mu, x_n)$$

converges to a certain fixed point x^* for $\mu < \mu_k$, including all the μ_i with $i < k$. Let

$$x_n = x^* + \epsilon_n,$$

as we did in Section 2.3.1 and, as usual,

$$\epsilon_{n+1} = F'(p, \mu, x^*)\, \epsilon_n, \tag{7.3}$$

when μ is close to μ_k. The convergence condition $|F'| < 1$ breaks down at μ_k, where

$$|F'(p, \mu_k, x^*)| = 1. \tag{7.4}$$

Assume that $|\epsilon_n|$ diminishes as $e^{-n/\tau}$, then it follows immediately from (7.3) that

$$\tau = -\left(\log |F'(p, \mu, x^*)|\right)^{-1} = -\left(p \log \mu + \sum_{i=1}^{p} \log f'(x_i^*)\right)^{-1},$$

where we have used the chain rule of differentiation and the x_i^* are points forming the p-cycle. Expanding $\log \mu$ near μ_k, i.e., writing

$$\log \mu = \log[\mu_k + (\mu - \mu_k)] = \log \mu_k - \frac{\mu - \mu_k}{\mu_k} + \cdots,$$

and taking into account

$$p \log \mu_k + \sum_{i=1}^{p} f'(x_i^*) = 0,$$

we have, by (7.4),

$$\tau = \frac{\mu_k}{p(\mu_k - \mu)}.$$

Comparing this value with the critical slowing down exponent Δ, defined in the conventional theory of critical dynamics

$$\tau \propto |\mu - \mu_c|^{-\Delta},$$

we see that the exponent takes the simple, yet universal value (Hao, 1981; cf. Franaszek and Pieranski, 1985)

$$\Delta = 1.$$

It is not surprising that Δ coincides with the mean-field theory value in critical dynamics, since we have started from the assumption of exponential decay for ϵ_n. It would be much more interesting to look for deviations of Δ from 1. However, the answer can be anticipated from the following consideration.

What we have called τ is nothing but the inverse of the Lyapunov number λ [cf. (6.56)]. In other words, we have calculated the way by which λ approaches zero at a k-th order period-doubling point. Nothing new can be expected, so long as k remains finite. However, from scaling considerations, the behaviour of the Lyapunov exponent near the accumulation point μ_∞ has been shown to be (Huberman and Rudnick, 1980; Shraiman, Wayne and Martin, 1981; the noise strength σ has been put to zero):

$$\lambda(\mu_\infty - \mu, 0) \propto (\mu_\infty - \mu)^t,$$

where the exponent t is given by

$$t = \frac{\log 2}{\log \delta} = 0.44980\ldots .$$

According to what has been said above, we expect the slowing down exponent Δ at the accumulation point to be given by the reciprocal of t, i.e.,

$$\Delta = \frac{1}{t} = 2.223\ldots . \tag{7.5}$$

The discontinuity between $\Delta = 1$ and $2.223\ldots$ should be resolved in analogy to what we have done with the operational dimension of one-dimensional mappings (see Section 6.7).

7.2   Transient Precursor of Bifurcations

The long transients in systems close to a bifurcation point manifest themselves most clearly in power spectra. This can be simply explained in the example of a period-doubling bifurcation in a mapping (or in the Poincaré map of a continuous dynamical system). Close to, but below, the bifurcation point, the system will eventually settle down to an orbit of the old period n (or period 1, if looking at the n-th iterate $f^{(n)}$), if one waits for a sufficiently long time to allow the transients to die out. Figures 7.3 (a) and (b) show the two different ways along which the transients converge to the fixed point x^*. When the slope

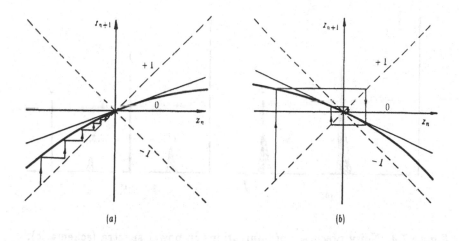

Figure 7.3: Convergence of the transients toward a fixed point. (a) The slope is between 1 and 0; (b) The slope is between 0 and -1.

at x^* is between 1 and 0, the transient points approach x^* from one side only, and there will be no new frequency component in the spectrum. As the slope changes to the interval $(0, -1)$, the transients fall, in turn, at two opposite sides of x^*. A point will come close to a previous point every two steps, thus giving rise to a new period 2 (or frequency 1/2) in the spectrum. In fact, we have used this property in Section 6.7 in the calculation of the box-counting dimension for islands formed by transients.

The described effect becomes more perceptible in the presence of noise and close to a bifurcation point, because, due to the critical slowing down, any fluctuation in the system will live longer and decay much more slowly. If an impatient experimentalist had taken the transient regime for a stationary state, he would see the type of spectra sketched in Figs. 7.4. In case (a), there will be a hump at the location of the fundamental frequency, without a sharp peak; in case (b), a hump at the 1/2 subharmonic frequency, again without a sharp

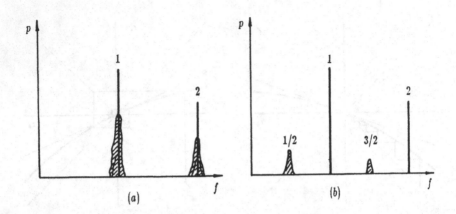

Figure 7.4: Noisy precursor of bifurcations in power spectra (schematic).
(a). The slope is close to +1; (b) The slope is close to −1, i.e., to a
period-doubling.

peak. The latter case makes it difficult to determine experimentally with high
precision the period-doubling point.

This kind of noisy precursors to bifurcations were analysed in detail by
Wiesenfeld (1985 a and b) and observed in experiments with a periodically
driven $p - n$ junction in the vicinity of a dynamical instability (Jeffries and
Wiesenfeld, 1985). Wiesenfeld's analysis employs the Floguet theory of ordinary differential equations with periodic coefficients and standard bifurcation
theory. The foregoing discussion on the example of one-dimensional mappings essentially covers the two cases, when a Floguet multiplier crosses +1
or −1. The more interesting case of a pair of complex Floguet multipliers[2]
crossing the unit circle (leading to a Hopf bifurcation) cannot be explored in
one-dimensional mappings, but may be observed in differential equations. Figure 7.5 shows schematically one episode in its development towards the next
period-doubling. The imaginary part of the pair gives rise to two noisy peaks
moving towards each other with a varying parameter (indicated by arrows in
the figure). They meet at the place of the new subharmonic, and then the

[2] See footnote on page 389.

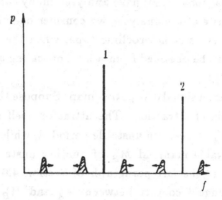

Figure 7.5: Noisy precursor to a Hopf bifurcation.

process evolves in a similar manner to what has been shown in Fig. 7.4(b).

A related case is the precursor near a symmetry-breaking bifurcation (Section 3.8.1). In the symmetric regime, only odd Fourier components are present in the spectrum, due to the antisymmetry of the equation. Close to the transition, noisy peaks may appear at the places of the even Fourier components. The situation resembles that shown in Fig. 7.4(b), with the difference that the sharp peaks correspond to odd frequency components. In addition, the phenomenon shows off only in certain periodic windows at the first stage of the period-doubling sequence, and does not repeat itself at the subsequent doublings. This is, of course, a consequence of the selection rules, that we have formulated in the language of symbolic dynamics (Section 3.8.2).

7.3   Chaotic Transients

Transient phenomena considered in the two previous sections are associated with bifurcations into periodic orbits and the transient behaviour itself is not chaotic. However, chaotic transients are also quite common and they are much more important from an experimental point of view, since they may be mistaken as true chaotic behaviour. As a rule, chaotic transients appear after a chaotic attractor collides with an unstable object and ceases to exist. Grebogi, Ott and Yorke (1983a, 1985, 1986) have analyzed many cases of such collisions. In order to have a taste of the analysis, we consider one of the simplest cases, namely, a boundary crisis of heteroclinic type, when the critical exponent of chaotic transients may be obtained from a kind of scaling arguments (Grebogi, Ott and Yorke, 1986).

Consider a two-dimensional dissipative map. Suppose that there is a chaotic attractor with a basin of attraction. The attractor itself is the closure of the unstable manifold M_A^u of a certain unstable period A, while the basin boundary is outlined by the stable manifold M_B^s of another unstable period B (recall what has been said in the last paragraph of Section 4.6.3). When we vary a parameter μ, a tangent contact between M_A^u and M_B^s occurs at $\mu = \mu_c$. This is a heteroclinic tangency. When $\mu > \mu_c$ the tangency develops into a small meniscus-shaped loop with two intersection points. Locally, this loop may be described by a parabola with characteristic sizes, proportional to r and $r^{1/2}$, where $r = \mu - \mu_c$. All points within this loop will be first attracted towards B along M_B^s and then get away along M_B^u. If we iterate these points backwards n steps, then they will approach the very neighbourhood of A. For sufficiently large n, this neighbourhood will be so small that the iteration may be well described by the linearized map at point A. The linearized map has two eigenvalues: $\alpha_1 > 1$ and $\alpha_2 < 1$, corresponding to the unstable and stable directions, respectively. After iterating back, the original linear sizes r and $r^{1/2}$ become

$$r \to r/\alpha_2^n, \quad r^{1/2} \to r^{1/2}/\alpha_1^n.$$

These two lengths characterize a small area, in which all points will get into the abovementioned meniscus-shaped region after n iterations and then escape. Therefore, the probability $p(r)$ that initial points fall in this region will be proportional to the probability of escape, which, in turn, measures the reciprocal of average lifetime of chaotic transients. Assume that

$$p(r) \propto r^\gamma, \tag{7.6}$$

where γ is an exponent characterizing the chaotic transients, we have

$$p(r) \propto \frac{r}{\alpha_2^n} \times \frac{r^{1/2}}{\alpha_1^n}.$$

Now, rescale r to $\alpha_2 r$ and iterate backwards $n + 1$ steps, we find

$$p(\alpha_2 r) \propto \frac{r}{\alpha_2^n} \times \frac{(\alpha_2 r)^{1/2}}{\alpha_1^{n+1}}.$$

Taking the ratio of these two probabilities, we arrive at

$$\frac{p(\alpha_2 r)}{p(r)} = \alpha_2^\gamma = \frac{\alpha_2^{1/2}}{\alpha_1},$$

which expresses the exponent γ through the two eigenvalues at A (Grebogi, Ott and Yorke, 1986):

$$\gamma = \frac{1}{2} + \frac{\log|\alpha_1|}{|\log|\alpha_2||}.$$

Critical exponent of chaotic transients for boundary crisis of homoclinic type may be obtained by similar arguments (Grebogi, Ott and Yorke, 1986).

7.4 Escape Rate from Strange Repellers

An unstable orbit in the basin of attraction of another attractor may act as a repeller. If one has managed to sit on it with infinitely high precision, one may stay there for ever. However, this never happens in practice, since infinitesimal deviations from the orbit, due to noise in a real experiment or round-off errors in a computer experiment, will be amplified and the point eventually escapes from the orbit, as if the latter repels. Unstable periodic orbits provide examples of trivial repellers. There may be *strange repellers* that have Cantor set structure, and the escape from them may be characterized by a well-defined escape rate (Kadanoff and Tang, 1984; Kadanoff, 1985).

The simplest example of strange repellers is given by those $\rho - \lambda$ orbits that coexist with attracting periods in one-dimensional mappings. We have seen in Section 3.6.7 that in one-dimensional mappings there are different types of chaotic orbits described by symbolic sequences of the $\rho\lambda^\infty$ type. One type that corresponds to a crisis, a homoclinic point, and a band-merging point, occurs at well-defined values of the parameter, that may be calculated, in principle, by the word-lifting technique (Section 3.2.1). These are the observable chaotic orbits or attractors. Another type of orbit may exist on some interval of the parameter, for example, in the period 3 window of the logistic map, but only has zero measure in the continuum of initial points. Nevertheless, they may show off as chaotic transients.

In higher dimensional systems, there may exist even more types of repellers. If they repel in certain directions and attract in others, a term *semi-attractor* has been coined for them (Kantz and Grassberger, 1985).

In order to introduce the notion of *escape rate*, let us consider a region $\mathcal{R}$, containing the repeller. The boundary of this region are regular, nonfractal surfaces or curves by construction. Consider a map f acting on points in this region. Suppose N_0 initial points were chosen which were distributed homogeneously in $\mathcal{R}$. After applying the map f^n to each initial point, we count the number N_n of points which still remain in $\mathcal{R}$. The *staying ratio* Γ_n is defined as

$$\Gamma_n = N_n/N_0,$$

when N_0 goes to infinity with n fixed. If Γ_n remains a positive number when $n \to \infty$, then there is an attractor in $\mathcal{R}$. If $\mathcal{R}$ contains an invariant set which repels, then Γ_n vanishes exponentially, i.e., $\Gamma_n \propto e^{-\alpha n}$ (Kadanoff, 1985). In this case the escape rate α is defined as

$$\alpha = -\lim_{n \to \infty} \frac{\log \Gamma_n}{n}.$$

Although a transient chaotic orbit approaches a non-chaotic attractor in the infinite time limit, it is possible to calculate Lyapunov exponents and various dimensions for the sufficiently long chaotic segment. Intuitively, the escape rate should have some relation to these exponents and dimensions. Indeed, Kantz and Grassberger (1985) have shown for one-dimensional mappings that

$$\alpha = (1 - D_1)\lambda,$$

where D_1 is the information dimension and λ the Lyapunov exponent. They have conjectured similar relation for higher dimensional systems, namely,

$$\alpha = \sum_{\{i,\, \lambda_i > 0\}} (1 - D_1^{(i)})\lambda_i,$$

where the sum runs over unstable directions, i.e., those having positive exponents, and $D_1^{(i)}$ is the partial (information) dimension[3] along the i-th direction.

---

[3]For partial dimensions see, e.g., Eckmann and Ruelle (1985).

References

Books, Conference Proceedings, and Collection of Papers

Entries in this part are ordered by their years of publication and are referred to as, for example, Poincaré (B1899). A more complete list will appear in the updated edition of Hao (B1984).

1. Poincaré H (1899), *Les Méthodes Nouvelles de la Mécanique Céleste*, tom 3, Gauthier-Villars.

2. Arnold V I, and Avez A (1968), *Ergodic Problems of Classical Mechanics*, Addison-Wesley.

3. Moser J (1973), *Stable and Random Motions in Dynamical Systems*, Princeton University Press.

4. Casati G, and Ford J (1977), eds., *Stochastic Behaviour in Classical and Quantum Hamiltonian Systems*, Lecture Notes in Physics **93**, Springer-Verlag.

5. Mori H (1978), ed., *Nonlinear Nonequilibrium Statistical Mechanics*, Proceedings of the Oji Seminar, Suppl. Progr. Thoer. Phys. **64**.

6. Collet P, and Eckmann J-P (1980), *Iterated Maps on the Interval as Dynamical Systems*, Birkhäuser.

7. Laval G, and Gresillon D (1980), eds., *Intrinsic Stochasticity in Plasmas*, Les Editions de Physique, Courtboeuf, Orsay.

8. Swinney H L, and Gollub J P (1981), eds., *Hydrodynamical Instabilities and the Transitions to Turbulence*, Springer-Verlag.

9. Haken H (1981), ed., *Chaos and Order in Nature*, Springer Series in Synergetics **11**, Springer-Verlag.

10. Sparrow C (1982), *The Lorenz Equations: Bifurcations, Chaos, and Strange Attractors*, Springer-Verlag.

11. Haken H (1982), ed., *Evolution of Order and Chaos*, Springer Series in Synergetics **17**, Springer-Verlag.

429

12. Bishop A R, Campbell D K, and Nocolaenko B (1982), eds., *Nonlinear Problems: Present and Future*, North-Holland.

13. Lichtenberg A J, and Liberman M A (1983), *Regular and Stochastic Motion*, Springer-Verlag.

14. Guckenheimer J, and Holmes P (1983), *Nonlinear Oscillations, Dynamical Systems and Bifurcations of Vector Fields*, Springer-Verlag.

15. Abraham R H, and Shaw C D (1983-1985), *Dynamics: the Geometry of Behavior*. Part I. *Periodic Behavior*; Part II. *Chaotic Behavior*; Part III. *Global Behavior*; Part IV. *Bifurcation Behavior*, Aerial Press.

16. Horton W, Reichl L, and Szebehely V (1983), eds., *Long-time Prediction in Nonlinear Dynamics*, Wiley.

17. Iooss G, Helleman R H G, and Stora R (1983), eds., *Chaotic Behaviour of Deterministic Systems*, Les Houches 1981, North-Holland.

18. Campbell D, and Rosen H (1983), eds., *Order in Chaos*, Proceedings of the International Conference on Order in Chaos held at the Center for Nonlinear Studies, Los Alamos, 24-28 May 1982, *Physica* **7D** and separate book, North-Holland.

19. Schuster H G (1984), *Deterministic Chaos. An Introduction*, 2nd edition, 1988, Physik-Verlag.

20. Bergé P, Pomeau Y, and Vidal Ch (1984), *L'ordre dans le Chaos, vers une approche deterministe de la turbulence*, Hermann Editions des Sciences. English translation (1986): *Order within Chaos, towards a deterministic approach to turbulence*, Wiley.

21. Hao Bai-lin (1984), ed., *Chaos*, An introduction and reprints volume, World Scientific.

22. Cvitanovic P (1984), ed., *Universality in Chaos*, An introduction and reprints volume, Adam Hilger.

23. Kuramoto Y (1984), ed., *Chaos and Statistical Methods*, Springer Series in Synergetics **24**, Springer-Verlag.

24. Chanda J (1984), ed., *Chaos in Nonlinear Dynamical Systems*, SIAM.

25. Horton C W, and Reichl L (1984), eds., *Statistical Physics and Chaos in Fusion Plasmas*, Wiley.

26. Shiraiwa K (1985), *Bibliography for Dynamical Systems*, Preprint Series No.1, 1985, Department of Mathematics, Nagoya University, Japan.

27. Velo G, and Wightman A S (1985), eds., *Regular and Chaotic Motions in Dynamic Systems*, Plenum.

28. Fischer D, and Smith W R (1985), eds., *Chaos, Fractals, and Dynamics*, Dekka.

29. Devaney R L (1986), *An Introduction to Chaotic Dynamical Systems*, Benjamin/Cummings.

30. Thompson J M T, and Stewart H B (1986), *Nonlinear Dynamics and Chaos. Geometrical methods for engineers and scientists*, Wiley.

31. Kaneko K (1986), *Collapse of Tori and Genesis of Chaos in Dissipative Systems*, World Scientific.

32. Steeb W-H, and Louw J A (1986), *Chaos and Quantum Chaos*, World Scientific.

33. Peitgen H-O, and Richter P H (1986), *The Beauty of Fractals. Images of Complex Dynamical Systems*, Springer-Verlag.

34. Holden A V (1986), ed., *Chaos*, Manchester University Press.

35. Mayer-Kress G (1986), ed., *Dimensions and Entropies in Chaotic Systems. Quantification of Complex Behaviour*, Springer Series in Synergetics, **32**, Springer-Verlag.

36. Bishop A R, Gruener A R, and Nicolaenko B (1986), eds., *Spatio-Temporal Coherence and Chaos in Physical Systems*, Physica, **23D**, North-Holland.

37. Nicolis J S (1987), *Chaotic Dynamics Applied to Biological Information Processing*, Akademie-Verlag Berlin.

38. Gleick J (1987), *Chaos: Making a New Science*, Viking.

39. Hao Bai-lin (1987), ed., *Directions in Chaos*, vol.1, World Scientific.

40. Procaccia I, and Shapiro M (1987), eds., *Chaos and Related Nonlinear Phenomena*, Plenum.

41. MacKay R S, and Meiss J D (1987), eds., *Hamiltonian Dynamical Systems*, a reprint selection, Adam Hilger.

42. Pike E R, Lugiato L (1987), eds., *Chaos, Noise and Fractals*, Adam Hilger.

43. Degn H, Holden A V, and Olsen L F (1987), eds., *Chaos in Biological Systems*, Plenum.

44. Glass L, and Mackey M C (1988), *From Clocks to Chaos. The Rhythms of Life*, Princeton University Press.

45. Kapitaniak T (1988), *Chaos in Systems with Noise*, World Scientific.

46. Hao Bai-lin (1988), ed., *Directions in Chaos*, vol. 2, World Scientific.

47. Livi R, Ruffo S, Ciliberto S, and Bniattl M (1988), eds., *Chaos and Complexity*, World Scientific.

48. Velarde M G (1988), ed., *Synergetics, Order and Chaos*, World Scientific.

49. Lundquist S, March N H, and Tosi M P (1988), eds., *Order and Chaos in Nonlinear Physical Systems*, Plenum Press.

Papers, including Reviews

Papers listed below are those which either have been referred to or have a closer relation to the text. More complete lists may be found in Shiraiwa (B1985), Hao (B1984) and the forthcoming updated edition of the latter.

1. Abraham N B, Gollub J P, and Swinney H L (1983) Testing nonlinear dynamics, *Physica*, 11D, 252.

2. Abraham N B, Albano A M, Das B, De Guzman G, Yong S, Gioggia R S, Puccioni G P, and Tredicce J R (1986) Calculating the dimension of attractors from small data sets, *Phys. Lett.*, 114A, 217.

3. Ahlers G, and Lücke M (1987) Some properties of an eight-mode Lorenz model for convection in binary fluids, *Phys. Rev.*, A35, 470.

4. Alekseev V M, and Yakobson M V (1981) Symbolic dynamics and hyperbolic dynamic systems, *Phys. Reports*, 75, 287.

5. Alfsen K H, and Frøyland J (1985) Systematics of the Lerenz model at $\sigma =10$, *Phys. Scripta*, 31, 15.

6. Alstrøm P (1986) Map dependence of the fractal dimension deduced from iterations of cirle maps, *Commun. Math. Phys.*, 104, 581.

7. Ananthakrishna G, Balakrishnan R, and Hao B-L (1987) Spatially chaotic spin patterns in a field-perturbed Heisenberg Chain, *Phys. Lett.*, 121A, 407.

8. Andrey L (1986) The relationship between entropy production and K-entropy, *Progr. Theor. Phys.*, 75, 1258.

9. Angelo P M, and Riela G (1981) A six-mode truncation of the Navier-Stokes equations on a two-dimensional torus: a numerical study, *Nuovo Cimento*, 64B, 207.

10. Arneodo A, Coullet P, and Tresser C (1979) A renormalization group with periodic behavior, *Phys. Lett.*, 70A, 74.

11. Arneodo A, Coullet P, and Tresser C (1982) Oscillators with chaotic bebehavior: an illustration of a theorem by Shil'nikov, *J. Stat. Phys.*, 27, 171.

12. Arnold V I (1965) Small denominators. I: Mappings of the circumference onto itself, *AMS Transl. Ser.* 2, 46, 213.

13. Aronson D G, Chory M A, Hall G R, and McGehee R P (1982) Bifurcations from an invariant circle for two-parameter families of maps of the plane: A computer-assisted study, *Commun. Math. Phys.*, 83, 303.

14. Aubry S (1983) Devil's staircase and order without periodicity in classical condensed matter physics, *J. Physique*, 44, 147.

15. Aubry S, and Le Daeron P Y (1983) The discrete Frenkel-Kontorova model and its extensions, *Physica*, 8D, 381.

16. Auerbach D, Cvitanovic P, Eckmann J-P, Gunaratne G and Procaccia I (1987) Exploring chaotic motion through periodic orbits, *Phys. Rev. Lett.*, 58, 2387.

17. Aurell E (1987) On the metric properties of the Feigenbaum attrractor, *J. Stat. Phys.*, 47, 439.

18. Avraham B-M, Procaccia I, and Grassberger P (1984) Characterization of experimental (noisy) strange attractors, *Phys. Rev.*, A29, 975.

19. Badii R, and Politi A (1985) Statistical description of chaotic attractors: the dimension function, *J. Stat. Phys.*, 40, 725.

20. Badii R, and Politi A (1987) Renyi dimensions from local expansion rates, *Phys. Rev.*, A35, 1288.

21. Bagley R J, Mayer-Kress J, and Farmer J D (1986) Mode locking, the Belousov-Zhabotinsky reaction, and one-dimensional mappings, *Phys. Lett.*, 114A, 419.

22. Baive D, and Franceschini V (1981) Symmetry breaking on a model of five-mode truncated Navier-Stokes equations, *J. Stat. Phys.*, 26, 471.

23. Bak P (1982) Commensurate phases ,incommensurate phases and the devil's staircase, *Repts. Progr. Phys.*, 45, 587.

24. Bak P, Bohr T, Jensen M H, and Christiansen P V (1984) Josephson junctions and circle maps, *Solid State Commun.*, 51, 231.

25. Bak P, Bohr T, and Jensen M H (1988) Circle maps, mode-locking and chaos, in Hao (B1988), p. 16.

26. Belair J, and Glass L (1985) Universality and self-similarity in the bifurcations of circle maps, *Physica*, 16D, 143.

27. Benedicks M, and Carleson L (1985) On iterations of $1 - ax^2$ on $(-1, 1)$, *Annals of Math.*, 122, 1.

28. Benettin G, Casartelli M, Galgani L, Giorgilli A, and Strelcyn (1978) On the reliability of numerical studies of stochasticity: I. Existence of time averages, *Nuovo Cimento*, 44B, 183.

29. Benettin G, Casartelli M, Galgani L, Giorgilli A, and Strelcyn (1979) On the reliability of numerical studies of stochasticity: II. Identification of time averages, *Nuovo Cimento*, 50B, 211.

30. Berre M L, Ressayre E, Gibbs H M, Kaplan D L, and Rose M H (1987) Conjecture on the dimensions of chaotic attractors of delayed-feedback dynamical systems, *Phys. Rev.*, A35, 4020.

31. Block L (1978) Homoclinic points of mappings of the interval, *Proc. AMS*, 72, 576.

32. Block L, Guckenheimer J, Misiurewicz M, and Young L-S (1979) Periodic points of one-dimensional maps, *Lect. Notes in Math.*, 819, 18.

33. Bohr T and Gunaratne G (1985) Scaling for supercritical circle maps: numerical investigation of the onset of bistability and period doubling, *Phys. Lett.*, 113A, 55.

34. Bohr T and Rand D (1987) The entropy function for characteristic exponents, *Physica*, 25D, 387.

35. Bohr T, Bak P, and Jensen M H (1984) Transition to chaos by interaction of resonances in dissipative systems. II. Josephson junctions, charge-density waves, and standard maps, *Phys. Rev.*, A30, 1970.

36. Bohr T, and Tél T (1988) The thermodynamics of fractals, in Hao (B1988), p. 194.

37. Boldrighini C, and Franceschini V (1979) A five-dimensional truncation of the plane incompressible Navier-Stokes equations, *Commun. Math. Phys.*, 64, 159.

38. Boyland P L (1986) Bifurcation of circle maps: Arnold tongues, bistability, and rotation intervals, *Commun. Math. Phys.*, 106, 353.

39. Brandstater A, Swift J, Swinney H L, Wolf A, Farmer J D, Jen (1983) Low-dimensional chaos in a hydrodynamic system, *Phys. Rev. Lett.*, 51, 1442.

40. Broomhead D S, McCreadie G, and Rowlands G (1981) On the analytic derivation of Poincare maps — the forced Brusselator problem, *Phys. Lett.*, 84A, 229.

41. Campanino M, and Epstein H (1981) On the existence of Feigenbaum's fixed point, *Commun. Math. Phys.*, 79, 261.

42. Casdagli M (1988) Rotational chaos in dissipative systems, *Physica*, 29D, 365.

43. Chang S-J, and McCown J (1984) Universal exponents and fractal dimensions of Feigenbaum attractors, *Phys. Rev.*, A30, 1149.

44. Chang S-J, and McCown J (1985) Universality behaviors and fractal dimensions associated with M-furcations, *Phys. Rev.*, A31, 3791.

45. Chen L-X, Li C-F, and Hong J (1984) Periodic and chaotic behaviours in optical bistability, *Chinese Phys. Lett.*, 1, 85.

46. Chen L-X, Li C-F, and Hong J (1985) Ikeda instability and degree of stability in optical bistability, *Acta Optica Sinica*, 5, 128 (in Chinese).

47. Chen R-X, and Chen S-G (1986) The topological entropy of one-dimensional unimodal maps, *Acta Physica Sinica*, 35, 1938 (in Chinese).

48. Chirikov B V (1979) A universal instability of many oscillator systems, *Phys. Reports*, 52, 265.

49. Cohen A, and Procaccia I (1985) Computing the Kolmogorov entropy from time signals of dissipative and conservative dynamical systems, *Phys. Rev.*, A31, 1872.

50. Coffman K, McCormick W D, and Swinney H L (1986) Multiplicity in a chemical reaction with one-dimensional dynamics, *Phys. Rev. Lett.*, 56, 999.

51. Collet P, Crutchfield J P, and Eckmann J-P (1983) Computing the topological entropy of maps, *Commun. Math. Phys.*, 88, 257.

52. Collet P, Eckmann J-P, and Koch H (1981) Period doubling bifurcations for families of maps on R^n, *J. Stat. Phys.*, 25, 1.

53. Conte R, and Dubois M (1988) Lyapunov exponents of experimental systems, in *Nonlinear Evolutions*, ed. J J P Leon, World Scientific, p. 767.

54. Coullet P, and Tresser C (1978) Iterations d'endomorphismes et groupe de renormalisation, *C. R. Acad. Sci. Paris*, 287, 577.

55. Cruchfield J P, Farmer J D, Packard N H, and Shaw R S (1980) Power spectral analysis of a dynamical system, *Phys. Lett.*, 76A, 1.

56. Crutchfield J P, Farmer J D, and Huberman B A (1982) Fluctuations and simple chaotic dynamics, *Phys. Reports*, 92, 45.

57. Crutchfield J P, Nauenberg M, and Rudnick J (1981) Scaling for external noise at the onset of chaos, *Phys. Rev. Lett.*, 46, 933.

58. Crutchfield J P, and Huberman B A (1980) Fluctuations and the onset of chaos, *Phys. Lett.*, 77A, 407.

59. Crutchfield J P, and Packard N H (1982) Symbolic dynamics of 1D maps: entropies, finite precision, and noise, *Int. J. Theor. Phys.*, 21, 433.

60. Crutchfield J P, and Packard N H (1983) Symbolic dynamics of noisy chaos, *Physica*, 7D, 201.

61. Curry J H (978) A generalized Lorenz system, *Commun. Math. Phys.*, 60, 193.

62. Curry J H (1979) On the Hénon transformation, *Commun. Math. Phys.*, 68, 129.

63. Curry J H (1980) An algorithm for finding closed orbit, *Lect. Notes in Math.*, 819, 111.

64. Curry J H (1981) On computing the entropy of the Hénon attractor, *J. Stat. Phys.*, 26, 683.

65. Cvitanović P, Jensen M H, Kadanoff L P, and Procaccia I (1985) Renormalization, unstable manifolds, and the fractal structure of mode locking, *Phys. Rev. Lett.*, 55, 343.

66. D'Humieres D, Beasley M R, Huberman B A, and Libchaber A (1982) Chaotic states and routes to chaos in the forced pendulum, *Phys. Rev.*, A26, 3483.

67. Da Costa L N, Knobloch E, and Weiss N O (1981) Oscillations in double diffusive convection, *J. Fluid Mech.*, 109, 25.

68. Daido H (1980) Analytic conditions for the appearance of homoclinic and heteroclinic points of a 2-dimensional mapping: the case of the Hénon mapping, *Progr. Theor. Phys.*, 63, 1190, 1831.

69. Daido H, and Tomita K (1979) Thermal fluctuation of a self-oscillating reaction system entrained by a periodic external force, *Progr. Theor. Phys.*, 61, 825.

70. De Gregorio S, Scoppola E, and Tirrozi B (1983) A rigorous study of periodic orbits by means of a computer, *J. Stat. Phys.*, 32, 25.

71. Derrida B, Gervois A, and Pomeau Y (1978) Iteration of endomorphisms on the real axis and representation of numbers, *Ann. Inst. Henri Poincaré*, 29A, 305.

72. Derrida B, Gervois A, and Pomeau Y (1979) Universal metric properties of bifurcations of endomorphisms, *J. Phys.*, A12, 269.

73. Dias de Deus J, Taborda Duarte J (1982) On the approach to the final aperiodic regime in maps of the interval, *Commun. Math. Phys.*, 84, 251.

74. Ding E-J (1986) Analytic treatment of periodic orbit: systametics for a nonlinear driven oscillator, *Phys. Rev.*, A34, 3547.

75. Ding E-J (1987a) Structure of parameter space for a prototype nonlinear oscillator, *Phys. Rev.*, A36, 1488.

76. Ding E-J (1987b) Scaling behavior in the supercritical sine circle map, *Phys. Rev. Lett.*, 58, 1059.

77. Ding E-J, and Hemmer P C (1987) Exact treatment of mode locking for a piecewise linear map, *J. Stat. Phys.*, 46, 99.

78. Ding M-Z, Hao B-L, and Hao X (1985) Power spectrum analysis and the nomenclature of periods in the Lorenz model, *Chinese Phys. Lett.*, 2, 1.

79. Ding M-Z, and Hao B-L (1988) Systematics of the periodic windows in the Lorenz model and its relation with the antisymmetric cubic map, *Commun. Theor. Phys.*, 9, 375.

80. Eckmann J-P (1981) Roads to turbulence in dissipative dynamical systems, *Rev. Mod. Phys.*, 53, 643.

81. Eckmann J-P, Kamphorst S O, Ruelle D, and Ciliberto S (1986) Liapunov exponents from time series, *Phys. Rev.*, A34, 4971.

82. Eckmann J-P, Thomas L E, and Wittwer P (1981) Intermittency in the presence of noise, *J. Phys.*, A14, 3153.

83. Eckmann J-P, and Ruelle D (1985) Ergodic theory of chaos and strange attractors, *Rev. Mod. Phys.*, 57, 617.

84. Eckmann J-P, and Wittwer P (1987) A complete proof of the Feigenbaum conjectures, *J. Stat. Phys.*, 46, 455.

85. Eidson J, Flynn S, Holm C, Weeks D , and Fox R F (1986) Elementary explanation of boundary shading in chaotic-attractor plots for the Feigenbaum map and the circle map, *Phys. Rev.*, A33, 2809.

86. El-Hamouly H, and Mira C (1982a) Lien entre les propriétés d'un endomorphisme et celles d'un difféomorphisme, *C. R. Acad. Sci. Paris*, sér.1, 293, 525.

87. El-Hamouly H, and Mira C (1982b) Singularites dues au feuilletage du plan des bifurcations d'un difféomorphisme bi-dimensionnel, *C. R. Acad. Sci. Paris*, sér.1, 293, 525.

88. Erber T, Johnson P, and Everett P (1981) Cebysev mixing and harmonic oscillator models, *Phys. Lett.*, 85A, 61.

89. Falk H (1984) Evolution of the density for a chaotic map, *Phys. Lett.*, 105A, 101.

90. Fan Y-S (1987) Periodic oscillations in the forced Brusselator, i*J. Shaanxi Normal U.*, Natur. Sci. Ser., No. 1, 1 (in Chinese).

91. Farmer J D (1981) Spectral broadening of period doubling bifurcation sequences, *Phys. Rev. Lett.*, 47, 179.

92. Farmer J D (1982a) Chaotic attractors of an infinite dimensional systems, *Physica*, 4D, 366.

93. Farmer J D (1982b) Information dimension and the probabilistic structure of chaos, *Z. Naturforsch.*, 37a, 1304.

94. Farmer J D (1985) Sensitive dependence on parameters in nonlinear dynamics, *Phys. Rev. Lett.*, 55, 351.

95. Farmer J D, Ott E, and Yorke J A (1983) The dimension of chaotic attractors, *Physica*, 7D, 153.

96. Farmer J D, and Satija I I (1985) Renormalization of the quasiperiodic transition to chaos for arbitrary winding numbers, *Phys. Rev.*, A31, 3520.

97. Farmer J D, and Sidorowich J J (1987) Predicting chaotic time series, *Phys. Rev. Lett.*, 59, 845.

98. Farmer J D, Crutchfield J P, Froehling H, Packard N H, Shaw R (1980) Power spectra and mixing properties of strange attractors, *Ann. N. Y. Acad. Sci.*, 357, 453.

99. Feigenbaum M J (1978) Quantitative universality for a class of nonlinear transformations, *J. Stat. Phys.*, 19, 25.

100. Feigenbaum M J (1979) The universal metric properties of nonlinear transformations, *J. Stat. Phys.*, 21, 69.

101. Feigenbaum M J (1980a) The onset spectrum of turbulence, *Phys. Lett.*, 74A, 375.

102. Feigenbaum M J (1980b) The transition to aperiodic behavior in turbulent systems, *Commun. Math. Phys.*, 77, 65.

103. Feigenbaum M J (1983) Universal behavior in nonlinear systems, *Physica*, 7D, 16.

104. Feigenbaum M J, Kadanoff L P, and Shenker S J (1982) Quasiperiodicity in dissipative systems: a renomalization group analysis, *Physica*, 5D, 370.

105. Feigenbaum M J, and Hasslacher B (1982) Irrational decimations and path intergrals for external noise, *Phys. Rev. Lett.*, 49, 605.

106. Feit S D (1978) Characterisic exponents snd strange attractors, *Commun. Math. Phys.*, 61, 249.

107. Franszek M, and Pieranski P (1985) Jumping particle model. Critical slowing down near the bifurcation point, *Canadian J. Phys.*, 63, 488.

108. Franceschini V (1980) A Feigenbaum sequence of bifurcations in the Lorenz model, *J. Stat. Phys.*, 22, 397.

109. Franceschini V (1983) Bifurcations of tori and phase locking in a dissipative system of differential equations, *Physica*, 6D, 285.

110. Franceschini V, and Russo L (1981) Stable and unstable manifolds of the Hénon mapping, *J. Stat. Phys.*, 25, 757.

111. Franceschini V, and Tebaldi C (1979) Sequqnces of infinite bifurcations and turbulence in a five-mode truncation of the Navier-Stokes equations, *J. Stat. Phys.*, 21, 707.

112. Franceschini V, and Tebaldi C (1981) A seven-mode truncation of the plane incompressible Navier-Stokes equations, *J. Stat. Phys.*, 25, 397.

113. Fraser A M, and Swinney H L (1986) Independent coordinates for strange attractors from mutual information, *Phys. Rev.*, A33, 1134.

114. Froehling H, Crutchfield J P, Farmer J D, Packard N H, Shaw R (1981) On determining the dimension of chaotic flows, *Physica*, 3D, 605.

115. Frøyland J (1983) Lyapunov exponents for multidimensional orbits, *Phys. Lett.*, 97A, 2.

116. Frøyland J, and Alfsen K H (1984) Lyapunov exponent spectra for the Lorenz model, *Phys. Rev.*, A29, 2928.

117. Fujisaka H, and Inoue M (1987) Statistical-thermodynamics formalism of self-similarity, *Progr. Theor. Phys.*, 77, 1334.

118. Fukuda W, and Katsura S (1986) Exactly solvable models showing chaotic behavior II, *Physica*, 136A, 588.

119. Gambaudo J-M, Glendinning P A, and Tresser C (1984) The rotation interval as a computable measure of chaos, *Phys. Lett.*, 105A, 97.

120. Gambaudo J-M, Procaccia I, Thomae S, and Tresser C (1986) New universal scenarios for the onset of chaos in Lorenz-type flows, *Phys. Rev. Lett.*, 57, 925.

121. Gao J-Y, Yuan J-M, and Narducci L M (1983) Instabilities and chaotic behavior in a hybrid bistable system with a short delay, *Opt. Commun.*, 44, 201.

122. Gaspard P (1983) Generation of a countable set of homoclinic flows through bifurcation, *Phys. Lett.*, 97A, 1.

123. Gaspard P, Kapral R, and Nicolis G (1984) Bifurcation phenomena near homoclinic systems:, *J. Stat. Phys.*, 35, 697.

124. Gaspard P, and Nicolis G (1983) What can we learn from homoclinic orbits in chaotic dynamics? *J. Stat. Phys.*, 31, 499.

125. Geisel T, Nierwetberg J, and Keller J (1981) Critical behavior of the Lyapunov number at the period-doubling onset of chaos, *Phys. Lett.*, 86A, 75.

126. Geisel T, and Nierwetberg J (1981) A universal fine structure of the chaotic region in period-doubling systems, *Phys. Rev. Lett.*, 47, 975.

127. Gibbon J D, and McGuinness M (1980) A derivation of the Lorenz equations for some unstable dispersive physical systems, *Phys. Lett.*, 77A, 295.

128. Gibbs H M, Hopf F A, Kaplan D L, and Shoemaker R L (1981) Observation of chaos in optical bistability, *Phys. Rev. Lett.*, 46, 474.

129. Glass L, Guevara M R, Belair J, and Shrier A (1984) Global bifurcations of a periodically forced biological oscillator, *Phys. Rev.*, A29, 1348.

130. Glass L, Guevara M R, and Shrier A, Perez R (1983) Bifurcation and chaos in a periodically stimulated cardiac oscillator, *Physica*, 7D, 89.

131. Glass L, and Perez R (1982) Fine structure of phase locking, *Phys. Rev. Lett.*, 48, 1772.

132. Glazier J A, Jensen M H, Libchaber A, and Stavans J (1986) Structure of Arnold tongues and the $f(\alpha)$ spectrum for period doubling: experimental results, *Phys. Rev.*, A34, 1621.

133. Gollub J P, and Benson S V (1980) Many routes to turbulent convection, *J. Fluid Mech.*, 100, 449.

134. Gong D-C, Qin G-R, Li R, Hu G, Mao J-Y,and Zhang L (1986) Experimental observation of the road from quasiperiodicity to chaos, *Kexue Tongbao (Sci. Bull. China)*, 31, 1601.

135. Gonzalez D L, and Piro O (1983) Chaos in a nonlinear driven oscillator with exact solution, *Phys. Rev. Lett.*, 50, 870.

136. Gonzalez D L, and Piro O (1985) Symmetric kicked self-oscillators: iterated maps, strange attractors, and symmetry of the phase-locking Farey hierarchy, *Phys. Rev. Lett.*, 55, 17.

137. Grassberger P (1981) On the Hausdorff dimension of fractal attractors, *J. Stat. Phys.*, 26, 173.

138. Grassberger P (1983a) On the fractal dimension of the Hénon attractor, *Phys. Lett.*, 97A, 224.

139. Grassberger P (1983b) Generalized dimensions of strange attractors, *Phys. Lett.*, 97A, 227.

140. Grassberger P, and Procaccia I (1983a) Estimation of the Kolmogorov entropy from a chaotic signal, *Phys. Rev.*, 28, 2591.

141. Grassberger P, and Procaccia I (1983b) Measuring the strangeness of strange attractors, *Physica*, 9D, 189.

142. Grassberger P, and Procaccia I (1983c) Characterization of strange attractors, *Phys. Rev. Lett.*, 50, 346.

143. Grassberger P, and Procaccia I (1983d) Estimation of the Kolmogorov entropy from a chaotic signal, *Phys. Rev.*, A28, 2591.

144. Grassberger P, and Procaccia I (1984) Dimensions and entropies of strange attractors from a fluctuating dynamics approach, *Physica*, 13D, 34.

145. Grebogi C, Ott E, Pelikan S, and Yorke J A (1984) Strange attractors that are not chaotic, *Physica*, 13D, 261.

146. Grebogi C, Ott E, and Yorke J A (1982a) Chaotic attractors in crisis, *Phys. Rev. Lett.*, 48, 1507.

147. Grebogi C, Ott E, and Yorke J A (1982b) Crises, sudden changes in chaotic attractors, and transient chaos, *Physica*, 7D, 181.

148. Grebogi C, Ott E, and Yorke J A (1983a) Fractal basin boundaries, long-lived chaotic transients, and unstable-unstable pair bifurcation, *Phys. Rev. Lett.*, 50, 935.

149. Grebogi C, Ott E, and Yorke J A (1983b) Are three-frequency quasiperiodic orbits to be expected in typical nonlinear dynamical systems? *Phys. Rev. Lett.*, 51, 339.

150. Grebogi C, Ott E, and Yorke J A (1985) Super persistent chaotic transients, *Ergod. Th. & Dynam. Sys.*, 5, 341.

151. Grebogi C, Ott E, and Yorke J A (1986) Critical exponent of chaotic transients in nonlinear dynamical systems, *Phys. Rev. Lett.*, 57, 1284.

152. Grebogi C, Ott E, and Yorke J A (1987) Unstable periodic orbits and the dimension of chaotic attractors, *Phys. Rev.*, A36, 3522.

153. Greenside H S, Wolf A, Swift J, and Pignataro T (1982) Impracticality of a box-counting algorithm for calculating the dimensionality of strange attractors, *Phys. Rev.*, A25, 3453.

154. Gregorio S, Scoppola E, and Tirozzi B (1983) A rigorous study of periodic orbits by means of a computer, *J. Stat. Phys.*, 32, 25.

155. Grossmann S, and Thomae S (1977) Invariant distributions and stationary correlation functions of the one-dimensional discrete processes, *Z. Naturforsch.*, 32a, 1353.

156. Gu Y (1987) Most stable manifolds and destruction of tori in dissipative dynamical systems, *Phys. Lett.*, 124A, 340.

157. Gu Y (1988) Most stable manifolds and transition to chaos in dissipative systems with competing frequencies, in Hao (B1988), p. 109.

158. Gu Y, Bandy D K, Yuan J-M, and Narducci L M (1985) Bifurcation routes in a laser with injected signal, *Phys. Rev.*, A31, 354.

159. Gu Y, Tung M, Yuan J-M, Feng D H, and Narducci L M (1984) Crises and hysteresis in coupled logistic maps, *Phys. Rev. Lett.*, 52, 701.

160. Gu Y, and Yuan J-M (1987) Classical dynamics and resonance structures in laser-induced dissociation of a Morse oscillator, *Phys. Rev.*, 36A, 3788.

161. Guckenheimer J (1977) Bifurcations of maps of the interval, *Inventiones Math.*, 39, 165.

162. Guckenheimer J (1980) Symbolic dynamics and relaxation oscillations, *Physica*, 1D, 227.

163. Guckenheimer J, and Williams R (1979) Structural stability of the Lorenz attractor, *Publ. Math. IHES*, 50, 307.

164. Guevara M R, and Glass L (1982) Phase locking, period-doubling bifurcations and chaos, *J. Math. Biol.*, 14, 1.

165. Gunaratne G H, and Procaccia I (1987) The organization of chaos, *Phys. Rev. Lett.*, 59, 1377.

166. Guo F, Pei L-Q, and Wu S-X (1986) Chaotic behaviour in a nonlinear circuit, *Acta Electronica Sinica*, 14, 29 (in Chinese).

167. Haken H (1975) Analogy between higher instabilities in fluids and lasers, *Phys. Lett.*, 53A, 77.

168. Haken H (1983) At least one Lyapunov exponent vanishes if the trajectory of an attractor does not contain a fixed point, *Phys. Lett.*, 94A, 71.

169. Halsey T C, Jensen M H, Kadanoff L P, Procaccia I, Shraiman B I (1986) Fractal measures and their singularities: the generalization of strange sets, *Phys. Rev.*, A33, 1141.

170. Hammel S M, Yorke J A, and Grebogi C (1987) Do numerical orbits of chaotic dynamical processes represent true orbits? *J. Complexity*, 3, 136.

171. Hao B-L (1981) Universal slowing-down exponent near period-doublihg bifurcation points, *Phys. Lett.*, 86A, 267.

172. Hao B-L (1982) Two kinds of entrainment-beating transitions in a driven limit cycle oscillator, *J. Theor. Biol.*, 98, 9.

173. Hao B-L (1983) Bifurcation, chaos, strange attractor, turbulence and all that, *Progress in Phys.*, 3, 329 (in Chinese).

174. Hao B-L (1985) Bifurcations and chaos in a periodically forced limit cycle oscillator, in *Advances in Science of China: Physics*, vol. 1, ed. Zhu Hong-yuan, Zhou Guang-zhou, and Fang Li-zhi, Science Press, p. 113.

175. Hao B-L (1986) Symbolic dynamics and systematics of periodic windows, *Physica*, 140A, 85.

176. Hao B-L (1987) Bifurcations and chaos in the periodically forced Brusselator, in *Collected Papers Dedicated to Professor Kazuhisa Tomita on the Occasion of his Retirement from Kyoto University*, Kyoto University, p.82.

177. Hao B-L (1988) Elementary symbolic dynamics, Chapter 14 in Lundqvist, March and Tosi, (B1988), p.387.

178. Hao B-L, Wang G-R, and Zhang S-Y (1983) U-sequences in the periodically forced Brusselator, *Commun. Theor. Phys.*, 2, 1075.

179. Hao B-L, and Ding M-Z (1988) Elementary symbolic dynamics in the study of chaos, in Velarde (B1988), p.706.

180. Hao B-L, and Zeng W-Z (1986) Information dimensions in unimodal mappings, in *Proceedings of the Sino-Japan Bilateral Workshop on Statistical Physics and Condensed Matter Theory*, ed. Xie Xi-de, World Scientific, p. 24.

181. Hao B-L, and Zeng W-Z (1987) Number of periodic windows in one-dimensional mappings, in *The XV International Colloquium on Group Theoretical Methods in Physics*, ed. R Gilmore, World Scientific, p. 199.

182. Hao B-L, and Zhang S-Y (1982a) Subharmonic stroboscopy as a method to study period-doubling bifurcations, *Phys. Lett.*, 87A, 267.

183. Hao B-L, and Zhang S-Y (1982b) Hierarchy of chaotic bands, *J. Stat. Phys.*, 28, 769.

184. Hao B-L, and Zhang S-Y (1983) Subharmonic stroboscopic sampling method for study of period-doubling bifurcation and chaotic phenomena in forced non-linear oscillators, *Acta Physica Sinica*, 32, 198 (in Chinese).

185. Hao B-L, and Zheng W-M (1988) Symbolic dynamics of unimodal maps revisited, Preprint ASITP-88-033, *Int. J. Mod. Phys.* B, to appear.

186. Hauser P R, Tsallis C, and Curado E M F (1984) Criticality of the routes to chaos of the $1 - a|x|^z$ map, *Phys. Rev.*, A30, 2074.

187. He D-R, Yeh W J, and Kao Y H (1984) Transition from quasiperiodicity to chaos in a Josephson-junction analog, *Phys. Rev.*, B30, 172.

188. He D-R, Yeh W-J, and Kao Y H (1985) Studies of return maps, chaos, and phase-locked states in a current driven Josephson-junction simulator, *Phys. Rev.*, B31, 1359.

189. Hemmer P C (1984) The exact invariant density for a cusp-shaped return map, *J. Phys.*, A17, L247.

190. Hénon M (1976) A two-dimensional mapping with a strange attractor, *Commun. Math. Phys.*, 50, 69.

191. Hénon M (1982) On the numerical computation of Poincaré maps, *Physica*, 5D, 412.

192. Hénon M, and Pomeau Y (1977) Two strange attractors with a simple structure, *Lect. Notes in Math.*, 565, 29.

193. Hentschel H G E, and Procaccia I (1983) The infinite number of generalized dimensions of fractals and strange attractors, *Physica*, 8D, 435.

194. Herman M R (1977) Measure de Lebesque et nombre de rotation, *Lect. Notes in Math.*, 597, 271.

195. Hirsch J E, Huberman B A, and Scalapino D J (1982) A theory of intermittence, *Phys. Rev.*, A25, 519.

196. Hirsch J E, Nauenberg M, and Scalapino D J (1982) Intermittency in the presence of noise: a renormalization group formulation, *Phys. Lett.*, 87A, 391.

197. Holmes P (1979) A nonlinear oscillator with a strange attractor, *Phil. Trans. Roy. Soc.*, A292, 419.

198. Hsu C S, and Kim M C (1985) Construction of maps with generating partitions for entropy evaluation, *Phys. Rev.*, A31, 3253.

199. Hu B (1982a) Introduction to real space renormalization group methods in critical and chaotic phenomena, *Phys. Reports*, 91, 233.

200. Hu B (1982b) A two dimensional scaling theory of intermittency, *Phys. Lett.*, 91A, 375.

201. Hu B (1983) A simple derivation of the stochastic eigenvalue equation in the transition from quasiperiodicity to chaos, *Phys. Lett.*, 98A, 79.

202. Hu B, and Mao J-M (1982) Period doubling: universality and critical-point order, *Phys. Rev.*, A25, 3259.

203. Hu B, and Rudnick J (1982a) Exact solutions to the Feigenbaum renormalization gruop equations for intermittency, *Phys. Rev. Lett.*, 48, 1645.

204. Hu B, and Rudnick J (1982b) Exact solutions to the renormalization-group fixed-point equations for intermittency in two-dimensional maps, *Phys. Rev.*, A26, 3035.

205. Hu B, and Rudnick J (1986) Differential-equation approach to functional equations: exact solutions for intermittency, *Phys. Rev.*, A34, 2453.

206. Hu B, and Satija I I (1983) A spectrum of universality classes in period doubling and period tripling, *Phys. Lett.*, 98A, 143.

207. Hu G (1986) The invariat distribution of non-fully developed chaos, *Chinese Phys. Lett.*, 3, 357.

208. Hu G, and Hao B-L (1983) A scaling relation for the Hausdorff dimension of the limiting sets in one-dimensional mappings, *Commun. Theor. Phys.*, 2, 1473.

209. Huang Y-N (1985) Determination of the stable periodic orbits for the Hénon map by analytical method, *Chinese Phys. Lett.*, 2, 98.

210. Huang Y-N (1986) An algebraic analytical method for exploring periodic orbits of the Hénon map, *Scientia Sinica* (Series A), 29, 1302.

211. Huberman B A, and Rudnick J (1980) Scaling behavior of chaotic flows, *Phys. Rev. Lett.*, 45, 154.

212. Huberman B A, and Zisook A B (1981) Power spectra of strange attractors, *Phys. Rev. Lett.*, 46, 626.

213. Hudson J L, and Mankin J C (1981) Chaos in the Belousov-Zhabotinskii reaction, *J. Chem. Phys.*, 74, 6171.

214. Huppert H E, and Moore D R (1976) Nonlinear double-diffusive convection, *J. Fluid Mech.*, 78, 821.

215. Ikeda K (1979) Multiple-valued stationary state and its instability of the transmitted light by a ring cavity system, *Opt. Commun.*, 30, 257.

216. Ikeda K, Daido H, and Akimoto O (1980) Optical turbulence: chaotic behavior of transmitted light from a ring cavity, *Phys. Rev. Lett.*, 45, 709.

217. Ikeda K, and Akimoto O (1982) Instability leading to periodic and chaotic self-pulsations in a bistable optical cavity, *Phys. Rev. Lett.*, 48, 617.

218. Ikeda K, and Kondo K (1982) Successive higher-harmonic bifurcations in systems with delayed feedback, *Phys. Rev. Lett.*, 49, 1467.

219. Ikeda K, and Matsumoto K (1987) High-diemnsional chaotic behavior in systems with time-delayed feedback, *Physica*, 29D, 223-235.

220. Ito R (1981) Rotation sets are closed, *Math. Proc. Camb. Phil. Soc.*, 89, 107.

221. Jakobson M V (1981) Absolutely continuous invariant measure for one parameter families of one dimensional maps, *Commun. Math. Phys.*, 81, 39.

222. Jeffries C, and Perez J (1982) Observation of a Pomeau-Manneville intermittent route to chaos in a nonlinear oscillator, *Phys. Rev.*, A26, 2117.

223. Jeffries C, and Perez J (1983) Direct observation of crises of the chaotic attractor in a nonlinear osillator, *Phys. Rev.*, A27, 601.

224. Jeffries C, and Wiesenfeld K (1985) Observation of noisy precursors of dynamical instabilities, *Phys. Rev.*, A31, 1077.

225. Jensen M H, Bak P, and Bohr T (1983) Complete devil's staircase, fractal dimension, and universality of mode-locking structure in the circle map, *Phys. Rev. Lett.*, 50, 1637.

226. Jensen M H, Bak P, and Bohr T (1984) Transition to chaos by interaction of resonances in dissipative systems. I. circle maps, *Phys. Rev.*, A30, 1960.

227. Jensen M H, Kadanoff L P, Libchaber A, Procaccia I, and Stavans J (1985) Global universality at the onset of chaos: results of a forced Rayleigh-benard experiment, *Phys. Rev. Lett.*, 55, 2798.

228. Jensen R, and Myers C R (1985) Images of the critical points of nonlinear maps, *Phys. Rev.*, A32, 1222.

229. Jetschke G, and Stiewe Ch (1985) Intermittency for tent maps is exactly calculable, *Phys. Lett.*, 112A, 265.

230. Jiang L-Y, and Peng S-L (1987) Rigorous bounds on the power spectra of arbitrary prime η-order renormalization group equations, *J. Phys.*, A20, 2325.

231. Kadanoff L P (1985) Applications of scaling ideas to dynamics, in Velo and Wightman (1985B).

232. Kadanoff L P, and Tang C (1984) Escape from strange repellers, *Proc. Natl. Acad. Sci. USA*, 81, 1276.

233. Kai T (1981) Universaility of power spectra of a dynamical system with an infinite sequence of period-doubling bifurcations, *Phys. Lett.*, 86A, 263.

234. Kai T (1982) Lyapunov number for a noisy 2^n Cycle, *J. Stat. Phys.*, 29, 329.

235. Kai T, and Tomita K (1979) Stroboscopic phase portrait of a forced nonlinear oscillator, *Progr. Theor. Phys.*, 61, 54.

236. Kaneko K (1982) On the period-adding phenomena at the frequency locking in a one-dimensional mapping, *Progr. Theor. Phys.*, 68, 669.

237. Kaneko K (1983a) Similarity structure and scaling property of the period-adding phenomena, *Progr. Theor. Phys.*, 69, 403.

238. Kaneko K (1983b) Doubling of torus, *Progr. Theor. Phys.*, 69, 1806.

239. Kaneko K (1983c) Transition from torus to chaos accompanied by frequency lockings with symmetry breaking, *Progr. Theor. Phys.*, 69, 1427.

240. Kaneko K (1984a) Fates of three-torus I. double devil's staircase in lockings, *Progr. Theor. Phys.*, 71, 282.

241. Kaneko K (1984b) Oscillation and doubling of torus, *Progr. Theor. Phys.*, 72, 202.

242. Kantz H, and Grassberger P (1985), Repellers, semi-attractors, and long-lived chaotic transients, *Physica*, 17D, 75.

243. Kaplan H (1983) New method for calculating stable and unstable periodic orbits, *Phys. Lett.*, 97A, 365.

244. Kaplan J L, and Yorke J A (1979a) Preturbulence: a regime observed in a fluid flow model of Lorenz, *Commun. Math. Phys.*, 67, 93.

245. Kaplan J L, and Yorke J A (1979b) The onset of chaos in a fluid flow model of Lorenz, *Ann. N. Y. Acad. Sci.*, 316, 400.

246. Katsura S, and Fukuda W (1985) Exactly solvable models showing chaotic behavior, *Physica*, 130A, 597.

247. Katzen D, and Procaccia I (1987) Phase transitions in thermodynamic formalism of multifractals, *Phys. Rev. Lett.*, 58, 1169.

248. Kawai H, and Tye S-H H (1984) Approach to chaos: universal quantitative properties of one-dimensional maps, *Phys. Rev.*, A30, 2005.

249. Keener J P, and Glass L (1984) Global bifurcations of a periodically forced nonlinear oscillator, *J. Math. Biol.*, 21, 175.

250. Keolian R, Putterman S J, Turkevich L A, Rudnick I, Rudnick J (1981) Subharmonic sequences in the Faraday experiment: departures from period-doubling, *Phys. Rev. Lett.*, 47, 1133.

251. Knobloch E, and Proctor M R E (1981) Nonlinear periodic convection in double diffusive systems., *J. Fluid Mech.*, 108, 291.

252. Knobloch E, and Weiss N O (1981) Bifurcations in a model of double-difusive convection, *Phys. Lett.*, 85A, 127.

253. Kumar K, Agarwal A K, Bhattacharjee J K, and Banerjee K (1987) Precursor transition in dynamical systems undergoing period doubling, *Phys. Rev.*, A35, 2334.

254. Lanford O E (1977) Computer pictures of the Lorenz attractor, Appendix to Williams (1977), *Lect. Notes in Math.*, 615, 113.

255. Lanford O E (1982) A computer-assisted proof of the Feigenbaum conjectures, *Bull. Amer. Math. Soc.*, 6, 427.

256. Lanford O E (1983) Period doubling in one and several dimensions, *Physica*, 7D, 124.

257. Li J-B, and Liu Z-R (1985) Chaotic properties of several forced nonlinear oscillator systems, *Acta Math. Scientia*, 2, 195 (in Chinese).

258. Li J-N (1985) Period-doubling bifurcation for a delay-differential equation related to optical bistability, *Chinese Phys. Lett.*, 2, 497.

259. Li J-N, and Hao B-L (1985) Bifurcation spectrum in a delay-differential system related to optical bistability, Preprint ASITP-85-021, *Commun. Theor. Phys.*, to appear.

260. Li T Y, Misiurewicz M, Pianigiani G, and Yorke J A (1982) Odd chaos, *Phys. Lett.*, 87A, 271.

261. Li T Y, and Yorke J A (1975) Period three implies chaos, Am. *Math. Monthly*, 82, 985.

262. Libchaber A, Laroche C, and Fauve S (1982) Period doubling cascade in mercury: quantitative measurement, *J. Physique Lett.*, 43, L211.

263. Ling F-H (1985) A numerical method for determining bifurcation curves of mappings, *Phys. Lett.*, 110A, 116.

264. Ling F-H (1988) Bifurcation curves of the Hénon map determined by a multiple shooting technique, *Chinese Phys. Lett.*, 5, 121.

265. Ling F-H, and Bao G-W (1987) A numerical implementation of Melnikov's method, *Phys. Lett.*, 122A, 413.

266. Liu K L, Lo W S, and Young K (1984) Generalized renormalization group equation for period-doubling bifurcations, *Phys. Lett.*, 105A, 10..

267. Liu K L, Lo W S, and Young K (1987) Entrainment and chaos in a discrete map with commensurate external forcing, *Nuovo Cimento*, 97, 170.

268. Liu K L, and Young K (1985) Stability of forced nonlinear oscillators via Poincaré map, *J. Math. Phys.*, 27, 502.

269. Liu K L, and Young K (1987) Generalized renormalization group analysis of period-doubling bifurcations, in Hao (B1987), p. 91.

270. Lorenz E N (1963) Deterministic nonperodic flow, *J. Atmos. Sci.*, 20, 130.

271. Lorenz E N (1980) Noisy periodicity and reverse bifurcation, *Ann. N. Y. Acad. Sci.*, 357, 282.

272. Lozi R (1978) Un attracteur etrange (?) du type attracteur de Hénon, *J. Physique*, 39, Coll. C5-9.

273. MacKay R S, and Tresser C (1986) Transition to topological chaos for circle maps, *Physica*, 19D, 206.

274. Machlup S, and Sluckin T J (1980) Driven oscillations of a limit-cycle oscillator, *J. Theor. Biol.*, 84, 119.

275. Malraison B, Atten P, Berge P, and Dubois M (1983) Dimension of strange attractors: an experimental determination for the chaotic regime of two convective systems, *J. Physique Lett.*, 44, 897.

276. Manneville P (1980) Intermittency in dissipative dynamical systems, *Phys. Lett.*, 79A, 33.

277. Manneville P, and Pomeau Y (1979) Intermittency and the Lorenz model, *Phys. Lett.*, 75A, 1.

278. Mao J-M, and Hu B (1987) Corrections to scaling for period doubling, *J. Stat. Phys.*, 46, 111.

279. Mao J-M, and Hu B (1988) Multiple scaling and the fine structure of period doubling, *Int. J. Mod. Phys.*, B2, 65.

280. Markus M, Kuschmitz D, and Hess B (1984) Chaotic dynamics in yeast glycolysis under periodic substrate input flux, *FEBS Lett.*, 172, 235.

281. Markus M, Müller S C, and Hess B (1985) Observation of entrainment, quasiperiodicity and chaos in glycolyzing yeast extracts under periodic glucose input, *Ber. Bunsenges. Phys. Chem.*, 89, 651.

282. Marotto F R (1978) Snap-back repellers imply chaos in R^n, *J. Math. Anal. Appl.*, 63, 199.

283. Marotto F R (1979) Chaotic behavior in the Hénon mapping, *Commun. Math. Phys.*, 68, 187.

284. Martin P C (1976) Instabilities, oscillations, and chaos, *J. Physique*, 37, Coll. C1-57.

285. May R M (1976) Simple mathematical models with very complicated dynamics, *Nature*, 261, 459.

286. May R M (1979) Bifurcations and dynamic complexity in ecological systems, *Ann. N. Y. Acad. Sci.*, 316, 517.

287. Mayer-Kress G (1987) Application of dimension algorithms to experimental chaos, in Hao (B1987), p. 122.

288. Mayer-Kress G, and Haken H (1981a) Intermittent behavior of the logistic system, *Phys. Lett.*, 82A, 151.

289. Mayer-Kress G, and Haken H (1981b) The influence of noise on the logistic model, *J. Stat. Phys.*, 26, 149.

290. Mayer-Kress G, and Haken H (1984) Attractors of convex maps with positive Schwarzian derivative in the presence of noise, *Physica*, 19D, 329.

291. McCreadie G A, and Rowlands G (1982) An analytical approximation to the Lyapunov number for 1D maps, *Phys. Lett.*, 91A, 146.

292. McDonald S W, Grebogi C, Ott E, and Yorke J A (1985) Fractal basin boundaries, *Physica*, 17D, 125.

293. McGuinness M J (1983) The fractal dimension of the Lorenz attractor, *Phys. Lett.*, 99A, 5.

294. Metropolis N, Stein M L, and Stein P R (1973) On finite limit sets for transformations on the unit interval, *J. Comb. Theor.*, A15, 25.

295. Misiurewicz M (1980) Strange attractors for the Lozi mapping, *Ann. N. Y. Acad. Sci.*, 357, 348.

296. Misiurewicz M, and Szlenk W (1980) Entropy of piecewise monotone mappings, *Studia Mathematica*, 67, 45.

297. Moore D R, Toomre J, Knobloch E, and Weiss N O (1983) Period doubling and chaos in partial differential equations for thermosolutal convection, *Nature*, 303, 663.

298. Mori H (1980) Fractal dimensions of chaotic flows of autonomous dissipative Systems, *Progr. Theor. Phys.*, 63, 1044.

299. Mori H, Okamoto H, and Ogasawara M (1984) Self-similar cascades of band splittings of linear mod 1 maps, *Progr. Theor. Phys.*, 71, 499.

300. Nauenberg M (1982) On the fixed points for circle maps, *Phys. Lett.*, 92A, 7.

301. Nauenberg M, and Rudnick J (1981) Universality and the power spectrum at the onset of chaos, *Phys. Rev.*, B24, 493.

302. Newhouse S E, Palis J, and Takens F (1983) Bifurcations and stability of families of diffeomorphisms, *Publ. Math. IHES*, No. 57, 5.

303. Newhouse S E, Ruelle D, and Takens F (1978) Occurrence of strange axiom A attractors near quasi-periodic flows on T^m ($m = 3$ or more), *Commun. Math. Phys.*, 64, 35.

304. Ni W-S, and Wei R-J (1985) Bifurcation and chaos in forced vibration systems containing a square nonlinear term, *Acta Physica Sinica*, 34, 503 (in Chinese).

305. Ni W-S, Tong P-I, and Hao B-L (1988) Homoclinic and heteroclinic intersections in the periodiocally forced Brusselator, Preprint ASITP-88-042, *Int. J. Mod. Phys. B*, to appear.

306. Oseledec V I (1968) A multiplicative ergodic theorem: Lyapunov characteristic numbers for dynamical systems, *Trans. Moscow Math. Soc.*, 19, 197.

307. Ostlund S, Rand D, Sethna J, and Siggia E (1983) Universal properties of the transition from quasiperiodicity to chaos in dissipative systems, *Physica*, 8D, 303.

308. Ostlund S, and Kim S H (1985) Renormalization of quasiperiodic mapping, *Phys. Scripta*, T9, 193.

309. Ott E (1981) Strange attractors and chaotic motions of dynamical systems, *Rev. Mod. Phys.*, 53, 655.

310. Ott E, Withers W D, and Yorke J A (1984) Is the dimension of chaotic attractors invariant under coordinate change? *J. Stat. Phys.*, 36, 687.

311. Packard N H, Crutchfield J P, Farmer J D, and Shaw R S (1980) Geometry from a time series, *Phys. Rev. Lett.*, 45, 712.

312. Pawelzik K, and Schuster H G (1987) Generalized dimensions and entropies from a measured time series, *Phys. Rev.*, 35, 481.

313. Pei L-Q, Guo F, Wu S-X, and Chua L O (1986) Experimental confirmation of the period-adding route to chaos in a nonlinear circuit, *IEEE Trans. Circuits Systems*, CAS-33, 438.

314. Peng S-L, and Qu C-C (1987) Existence of non-bijective solution for generalized Feigenbaum's functional equation, *Kexue Tongbao (Sci. Bull. China)*, 32, 371.

315. Perez R, and Glass L (1982) Bistability, period doubling bifurcations and chaos in a periodically forced oscillator, *Phys. Lett.*, 90A, 441.

316. Pomeau Y, and Manneville P (1980) Intermittent transition to turbulence in dissipative dynamical systems, *Commun. Math. Phys.*, 74, 189.

317. Procaccia I, Thomae S, and Tresser C (1987) First return maps as a unified renormalization scheme for dynamical systems, *Phys. Rev.*, A35, 1884.

318. Qian M, and Yan Y (1986) Transversal heteroclinic cycle and its application to Hénon mapping, *Kexue Tongbao (Sci. Bull. China)*, 31, 10.

319. Qin G-R, Gong D-C, Yang C-Y, Hu G, Mao J-Y, and Zhang L (1985) Division of frequency and chaos, *Chinese Phys. Lett.*, 2, 35.

320. Qiu X-M, and Wang X-G (1986) A new strange attractor in MHD flow in sheared magnetic field, *Chinese Phys. Lett.*, 3, 105.

321. Qiu X-M, and Wang X-G (1987) Chaotic attractor of MHD flow in a sheared magnetic field, *Chinese Phys. Lett.*, 4, 49.

322. Rand D, Ostlund S, Sethna J, and Siggia E D (1982) A universal transition from quasi-periodicity to chaos in dissipative systems, *Phys. Rev. Lett.*, 49, 132.

323. Richetti P, Roux J L, Argoul F, and Arneodo A (1987) From quasiperiodicity to chaos in the Belousov-Zhabotinskii reaction II: modelling and theory, *J. Chem. Phys.*, 86, 3339.

324. Riela G (1982) A new six-mode truncation of the Navier-Stokes equations on a two-dimensional torus: a numerical study, *Nuovo Cimento*, 69B, 295.

325. Robbins K A (1979) Periodic solutions and bifurcation structure at high R in the Lorenz model, *SIAM J. Appl. Math.i*, 36, 457.

326. Rössler O E (1979a) Chaotic oscillations — an example of hyperchaos, *Lect. Notes in Appl. Math.*, vol. 17.

327. Rössler O E (1979b) An equation for hyperchaos, *Phys. Lett.*, 71A, 155.

328. Ruelle D (1979) Sensitive dependence on initial condition and turbulent behavior of dynamical systems, *Ann. N. Y. Acad. Sci.*, 316, 408.

329. Ruelle D (1981) Small random perturbations of dynamical systems and the definition of attractors, *Commun. Math. Phys.*, 82, 137.

330. Ruelle D (1983) Five turbulent problems, *Physica*, 7D, 40.

331. Ruelle D, and Takens F (1971) On the Nature of turbulence, *Commun. Math. Phys.*, 20, 167; 23, 343.

332. Russell D A, Hanson J D, and Ott E (1981) Dimension of strange attractors, *Phys. Rev. Lett.*, 45, 1175.

333. Saltzman B (1962) Finite amplitude convection as an initial value problem. I, *J. Atmos. Sci.*, 19, 329.

334. Sano M, and Sawada Y (1985) Measurement of the Lyapunov spectrum from a chaotic time series, *Phys. Rev. Lett.*, 55, 1082.

335. Sarkovskii A N (1964) Coexistence of cycles of a continuous map of a line into itself, *Ukranian Math. J.*, 16, 61.

336. Sato S, Sano M, and Sawada Y (1983) Universal scaling property in bifurcation structure of Duffing's and of generalized Duffing's equations, *Phys. Rev.*, A28, 1654.

337. Schmitz R A, Graziani K R, and Hudson J L (1977) Experimental evidence of chaotic states in the Belousov-Zhabotinskii reaction, *J. Chem. Phys.*, 67, 3040.

338. Schmutz M, and Rueff M (1984) Bifurcation schemes of the Lorenz model, *Physica*, 11D, 167.

339. Schreiber I, and Marek M (1982) Strange attractors in coupled reaction-diffusion cells, *Physica*, 5D, 258.

340. Sethna J P, and Siggia D E (1984) Universal transition in a dynamical system forced at two incommensurate frequencies, *Physica*, 11D, 193.

341. Shaw R S (1981) Strange attractors, chaotic behavior and information flow, *Z. Naturforsch.*, 36A, 80.

342. Shenker S J (1982) Scaling behavior in a map of a circle onto itself: empirical results, *Physica*, 5D, 405.

343. Shimada T, and Morioka N (1978) Chaos and limit cycles in the Lorenz model, *Phys. Lett.*, 66A, 182.

344. Shimada T, and Nagashima T (1979) A numerical approach to ergodic problem of dissipative systems, *Progr. Theor. Phys.*, 61, 1605.

345. Shraiman B I (1984) Transition from quasiperiodicity to chaos: a perturbative renormalization-group approach, *Phys. Rev.*, A29, 3464.

346. Shraiman B, Wayne C E, and Martin P C (1981) A scaling theory for noisy period-doubling transitions to chaos, *Phys. Rev. Lett.*, 46, 935.

347. Silnikov L (1969) On a new type of bifurcation of multidimensional dynamical systems, *Sov. Math. Dokl.*, 10, 1368; Russ. Orig. Doklady 189, 59.

348. Simm C W, Sawley M L, Skiff F, and Pochelon A (1987) On the analysis of experimental signals for evidence of deterministic chaos, *Helvetica Physica Acta*, 60, 510.

349. Simò C (1979) On the Hénon-Pomeau attractor, *J. Stat. Phys.*, 21, 465.

350. Simoyi R H, Wolf A, and Swinney H L (1982) One-dimensional dynamics in a multi-component chemical reaction, *Phys. Rev. Lett.*, 49, 245.

351. Sinai Ya., and Vul E B (1980) Discovery of closed orbits of dynamical systems with the use of computers, *J. Stat. Phys.*, 23, 27.

352. Singer D (1978) Stable orbits and bifurcations of maps of the interval, *SIAM J. Appl. Math.*, 35, 260.

353. Smale S (1967) Differentiable dynamical systems, *Bull. Am. Math. Soc.*, 13, 747.

354. Stefan P (1977) A theorem of Sarkovskii on the existence of periodic orbits of continuous endomorphisms of the real line, *Commun. Math. Phys.*, 54, 237.

355. Stein P, and Ulam S (1964) Nonlinear transformation studies on electronic computers, *Rozprawy Metamatyczne*, 339, 401.

356. Swift J W, and Wiesenfeld K (1984) Suppression of period doubling in symmetric systems, *Phys. Rev. Lett.*, 52, 705.

357. Swinney H L (1983) Observations of order and chaos in nonlinear systems, *Physica*, 7D, 3.

358. Szépfalusy P, and Tél T (1986), New approach to the problem of chaotic repellers, *Phys. Rev.*, A34, 2520.

359. Takens F (1981) Detecting strange attractors in turbulence, *Lect. Notes in Math.*, 898, 336.

360. Tedeschini-Lalli L (1982) Truncated Navier-Stokes equations: continuos transition from a five-mode to a seven-mode model, *J. Stat. Phys.*, 27, 365.

361. Tél T (1982a) On the construction of invariant curves of period-two points in two-dimensional maps, *Phys. Lett.*, 94A, 334.

362. Tél T (1982b) On the construction of stable and unstable manifolds of two-dimensional invertible maps, *Z. Phys.*, B49, 157.

363. Tél T (1983) Invariant curves, attractors, and phase diagram of a piecewise linear map with chaos, *J. Stat. Phys.*, 33, 195.

364. Tél T (1986) Characteristic exponents of chaotic repellers as eigenvalues, *Phys. Lett.*, 119A, 65.

365. Tél T (1987) Escape rate from strange sets as an eigenvalue, *Phys. Rev.*, A36, 1502.

366. Termonia Y (1984) Kolmogorov entropy from a time series, *Phys. Rev.*, A29, 1612.

367. Testa J, and Held G A (1982) Period doubling, bifurcations, chaos and periodic windows of the cubic map, *Phys. Rev.*, A28, 3085.

368. Tomita K (1982) Chaotic response of nonlinear oscillators, *Phys. Reports*, 86, 113.

369. Tomita K, and Kai T (1978) Stroboscopic phase portrait and strange attractors, *Phys. Lett.*, 66A, 91.

370. Tomita K, and Kai T (1979a) Stroboscopic phase portrait of a forced nonlinear oscillator, *Progr. Theor. Phys.*, 61, 54.

371. Tomita K, and Kai T (1979b) Chaotic response of a nonlinear oscillator, *J. Stat. Phys.*, 21, 65.

372. Tomita K, and Tsuda I (1980) Towards the interpretation of the global bifurcation structure of the Lorenz system: a simple one-dimensional model, *Progr. Theor. Phys. Suppl.*, 69, 185.

373. Tresser C, Coullet P, and Arneodo A (1980) Topological horseshoe and numerically observed chaotic behavior in the Hénon mapping, *J. Phys.*, A13, L123.

374. Tresser C, and Coullet P (1978) Iterations de endomorphismes et groupe de renormalisation, *C. R. Acad. Sci. Paris*, 287A, 577.

375. Turner J S, Roux J C, McCormick W D, and Swinney H L (1981) Alternating peiodic and chaotic regimes in a chemical reaction — experiment and theory, *Phys. Lett.*, 85A, 9.

376. Tyson J J (1973) Some further studies of nonlinear oscillations in chemical systems, *J. Chem. Phys.*, 58, 3919.

377. Ulam S M, and von Neumann J (1947) On combinations of stochastic and deterministic processes, *Bull. Amer. Math. Soc.*, 53, 1120.

378. Vastano J A, and Kostelich E J (1986) Comparison of algorithms for determining Lyapunov exponents from experimental data, in Mayer-Kress, (B1986).

379. Velarde M G (1981) Steady states, limit cycles and the onset of turbulence: a few model calculations and exercises, in *Nonlinear Phenomena at Phase Transitions and Instabilities*, ed. T. Riste, Plenum.

380. Wang B-R, Miao G-Q, and Wei R-J (1984) Experimental observation of subharmonic in liquid Nitrogen and water, *Acta Physica Sinica*, 33, 434 (in Chinese).

381. Wang G-R (1983) The period-doubling bifurcation sequences of the trimolecular model with forced osillation term, *Acta Physica Sinica*, 32, 960 (in Chinese).

382. Wang G-R, Chen S-G, and Hao B-L (1983) Intermittent chaos in the forced Brusselator, *Acta Physica Sinica*, 32, 1139 (in Chinese).

383. Wang G-R, Chen S-G, and Hao B-L (1984a) Kolmogoroff capacity and Lyapunoff dimension of strange attractors in the forced Brusselator, *Acta Physica Sinica*, 33, 1246 (in Chinese).

384. Wang G-R, Chen S-G, and Hao B-L (1984b) On the nonconvergence problem in computing the capacity of strange attractors, *Chinese Phys. Lett.*, 1, 11.

385. Wang G-R, Chen S-G, and Hao B-L (1985) Numerical calculation of the attractor dimensions, *Chinese J. Comput. Phys.*, 2, 47 (in Chinese).

386. Wang G-R, and Chen S-G (1986) Universal constants and universal functions of period-n-tupling sequences in one-dimensional unimodal mappings, *Acta Physica Sinica*, 35, 58 (in Chinese).

387. Wang G-R, and Hao B-L (1984) Transition from quasiperiodic regime to chaos in the forced Brusselator, *Acta Physica Sinica*, 33, 1321 (in Chinese).

388. Wang P-Y, Dai J-H, Zhang H-J (1985) Bifurcation, chaos and transient behavior in liquid crystal hybrid optical bistable devices, *Acta Physica Sinica*, 34, 581 (in Chinese).

389. Wang P-Y, Dai J-H, and Zhang H-J (1984) Bifurcation, chaos and transient behavior in liguid crystal hybrid optical bistable devices, *Acta Physica Sinica*, (in Chinese).

390. Wang Y-Q, and Chen S-G (1984) Metric properties of chaotic region in one-dimensional maps, *Acta Physica Sinica*, 33, 341 (in Chinese).

391. Wang Y-Q, and Chen S-G (1986) Universal measure of one dimensional unimodal maps, *Chinese J. Comput. Phys.*, 3, 387 (in Chinese).

392. Wang Y-X, He Y-S, and Peng S-L (1984) Frequency sweep method for study of period-doubling bifurcation and chaotic behaviour, *Acta Physica Sinica*, 33, 671 (in Chinese).

393. Wegmann K, and Rössler O E (1978) Differenct kinds of chaotic oscillations in the Belousov-Zhabotinskii reaction, *Z. Naturforsch.*, 33a, 1179.

394. Wei R-J, Tao Q-T, and Ni W-S (1986) Bifurcation and chaos of direct radiation loudspeaker, *Chinese Phys. Lett.*, 3, 469.

395. Wiesenfeld K (1985a) Virtual Hopf phenomenon: a new precursor of period-doubling bifurcations, *Phys. Rev.*, A32, 1744.

396. Wiesenfeld K (1985b) Noisy precursors of nonlinear instabilities, *J. Stat. Phys.*, 38, 1071.

397. Wiesenfeld K, Knobloch E, Miracky R F, and Clarke J (1984) Calculation of period-doubling in a Josephson circuit, *Phys. Rev.*, A29, 2102.

398. Williams R F (1977) The structure of Lorenz attractors, *Lect. Notes in Math.*, 615, 94.

399. Williams R F (1979) The structure of Lorenz attractors, *Publ. Math. IHES*, 50, 321.

400. Wolf A, Swift J B, Swinney H L, and Vastano J A (1985) Determining Lyapunov exponents from a time series, *Physica*, 16D, 285.

401. Wolf A, and Swift J (1981) Universal power spectra for the reverse bifurcation sequence, *Phys. Lett.*, 83A, 184.

402. Wu S-X, Pei L-Q, and Guo F (1985) U-sequence and period-tripling phenomena in a forced nonlinear oscillator, *Chinese Phys. Lett.*, 2, 213.

403. Xu J-H, and Li W (1986) The dynamics of large scale neuron-glia network and its relation to the brain function (I), *Commun. Theor. Phys.*, 5, 339.

404. Xu N, and Xu J-H (1988) The fractal dimensions of EEG as a physical measure of conscious human brain activities, *Bull. Math. Biology*, 50, 559.

405. Xu Y, Mao Z-M, Duan Y-L (1986) Mechanism of bifurcation and chaos in unijunction transistor second order circuit, *Acta Electronica Sinica*, 14, 123 (in Chinese).

406. Yang W-M, and Hao B-L (1987) How the Arnold tongues become sausages in a piecewise linear circle map, *Commun. Theor. Phys.*, 8, 1.

407. Yeh W-J, He D-R, and Kao Y H (1984) Fractal dimension and self-similarity of the devil's staircase in a Josephson-junction simulator, *Phys. Rev. Lett.*, 52, 480.

408. Yeh W-J, and Kao Y H (1982) Universal scaling and chaotic behavior of a Josephson-junction analog, *Phys. Rev. Lett.*, 49, 1888.

409. Yeh W-J, and Kao Y H (1983) Intermittency in Josephson junctions, Appl. *Phys. Lett.*, 42, 299.

410. Yorke J A, and Alligood K T (1985) Period doubling cascades of attractors: a prerequisite for horseshoes, *Commun. Math. Phys.*, 101, 305.

411. Yorke J A, and Yorke E D (1979) Metastable chaos: the transition to sustained chaotic behavior in the Lorenz model, *J. Stat. Phys.*, 21, 263.

412. Yoshida T, Mori H, and Shigematsu H (1983) Analytic study of chaos of the tent map: band structures, power spectra and critical behavior, *J. Stat. Phys.*, 31, 279.

413. Young K (1985) Further regularities in period-doubling bifurcations, *Phys. Lett.*, 111A, 161.

414. Young L-S (1981) Capacity of attractors, *Ergod. Th. & Dynam. Sys.*, 1, 381.

415. Young L-S (1982) Dimension, entropy and Lyapunov exponents, *Ergod. Th. & Dynam. Sys.*, 2, 109.

416. Young L-S (1984) Dimension, entropy and Lyapunov exponents in differentiable dynamical systems, *Physica*, 124A, 639.

417. Yu X-L, Jin H-G, Yan G-H, Wang G-R, and Chen S-G (1987) The forced Brusselator and the circle map, preprint, submitted to *Acta Physica Sinica* (in Chinese).

418. Yuan J-M, Tung M, Feng D H, and Narducci L M (1983) Instability and irregular behavior of coupled logistic eguations, *Phys. Rev.*, A28, 1662.

419. Zeng W-Z (1985) A recursion formula for the number of stable orbits in the cubic map, *Chinese Phys. Lett.*, 2, 429.

420. Zeng W-Z (1987) On the number of stable cycles in the cubic map, *Commun. Theor. Phys.*, 8, 273.

421. Zeng W-Z, Ding M-Z, and Li J-N (1985) Symbolic description of periodic windows in the antisymmetric cubic map, *Chinese Phys. Lett.*, 2, 293.

422. Zeng W-Z, Ding M-Z, and Li J-N (1988) Symbolic dynamics for one dimensional mappings with multiple critical points, *Commun. Theor. Phys.*, 9, 141.

423. Zeng W-Z, Hao B-L, Wang G-R, and Chen S-G (1984) Scaling property of period-n-tupling sequences in one-dimensional mappings, *Commun. Theor. Phys.*, 3, 283.

424. Zeng W-Z, and Hao B-L (1986) Dimensions of the limiting sets of period-n-tupling sequences, *Chinese Phys. Lett.*, 3, 285.

425. Zeng W-Z, and Hao B-L (1987) The derivation of a sum rule detrmining the q-th order information dimension, *Commun. Theor. Phys.*, 8, 295.

426. Zhang H-J, Dai J-H, Wang P-Y, Jin C-D, and Hao B-L (1987) Analytical study of a bimodal map related to optical bistability, *Commun. Theor. Phys.*, 8, 281.

427. Zhang H-J, Wang P-Y, Dai J-H, Jin C-D, and Hao B-L (1985) Analytical study of a bimodal mapping related to a hybrid optical bistable device using liquid crystal, *Chinese Phys. Lett.*, 2, 5.

428. Zheng W-M (1986a) Derivation of the spectrum for the standard mapping: a simple renormalization group procedure, *Phys. Rev.*, A33, 2850.

429. Zheng W-M (1986b) Simple renormalization procedure for quasiperiodic maps, *Phys. Rev.*, A34, 2336.

430. Zheng W-M (1987a) Global scaling properties of the spectrum for the Fibonacci chains, *Phys. Rev.*, A35, 1467.

431. Zheng W-M (1987b) 'Momentum-space' suppression technique for Josephson junctions, *Phys. Lett.*, 121A, 451.

432. Zheng W-M (1988a) Construction of median itineraries without using the anti-harmonic, Preprint ASITP-88-006.

433. Zheng W-M (1988b) Generalized composition law for symibolic sequences, Preprint ASITP-88-007, *J. Phys.* A, to appear.

434. Zheng W-M (1988c) The W-sequence for circle maps, Preprint ASITP-88-010.

435. Zheng W-M (1988d) Construction of median itineraries for the antisymmetric cubic map, Preprint ASITP-88-034.

436. Zheng W-M, and Hao B-L (1988) Symbolic dynamics analysis of symmetry breaking and restoration, Preprint ASITP-88-035.

437. Zhong W-Y, and Yang P-C (1986) The transition of a multi-dimensional Lorenz system, *Advances in Atmos. Sci.*, 3, 289.

438. Zhu C-X (1984) Chaotic phenomena in nonlinear dynamics, *Progress in Mech.*, 14, 129 (in Chinese).

439. Zisook A B (1981) Universal effects of dissipation in two-dimensional mappings, *Phys. Rev.*, A24, 1640.

440. Zong Y-T, and Wei R-J (1984) Bifurcation in air acoustic systems, *Chinese Phys. Lett.*, 1, 61.

441. Zou C-Z (1986) Amalgamation of periodic and chaotic solutions in a double parallel-connected Lorenz system, *Chinese Phys. Lett.*, 3, 161.

442. Zou C-Z, Zhou X-J, and Yang P-C (1985) The statistical structure of Lorenz strange attractors, *Advances in Atmos. Sci.*, 2, 215.

443. Zou C-Z, Yang P-C, and Zhou X-J (1986) The effect of aspect ratio on the bifurcation properties of a double parallel-connected Lorenz system, *Advances in Atmos. Sci.*, 3, 406.

Subject Index